S0-CAH-053

THERMODYNAMIC PROPERTIES OF AIR

NATIONAL STANDARD REFERENCE DATA SERVICE OF THE USSR: A Series of Property Tables

1. Thermodynamic Properties of Helium
2. Thermodynamic Properties of Nitrogen
3. Thermodynamic Properties of Methane
4. Thermodynamic Properties of Ethane
5. Thermodynamic Properties of Oxygen
6. Thermodynamic Properties of Air
7. Thermodynamic Properties of Ethylene
8. Thermophysical Properties of Freons, Part 1
9. Thermophysical Properties of Freons, Part 2
10. Thermophysical Properties of Neon, Argon, Krypton, and Xenon

In Preparation

Thermodynamic Properties of n-Hexane
Thermodynamic Properties of Propane
Thermodynamic Properties of Hydrogen

THERMODYNAMIC PROPERTIES OF AIR

V. V. Sychev
A. A. Vasserman
A. D. Kozlov
G. A. Spiridonov
V. A. Tsymarny

Theodore B. Selover, Jr.
English-Language Edition Editor

HEMISPHERE PUBLISHING CORPORATION
A subsidiary of Harper & Row, Publishers, Inc.
Washington New York London

DISTRIBUTION OUTSIDE NORTH AMERICA
SPRINGER-VERLAG
Berlin Heidelberg New York London Paris Tokyo

Thermodynamic Properties of Air

Originally published by Standards Publishers, Moscow, 1978 as Termodinamicheskiye Svoystva Vozdukha in the Monograph Series of the National Standard Reference Data Service of the USSR; State Committee on Standards of the Council of Ministers of the USSR.

English translation by G. E. Slark.

1 2 3 4 5 6 7 8 9 0 B C B C 8 9 8 7 6

This book was set in English Times by Hemisphere Publishing Corporation. The cover designer was Sharon Martin DePass and the typesetter was Sandra F. Watts.
BookCrafters, Inc. was printer and binder.

Library of Congress Cataloging in Publication Data

Termodinamicheskie svoĭstva vozdukha. English.
Thermodynamic properties of air.

(National standard reference data service of the USSR)
Translation of: Termodinamicheskie svoĭstva vozdukha.
Bibliography: p.
1. Air—Thermal properties. I. Sychev, V. V. (Vîacheslav Vladimirovich), date. II. Selover, Theodore B., date. III. Slark, G. E. IV. Title. V. Series.
QD163.T4713 1987 551.5 87-122
ISBN 0-89116-610-6 Hemisphere Publishing Corporation

DISTRIBUTION OUTSIDE NORTH AMERICA:
ISBN 3-540-17640-3 Springer-Verlag Berlin

CONTENTS

PREFACE TO THE SERIES

This treatise is part of a continuing series on thermodynamic properties of technologically important fluids. These very important contributions by scientists and engineers working through the aegis of the Soviet National Service for Standard Reference Data have been released to Hemisphere Publishing Corporation for translation to make them available to the English reading technical community. The authors are Soviet experts in the field.

While a team of translators was involved in producing the English versions, the overall series is being published under the technical editorship of T. B. Selover, Jr.

Each volume presents a comprehensive survey of the world's literature up to its publication date. A special effort has been made to give a thorough presentation of Russian as well as other work. Many studies not previously known to Western counterparts are included. The results have been to broaden the range of applicability of data and to improve upon equations of state to provide accurate computation methods for generating smoothed tables of properties.

For some volumes there are no equivalent comprehensive surveys available in English. Thus, a valuable service has been fulfilled to workers in the fields of process design, equipment development, custody transfer, and safety.

Each volume is set up in the same way with Part I dealing with a study of all necessary aspects of experimental data interpretation and analysis. Then in Part II the fundamental constants, symbols with units, and data tables can be found. The use of SI units is consistent throughout.

The section on experimental data is particularly important because it covers, in detail, the key studies. Possible errors in measurement or data analysis that could have led to inaccurate tabular results in publications are pointed out.

Methods of constructing the equation of state and procedures for computing the data tables are covered thoroughly. There is a detailed error analysis of data generated from the equation of state relative to literature values.

Properties of each fluid in phase equilibria are treated at the freezing curve and the saturation curve with tables given for both temperature and pressure dependence. Properties cover the range of triple point to critical point in the condensed phase. Single phase properties in each volume cover a range of temperature and pressure wide enough to include most practical applications. Ideal gas property equations and calculation methods are included. The scope of properties generated and covered in the tables is more comprehensive than is typically found in any one English-language treatise. This gives the advantage of internal consistency.

An extensive bibliographic listing is included with each volume. All Russian citations have been translated. In some cases the availability in English of translated Russian sources is given.

Frequent reference to key English-language references has been made to help in providing correct translations. In some cases the original symbols have been changed to avoid confusion within the text or to avoid misuse of terms commonly found in English. Where mistakes in the original Russian text have been found, they have been corrected and so noted as editor's changes. Added descriptive clarification has been used in some places for tables, figures, and text to clarify meaning.

Although we recognize that key review papers have been published in 1986 for some of the fluids in this set, the series stands on its own merit. It represents a vast accumulation of knowledge never before available to Western countries. Through careful study of these volumes, workers in the field will develop an appreciation for both the scope of studies and where differences exist. Hopefully, this will lead to more dialogue between the authors and their Western counterparts.

Theodore B. Selover, Jr.

PREFACE

This book is the second in a series published by the State Service for Standard Reference Data, which is concerned with thermophysical properties of gases and liquids of engineering importance. Studies of thermophysical properties of substances have been performed according to the program of the Soviet National Committee on the Gathering and Assessment of Numerical Data for Science and Technology of the Presidium of the USSR Academy of Sciences and the Commission of the USSR Academy of Sciences on Thermodynamic Tables. This volume has been prepared by a task force on the properties of atmospheric gases and contains detailed tables of thermodynamic properties of air at temperatures between 70 and 1500 K and pressures between 0.01 and 100 MPa. The tabulated data on thermodynamic properties of air are needed for designing air-separation facilities, power plants, and chemical engineering equipment.

Experimental studies of thermodynamic properties of air were started at the end of the past century and have since been performed by many investigators in major laboratories all over the world. The thermal properties of air have been rather fully investigated over a wide range of temperatures and pressures. However, the calorific and acoustic properties of air have been investigated much less thoroughly.

As experimental data became available, tables of thermodynamic properties of air were compiled on the basis of correlation and critical analysis of available experimental data. However, even the most recent Soviet and Western studies do not utilize the results of all the experimental investigations performed up to now, these include experimental studies of the density of gas at high temperatures and the density of liquid under pressure which were performed by us. It is also important that the error of tabulated quantities in all the previous publications has not been analyzed with sufficient rigor. The analysis is usually based on comparison of analytic and experimental data, and these are lacking for a number of properties.

In this book we present a wide range of tabulated quantities, based on experimental pvT data over virtually the entire range of parameters of state. The equations of state were compiled and the tables were calculated with consideration for the thermodynamic behavior of air as a mixture of gases. The use of the mathematical simulation method by computer developed by us made it possible to validate rather precisely the estimates of error in calculating the tabulated thermodynamic properties.

This publication is the result of investigation of thermophysical properties of gases and liquids of engineering importance. The work was performed at the All-Union Research Center of the State Service for Standard Reference Data, Moscow Energetics Institute, and the Odessa Institute of Marine Engineers.

The authors are thankful to engineering programmers A. Ya. Kreizerova, Yu. I. Kas'yanov, and L. R. Malova and to engineers N. V. Knyazeva and N. A. Kochetova of the staffs of the abovementioned institutions for assistance in processing the experimental data and in performing the calculations.

V. V. Sychev
A. A. Vasserman
A. D. Kozlov
G. A. Spiridonov
V. A. Tsymarny

FOREWORD

The acceleration of scientific and technological progress is one of the major tasks of the tenth Five-Year Plan, promulgated by the Twenty-fifth Congress of the Communist Party of the Soviet Union. In light of this task, it is important to develop and expand the sphere covered by standardization, since such expansion would allow accumulation of the latest scientific and technological achievements, organically combine pure and applied sciences, and promote rapid and practical implementation of scientific achievements.

The State Service for Standard Reference Data oversees one of the new trends in standardization, that of standardization of the most reliable data available on the physical constants and properties of materials and substances.

The development of standard reference data is a multifaceted scientific and technological task that must be based on the Soviet and worldwide practice of selection and estimation of the reliability of data. An important stage in this work consists of preparing standard reference publications that contain not only data with an estimated reliability but also modern methods of obtaining such estimates. It is precisely these such publications that should serve as a methodological basis for the development of official tables of standard reference data on the properties of materials and substances.

The experience accumulated in working out tables of data on the thermodynamic properties of gases and liquids is of great interest from this point of view. This work has been particularly developed in the USSR. The work of Soviet investigators in thermophysics, which has established the basis for Soviet and international tables of thermophysical properties, has been acclaimed worldwide. Currently, this work is carried out by over 40 research organizations in the USSR under the auspices of the scientific program of the

State Service of Standard Reference Data and the Working Group on Thermodynamic Tables of the Soviet National Committee on Numerical Scientific and Technological Data of the Presidium of the USSR Academy of Sciences.

The publication of this series of books will supply reliable reference information to a wide circle of engineers and scientists and will serve as a basis for the development and practical utilization of pertinent official tables of standard reference data. The use of such data, obtained on the basis of the most up-to-date and accurate methods of study, is one of the conditions necessary for improving the level of scientific and developmental work. It ensures effective monitoring of industrial processes and quality of industrial output and promotes efficient utilization and accounting of the consumption of raw and finished materials, fuel, and energy. It is thus obvious that supplying reliable data on the properties of substances and materials is an important economic activity. The State Service of Standard Reference Data has been established primarily to supply this need.

V. V. Boytsov

PART ONE

CHAPTER

ONE

EXPERIMENTAL DATA ON THE THERMODYNAMIC PROPERTIES OF AIR

The compressibility of air was investigated experimentally for the first time as early as the end of the nineteenth century. Subsequently, data on the thermodynamic properties of air were frequently supplemented and refined. The last studies were performed rather recently, and their results will help improve the accuracy and reliability of the tables.

It was noted by many investigators that air can be regarded, for practical purposes, as a binary mixture of nitrogen and oxygen. For this reason it had been possible to supplement data on its properties by experimental results obtained in investigating such mixtures. Nevertheless, in this book we used only experimental data on the properties of air. The pertinent data on the properties of mixtures are employed only for comparison since they cannot be used for determining the properties of air with an accuracy comparable to that of experimental work.

1.1. THERMAL PROPERTIES OF SINGLE-PHASE AIR

A listing and a brief description of the principal studies containing original experimental data on the density (or compressibility) of single-phase air are given in Table 1.1. The table does not mention the familiar studies of Holborn

Table 1.1 Experimental studies of the density of single-phase air

Year	Author	Reference	Range of parameters		Number of points	Phase	Method
			ΔT, K	Δp, MPa			
1880	Amagat	[25]	273	3–40	13	G	VV
1888	Amagat	[26]	288	76–304	7	G	VV
1893	Amagat	[27]	273–473	0.1–304	152	G	VV
1896	Witkowski	[110]	128–373	0.1–13.2	176	G	CV
1908	Koch	[77]	194; 273	2.5–20.2	45	G	CV
1913	Occhialani and Bodareu	[87]	287-293	6.5–33.8	11	G	B*
1915	Holborn and Schultze	[67]	273–473	2–10	42	G	CV
1923	Penning	[91]	128–293	2.5–6.2	62	G	VV
1936	Burnett	[37]	303	0–6.3	20	G	B
1945	Kiyama	[73]	303	9.8–455	20	G	—
1953	Michels et al.	[85]	273–348	0.6–227	157	G	VV
1954	Michels et al.	[84]	118–248	0.6–101	199	G	CV
1960	Rogovaya Kaganer	[18]	173–273	2.7–10.9	23	G	CV
1968	Vukalovich et al.	[10]	288–873	1.4–72.2	349	G	CV
1971	Romberg	[96]	84–122	0.02–1.95	124	G	B
1973	Blanke	[34]	61–170	0.01–4.9	117	G, L	CV
1976	Vasserman et al.	[6]	78–199	1–59	109	L	CV

VV, Variable volume; CV, constant volume; B, Burnett; B*, modification of Burnett method (the second vessel is not evacuated during the experiments); G, gas; L, liquid.

and Otto [64–66] since these are simply a workup of previously obtained [67] experimental data. It also does not include studies in which the measurements were performed near atmospheric pressures. These are mentioned, among others, in the bibliographical survey by Hall [57].

In addition to data for single-phase air, some investigators also list information on the density at saturation and for two-phase air. For example, Michels et al. [84] list, in addition to 199 points for the single-phase region, data for two-phase air. Of the 423 points listed by Blanke [34] only 117 were obtained in the single-phase region while the others are for the two- and three-phase regions.

The claim that the density of air has been investigated experimentally in a large number of studies does not hold up on close scrutiny.

As seen in Table 1.1, some of these studies were performed during the nineteenth century and are apparently only of historical interest. A detailed analysis of studies performed up to 1968 is given in the books by Vasserman et al. [7] and Vasserman and Rabinovich [9]. For this reason we present here only a digest, which describes the accuracy of experimental data that are of significance for the subsequent presentation.

The most thorough studies are those by Michels with his coworkers, which yielded highly accurate results and covered a wide range of variables. The reproducibility and accuracy in determining the compressibility pv is estimated by Michels et al. to be about 0.01%. At low temperatures the scatter of values of pv is 0.1% and becomes as high as 0.2% only in individual points. Unfortunately, only eight points are given in [84] for liquid air. The scatter of experimental data is also low in [67, 77, 91, 110], and is as high as 0.5% only in the study by Amagat [27]. At the same time, there is a lack of agreement among data on individual authors, which increases at higher pressures and is due to lack of allowance for systematic errors. In particular, the deviation of values in [27] from those in [84, 85] excess 1%. The high accuracy of data in [84, 85] makes it sensible to compare all the remaining data with those listed here in determining the weights of data used for compiling equations of state.

The data of Holborn and Schultze [67] agree with data in [84, 85] within 0.05%. The results obtained by Penning [91] and Witkowski [110] deviate on the average by 0.2–0.3%, with data in [110] being systematically on the high side of temperatures below 0 °C. The data of Koch [77] are approximately 0.2% on the low side and do not conform to the limit condition ($Z \rightarrow 1$ as $\omega \rightarrow 0$). The maximum deviations of data in [77, 110] from those in [84, 85] amount to 0.5%.

Up to 1968 the range of variables over which experimental measurements of the density of air were performed was not as wide as for other gases of engineering importance. No data were available for low temperatures, including the liquid phase, or at high pressures and temperatures.

The resolutions of the International Union of Pure and Applied Chemistry and corresponding resolutions of Soviet scientific organizations mentioned air as a substance for which tables of thermodynamic properties should be compiled on a priority basis. These resolutions stimulated a large number of experimental studies [6, 10, 34, 96] which have significantly supplemented information on thermodynamic properties of air. The data of all these studies taken together were not used in compiling equations and calculating tables and were not analyzed in detail. The only exceptions are data of Vukalovich et al. [10], which were used by Vasserman and Kreyzerova [8] and by Spiridonov et al. [19]. Hence, it is advisable to analyze the above-mentioned experimental studies in more detail.

Vukalovich et al. [10], who performed their study at the Moscow Energetics Institute, used a constant volume piezometer (Fig. 1) with a cold valve and a mercury separator which was located in the zone at room temperature. Such a design is simple and has a number of methodological advantages. The dead-space volume, which includes the volume of the capillary and separator, is

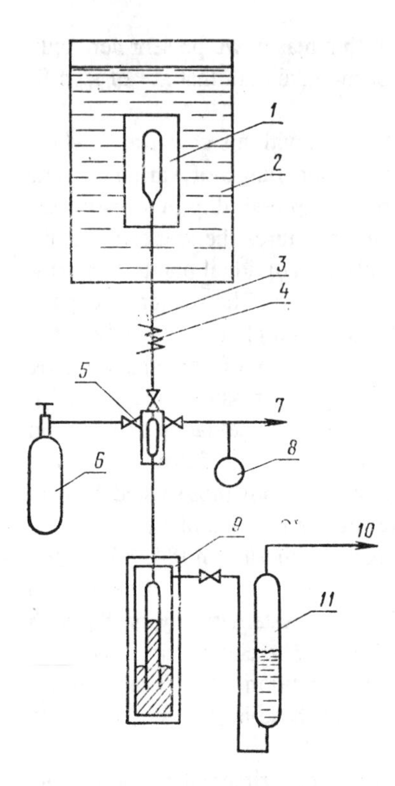

Figure 1 Schematic of the Moscow Energetics Institute's experimental facility for investigating the *pvT* behavior of gaseous air: (1)Piezometer; (2) thermostat; (3) capillary; (4) guard heater; (5) port for viewing the mercury level; (6) separating vessel for determining the mass of air used in the study; (7) line for connecting to the vacuum pump; (8) vacuum gauge; (9) mercury-water separator; (10) line connecting to the MP-600 [piston] manometer and hydraulic press; (11) water-oil trap.

equal to 0.4 cm^3 for a piezometer volume of $\sim$335 cm^3. This necessitated a relatively small correction.

The majority of experimental points in [10] were obtained along isotherms since it is very easy to discharge a part of the gas from the piezometer through the cold valve after performing measurements at a given pressure. This equipment can also be used for performing measurements along isochores. When gas is discharged from the piezometer, measurements can be performed at another density. The mass of the air was determined by the absolute weighting method. Weighing was performed using light and rather strong tanks. This permitted elimination of the use of adsorbents for sample trapping, resulting in a number of advantages [15].

The investigators employed atmospheric air which was compressed by a reciprocating compressor and was then purified by filters. After purification and during experiments, the air was periodically subjected to analysis. This confirmed constancy of composition, including constancy of the oxygen content following measurements at temperatures above 600 °C. This is very important in high-temperature measurements since it shows that there is no interaction between the air and the piezometer metal wall. The composition of air investigated by Vukalovich et al. [10] was (% by volume): N_2, 78.0 $\pm$ 0.2; O_2,

21.0 ± 0.05; inert gases (Ar), 0.93 ± 0.3; CO_2, 0.02–0.04, NO_x and C_xH_x, trace amounts; H_2O, CO, and H_2, not present.

It should be noted that the composition of air quoted by different investigators is somewhat different. Thus, the U.S. National Bureau of Standards team [104] and Baehr and Schwier [30] used the following values.

N_2	O_2	Ar	CO_2	
78,09	20,95	0,93	0,03	[104]
78,41	20.66	0.93	—	[30]

The majority of experimental studies do not list information on the composition of air they used. Information is given only on steps taken for removing water vapor and CO_2. This apparently suffices since the concentrations of water vapor and CO_2 are the least constant. Given the closeness of properties of oxygen and nitrogen, insignificant variations in their content cannot have a significant effect on the thermal properties of air.

In spite of the fact that the air used by Vukalovich et al. [10] contained CO_2, the difference of compositions was not considered when comparing with results in [85]. The correction in question is an order of magnitude smaller than the experimental error.

The temperature and pressure measurement error are estimated in [10] at 0.02–0.07 K and 0.05%, respectively, and the error of weighing at $(2–30) \cdot 10^{-4}$ g. The maximum possible error of density determination (including the reference error) is 0.15%.

Comparison of experimental data in [10] with data of [85] on the 50 °C isotherm showed that the differences in the specific volume do not exceed 0.05%. The deviation from data of Holborn and Schultze [67] at the 100, 150, and 200 °C isotherms is about 0.1%. Calculations performed by Vasserman with Krayzerova [8] show that the 0.15% error claimed by the authors of [10] is apparently an overestimate (i.e., the error is smaller).

Romberg [96] employed the familiar Burnett method, which allows determination of compressibility without directly measuring the volume of the piezometer and weighing the gas. He used pressure relieved measuring vessels in spite of the fact that the working pressures were relatively low. Romberg also notes that the volume of the connecting tubes remain virtually unchanged even under pressure. The test vessel was maintained at constant temperature in a liquid nitrogen–filled cryostat and subjected to elevated pressure. The uniformity of the temperature field was improved by using thick walled copper casing and stirrer. The device does not contain dead-space volumes which are at a temperature different from the experimental temperature. This was done by using a membrane-type differential manometer built into the bottom of one of the vessels. The pressure from the press of the piston-type manometer is transmitted to the membrane situated in the low working-temperature zone through helium.

All these measures made it possible to maintain the scatter of experimental data within limits not exceeding 10^{-3} bar and 10^{-2} K. The overall measuring error is estimated by Romberg at 0.03%.

Romberg addressed in detail the question of the composition of air used in the tests. He assumed, apparently on the basis of data in [104], that the starting air composition is 78.09% N_2, 20.95% O_2, 0.93% Ar, and 0.03% CO_2 (by volume). However, at the experimental temperature the CO_2 is in the crystal state, which results in a slight change in composition, i.e., 78.11% N_2, 20.96% O_2, and 0.93% Ar. Instead of purifying atmospheric air, Romberg synthesized it from the pure components. Analysis of the mixture he used showed that it was composed of 78.16% N_2 and 20.91% O_2. This discrepancy apparently did not significantly affect the measured value of compressibility.

The study by Romberg, which yielded a large body of information, was performed over a range of temperatures (from below 118 K up to the critical). This is important for compiling the equation of state since all the previously published tables contain only extrapolations of experimental data for this region. Given the extreme care in performing the different stages of the study, it seems a bit strange that Romberg used obsolete values of molecular weight and used a resistance thermometer calibrated to the 1948 temperature scale.

Blanke in his thorough investigation [34] attained the lowest temperatures, obtained data on the density of the gas and the liquid, also on the state of saturation, and in the two- and three-phase regions. Unfortunately, the maximum pressure attained by Blanke did not exceed 6 MPa.

He used a constant-volume piezometer with a membrane-type differential manometer which served as a separating device. The mass of air was determined by a gas meter through which the tested air was passed following the measurements. The volume of the gas meter, which consisted of 20 glass vessels, was selected in a manner that allowed reaching a final pressure not in excess of 0.1 MPa.

Air was sampled far from city limits, which ensured that it did not contain industrial gases. Water vapor and CO_2 were removed by caustic and phosphorus pentoxide. Without special analysis, Blanke assumed that the composition of the gas was that cited by Baehr and Schwier [30]. He estimated the accuracy of measurement of the principal experimental quantities (volume of piezometer and gas meter, temperature, pressure). He also calculated corrections for the effect of pressure and temperature on the piezometer volume and analyzed errors. Blanke analyzed the effect of various factors on the error, determined for different ranges of variables. This is very important due to the complex configuration of the thermodynamic surface over the range explored in his study, the different parts of which have different values of derivatives. The relative error of density of the gas $\delta\rho$ ranged from 0.05 to 0.13%, which corresponds to a pressure error $\delta\rho$ of 0.04–0.2%. For liquid at temperatures below 115 K the value of $\delta\rho$ is virtually constant and amounts ot 0.04–0.06%. The value of $\delta\rho$ increases to 0.4% upon approaching the critical region.

In spite of these new measurements, the density of liquid air remained insufficiently investigated. In order to supplement information over this range measurements of air density at temperatures to 78 K and pressures to 60 MPa were performed in the Odessa Institute of Marine Engineers [6]. This was done with a version of a constant-volume piezometer (Fig. 2). It had no dead-space volume, and its volume was independent of the pressure of the air in it. The first of these changes was attained by placing the valve and the differential manometer serving as separator in the working-temperatures zone. The second was done using a pressure-relieved thin-walled piezometer.

The differential manometer with contact indicator of membrane deflection, which serves as a separator, ensures constancy of the piezometer volume and recording of the time when the pressure to both sides of the membrane becomes equal. The fact that the valve was used only for filling the piezometer with the test substance and for discharging the latter into the weighing tank made possible constant volume measurements.

This arrangement requires only a single correction for the temperature

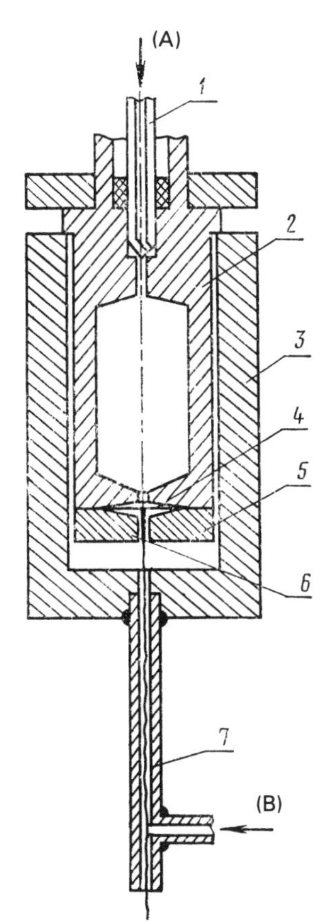

Figure 2 Design of piezometer employed in the Odessa Institute of Marine Engineers for investigating the *pvT* behavior of liquid air: (1) Charging-valve needle; (2) piezometer housing; (3) high-pressure chamber housing; (4) separating differential manometer membrane; (5) membrane travel limiter; (6) contact indicator of membrane deflection; (7) air-tight electric-cable inlet; (A) connection to air charging line; (B) high-pressure line, connecting the piezometer to the MP-600 piston-type manometer.

dependence of the piezometer volume. This correction is calculated for each test point from data on the coefficient of linear expansion of stainless steel, which is the piezometer material.

Particular attention was paid to the piezometer filling procedure. The liquid nitrogen–cooled piezometer was filled from a tank in which the air was maintained at 6–7 MPa, which is higher than the critical pressure. In order to prevent the possible condensation of oxygen which may occur under these circumstances and may change the composition of the air, the piezometer was filled through a long capillary. The low velocity at which the gaseous air entered the piezometer ensured that it would condense completely and the final composition would remain unchanged.

The air in the piezometer weighed between 28 and 37 g, determined by weighing the separating vessel. This vessel was cooled by liquid nitrogen when the air was discharged after the experiment. This reduced the mass of air remaining in the piezometer and in the connecting tubing to 0.05 g. Moreover, this mass of air was also determined experimentally.

The air in these experiments was taken from an air-separation facility, where moisture, acetylene, CO_2, and industrial gases had been removed from it. No special measurements of the air composition was performed. The composition of the air used in various runs was checked indirectly. For this purpose the results were compared with the rather scarce data on the range investigated by Vasserman et al. [6], including the region in [84]. In addition, the density of saturated liquid was determined in [6] by extrapolating the isochores to the incipient condensation curve, which was compared with the corresponding values from [34]. The fact that the agreement with data in [34, 84] was within several hundredths of a percentage point indicated that the data in [6] are reliable.

Vasserman et al. estimate the error of temperature and pressure measurements at 0.02 K and $(2\text{–}3)\cdot 10^{-3}$ MPa respectively. The maximum error of density measurement (including referencing errors) was 0.1%.

As seen from this survey, the experimental data of Vasserman et al. [6], Blanke [34], and Romberg [96] obtained during the early 1970s pertain to a low-temperature range, including the liquid phase, which is very important for cryogenics. This made it possible to compile tables of thermodynamic properties of air validated experimentally over a wider range of parameters than all the previous publications.

1.2 CALORIFIC AND ACOUSTIC PROPERTIES OF SINGLE-PHASE AIR

Studies of calorific properties of air, as for many other substances, are much less detailed than those of the *pvT* behavior. The largest group of studies is made up of those for measuring the Joule–Thomson effect, which stems from its applications in cryogenics. The specific heat at constant pressure and, even more so, at constant volume have been explored much less. Very little information is available on the heat of vaporization of air.

The significance of studies of calorific properties is quite varied. Some of these are of no current value due to their poor accuracy and the fact that they were performed quite long ago. Experimental studies of calorific and acoustic properties are represented in the corresponding tables in less detail than information on the experimental studies of pvT behavior given in Sec. 1.1. This is done to the fact that calorific data are used only for checking the reliability of the equation of state, whereas the pvT data are the main information used for compiling it. The tables do not mention studies which present data over a narrow range of temperatures at close to atmospheric pressure, containing only graphical information, or studies performed during the nineteenth century. The tables also make no mention of studies of mixtures (in particular, nitrogen–oxygen binary mixtures), the results of which will be used only in very special cases for approximations and comparisons.

A rather detailed list of studies performed up to and including 1968 is contained in [7, 9, 30, 46] and in the NBS survey [57]. Note that there are inaccuracies in [57] concerning the kind of study (experimental or analytic), the number of experimental points, and the nomenclature of investigated properties.

Following tradition, we shall first consider studies of the Joule–Thomson effect. They were stared by Joule and Thomson during the nineteenth century. At the end of that century Witkowski [111] also measured this effect at high pressures. A list of experimental studies of this effect performed during a 70-year span is given in Table 1.2. This table includes measurements of the differential and integral adiabatic Joule–Thomson effect as well as a single study of the isothermal effect.

A thorough survey of measurements performed at the start of the present century is given by Hoxton [68], who also presents his own original data. The most detailed studies of the Joule–Thomson effect are those by Roebuck [93, 94] and Hausen [59], performed over rather wide temperature and pressure ranges. Unfortunately, the studies by Roebuck suffer from a recurrent error due to incorrect calibration of the manometer. This error was detected in a subsequent study by Roebuck and Murrel [95], and they have recommended that the previous data be corrected (the measured pressures should be multiplied by 0.9677). Naturally, this requires correction of experimental values of α_i. The corrected values of α_i for air at atmospheric pressure are listed in [95], and these are 3.3% higher than the values listed in [93, 94]. Most likely the values not only of α_i, but also of p_m, the reference pressure should be corrected. Given the complex dependence of the adiabatic Joule–Thomson effect on the temperature and pressure, a 3% correction of p_m in the critical region can have a significant effect on the values of α_i. These considerations are reinforced by the fact that a 3.3% correction of α_i only decreases the differences between the data

Table 1.2 Studies of the Joule–Thomson effect

			Range of parameters			
Year	Author	Reference	ΔT, K	Δp, MPa	Number of points	Quantity measured
1907	Olszewski	[88]	397–532	1.62	8	I
1909	Dalton	[42]	273	4.0	—	I
1909	Bradley and Hale	[35]	153–273	20.5	54	I
1911	Vogel	[107]	283	14.7	—	D
1916	Noell	[86]	218–523*	14.7	—	D
1919	Hoxton	[68]	288–363	0.86	14	D
1925	Roebuck	[93]	273–553	21.6	63	I
1926	Hausen	[59]	98–283	19.6	87	I
1930	Roebuck	[94]	123–248	21.6	82	I
1943	Baker	[31]	273	5.6	2	I
1948	Brilliantov	[4]	94–156	0.2	27	D
1956	Ishkin and Kaganer	[12]	90–298	11.2	161	T
1974	Dawe and Snowdon	[44]	222–366	10.0	133	I

I, Integral adiabatic Joule–Thomson effect; D, differential adiabatic Joule–Thomson effect; T, isothermal Joule–Thomson effect.

*Temperature at high-pressure side.

of Roebuck [93] and Hausen [59] at low pressures, but increases them at high pressures.

To our knowledge, Ishkin and Kaganer [12] were the only ones to measure the isothermal Joule–Thomson effect. Upon analyzing the experimental methods for measuring this effect, they gave preference to the isothermal method. They point out that the main advantage of this method consists of eliminating transfer of heat between the air and the surroundings and improving the accuracy of measurements at low test-gas flow rates. A great deal of attention was given by Ishkin and Kaganer to analysis of errors of measurement of individual experimental values. However, they committed a systematic error on the order of 5% due to erroneous calibration of the gas meter.

In fact, measurement of the gas flow rate and elimination of heat losses or careful accounting for the latter are the main experimental difficulties in measuring the adiabatic Joule–Thomson effect. This makes it natural to take advantage of the isothermal technique. Nevertheless, all the complexities of calorimetric experimentation are apparently inherent to both methods of Joule–Thomson effect measurement.

It appears proper to present here the authoritative opinion of V. V. Altunin, who measured both the isothermal and adiabatic Joule–Thomson effect in numerous studies of CO_2. In his book [1], which gives a description of experimental techniques and facilities and presents analysis of data and of sources of error,

Altunin discusses in detail the difficulties arising in performing calorific experiments. He points out that "implementation of isothermal throttling [in measuring the Joule–Thomson effect] is more complicated than that of the adiabatic . . ." and simply notes that it is theoretically possible to attain higher experimental accuracy when measuring the isothermal effect.

Experimental data on the enthalpy of gaseous air were published by Dawe and Snowdon [44]. Actually they measured the integral Joule–Thomson effect during expansion of gas from an initial pressure of 0.4–10 MPa to atmospheric along five isotherms. They do not present direct results of measurements; rather they give information on the number of experimental points obtained along each isotherm and equations describing the experimental data along isotherms as a function of pressure with an error not exceeding 0.04 K. Dawe and Snowdon used these data for calculating the enthalpy, also represented by equations along the isotherms. In conclusion, in comparing their experimental data with the familiar analytic results of Din [46] and Vasserman with his coworkers [7], they noted that their results aer in excellent agreement with those in the latter study and disagree with data in [46]. The disagreement increases with rising pressures. It is characteristic that the same lack of agreement exists with Din's data on properties of nitrogen.

In this study, as in the majority of investigations of calorific properties, purely experimental data are intermingled with results found upon subsequent data reduction. In particular, the majority of values of specific heat c_p were obtained by working up experimental data on other properties. For example, Hausen [59] and Roebuck [93, 94] used data on the Joule–Thomson effect for calculating values of c_p. In the majority of handbooks these are regarded as experimental, without justification. These studies are included in Table 1.3, first, due to the small number of purely experimental studies. Second, the calorific properties calculated analytically from measurements of other calorific or acoustic quantities (for example, values of enthalpy determined from data on the Joule–Thomson effect or determination of the ratio $\varkappa = c_p/c_v$ from data on the speed of sound) have an accuracy not less than the results of direct measurements.

Many of the studies concerned with measuring the isobaric specific heat were performed at atmospheric pressure. Measurements of c_p at elevated pressures were performed only by Holborn and Jakob [63]. There are simply no data on specific heat c_p of liquid air at pressures above the saturation pressure.

The published information on c_p of compressed air was obtained primarily by working up experimental data on other thermodynamic properties. Such a study was performed first in 1923 by Jakob [72] using two different methods—on the basis (1) of thermal data and (2) of data on the Joule–Thomson effect. Subsequently, a number of Jakob's data were corrected [109]. As mentioned above, Roebuck's [93, 94] and Hausen's [59] data on c_p at high pressures were obtained on the basis of data on the Joule–Thomson effect.

Table 1.3 Studies of specific heat

Year	Author	Reference	Range of parameters ΔT, K	Δp, MPa	Number of points	Properties listed
1906	Cook	[40]	91–295	0.1	20	$\varkappa$
1908	Koch	[77]	194; 273	2.5–20.2	16	$\varkappa$
1911 1913	Scheel and Heuse*	[98, 99, 100]	90–295	0.1	3**	c_p
1916	Holborn and Jakob*	[63]	323	4.9–29.4	6	c_p
1921	Womersley†	[113]	273–2273	0.1	20	c_p
1923	Partington and Shilling	[90]	288–1073	0.1	9	w, c_p, c_v
1923	Jakob	[72]	193–523	0.1–19.6	—	c_p, c_v, $\varkappa$
1925	Giacomini	[55]	83–273	0.1	4	c_v
1925	Brinkworth	[36]	155–290	0.1	3**	c_p, $\varkappa$
1925	Roebuck	[93]	273–553	0.1–22.3	60	c_p
1926	Hausen	[59]	193–323	0.1–19.6	20	c_p
1928	Shilling and Partington	[102]	273–1573	0.1	—	c_p, c_v, $\varkappa$
1928	Eucken and Hauck*	[51]	138–165	5.5–16.2	16	c_v
1929	Eucken and von Lüde*	[52]	271–480	0.1	6	c_p
1930	Lourie	[82]	273–2273	0.1	21	c_p, c_v
1930	Roebuck	[94]	173–273	0.1–22.3	28	c_p
1931	Henry*	[61]	273–623	0.1	9	c_v
1937	Hubbard and Hodge	[69]	300	0.1–10.1	3	λ
1939	Kistiakowsky and Rice*	[75]	271–366	0.1	4	c_p
1940	Bennewitz and Schultze*	[33]	325–347	0.1	2**	c_p
1943	Dailey and Felsing‡	[41]	346–605	0.1	7	c_p
1953	Glassman and Bonilla	[56]	200–2500	0.1	—	c_p
1955	Andrussow	[28]	195–473	0.1	4	c_v
1965	Chashkin et al.***	[24]	128–138		57	c_v

*These articles list experimental data. The remaining studies employed indirect methods.

**Averaging of a large number of data points.

***The $\rho = 0.318$ g/cm^3 isochore.

†Quantities averaged over the interval of 0 °C — t.

‡The experimental values are listed in Dailey's thesis.

A survey of direct and indirect determinations of specific heat (c_p, c_v, and $\varkappa$) is given by Masi [83].

Direct measurements of isochoric specific heat were performed only by Eucken and Hauck [51], Henry [61], and relatively recently by Chashkin with his coworkers [24]. Unfortunately, no details are given in [51, 61] on the experimental technique. The study by Chashkin et al. [24] is concerned with investigating the effect of impurities on the behavior of the isochoric specific heat in the critical range. For this reason the air and nitrogen under study had a 1.2% impurity content. In addition, the data obtained in [24] pertain to the range of variables which must be described by the nonanalytic form of the equation.

A list of measurements of the speed of sound in air is given in Table 1.4. The overwhelming majority of measurements were performed for gaseous air at pressures not exceeding atmospheric. In some studies [71, 103] measurements were performed only at different frequencies in order to investigate the dispersion. Hodge [62] and Koch [76] were the only ones to perform measurements over a relatively wide range of pressures, and only van Itterbeek and van Dael [70] present data on the speed of sound in liquid air.

It should be noted that even during the past several years, in spite of

Table 1.4 Investigations of the speed of sound

Year	Author	Reference	Range of parameters		Number of points
			ΔT, K	Δp, MPa	
1906	Cook	[40]	91–295	0.1	20
1908	Koch	[76]	194; 273	0.1–20.3	18*
1921	Dixon et al.	[48]	273–973	0.1	8*
1924	Dixon and Greenwood	[49]	293–363	0.1	9
1928	Shilling and Partington	[102]	273–1273	0.1	12
1933	Kaye and Sherratt	[74]	291; 373	0.1	2**
1937	Hodge	[62]	300	0.1–10.1	11
1938	Colwell et al.	[39]	273	0.1	1**
1943	Tucker	[106]	292–377	0.1	4
1945	Quigley	[92]	92–259	0.1	29
1956	van Itterbeek and de Rop	[71]	229–313	0.1–1.33	44
1958	van Itterbeek and van Dael	[70]	77; 90	0.4–7.3	28

*Smoothed quantities.

**Averages of a large number of data points.

refinements of calorific experimentation techniques, no attempt were made to experimentally refine information on calorific properties of air, and these data remain up to now limited, variable and insufficiently reliable. For this reason data on calorific and acoustic properties of single-phase air cannot make a significant contribution to the compiling of the equation of state for a wide range of variables.

1.3 THERMODYNAMIC PROPERTIES AT THE PHASE-EQUILIBRIUM CURVES

The thermodynamic properties of air at the phase-equilibrium curves were not investigated in sufficient detail, and in addition, a part of these data are insufficiently accurate. Data on the pressure of saturated vapor and boiling liquid are most numerous. A number of investigators measured the density of saturated liquid and vapor but information on the heat of vaporization, specific heat, and speed of sound is very limited.

Since air is a mixture of gases, its behavior at phase equilibrium is more complicated than that of pure substances. The isobars and isotherms of two-phase air do not coincide, and the projection of each of the phase-transition curves on the pT curve is described by two curves. In particular, the liquid–vapor equilibrium curve is represented by incipient-boiling and incipient-condensation curves.

According to data of Walker et al. [108] and Blanke [34], liquid and vapor coexist at temperatures above 60 K. When the temperature is lowered, the crystal phase appears and at temperatures of 56 to 60 K and low pressures the three phases—crystal, liquid, and vapor—coexist. At lower temperatures only two phases—the crystal and the vapor—coexist. There also exist two-phase equilibrium curves, analogous to the incipient boiling (bubble point) and incipient condensation (dew point) curves. The curve corresponding to inception of vaporization from the crystal is termed by some [108] the sublimation curve.

1.3.1 The Incipient Solidification Curve

The solidification temperature of air at below-atmospheric pressures was measured by Walker et al. [108], who point out that the solidification starts and ends approximately between 60 and 56 K. Blanke, in his very carefully performed study [34], found that the incipient solidification temperature is 59.75 $\pm$ 0.07 K and obtained data in the three-phase region up to 54.3 K. An indirect confirmation of these values was obtained by Ruhemann with his coworkers [97], who present a melting diagram for the N_2–O_2 mixture. It is seen from the diagram that when the mixture has the same composition as air, the temperatures of inception and termination of solidification are approximately 60.3 and 56.4 K. According to data of Prikhod'ko and Yavnel' [17], who investigated phase transitions in solid N_2–O_2 mixtures at 20 to 65 K, the temperature range over which a mixture having a composition close to that of air solidifies is somewhat narrower (59.4–57.9 K).

This means that data on solidification temperatures of air and on N_2–O_2 mixtures of corresponding composition do not agree with one another. It is possible that the experimental results were significantly affected by features of the experimental technique. Thus, Blanke notes, upon analyzing the study by Ruhemann et al., that the compositions of liquid and crystalline air subjected to rapid cooling are significantly different, whereas when cooling occurs at a slow pace they succeed in equalizing. Naturally, the temperatures at the termination of solidification will then be different. The most reliable value of incipient solidification temperature is apparently that listed by Blanke (59.75 K).

The pressure dependence of the solidification temperature of air has not been investigated experimentally. Since the variation in the solidification temperature is not too significant when pressure is changed by 100 MPa (approximately 20 K for nitrogen and 9 K for oxygen), it can be assumed that the curve of incipient solidification of air at pressures to 100 MPa lies basically outside the range of parameters for which the tables here are compiled. The pressure must be limited to the 45 to 75 MPa range only at the 70 and 75 K isotherms, respectively, for the tables in this book covering the single-phase region.

1.3.2 Pressure of Saturated Vapor and Boiling Liquid: Parameters of Critical Points

As noted, the liquid–vapor phase-equilibrium curve for air consists of two connecting branches. These correspond to incipient boiling, or the bubble point line, and incipient condensation, or the dew point line. In pT coordinates the first curve lies above the second.

A list of studies containing experimental values of saturated-vapor pressure and boiling-liquid pressure is given in Table 1.5. The majority of these values were obtained by determining points of inflection of quasiiochores. Furukawa and McCoskey [54] performed their experimental work along isotherms. The pressure was changed by adding a certain quantity of air to the piezometer, and the nature of this change was different in the single- and two-phase regions. This made it possible to determine the incipient condensation pressure from the location of the point of intersection of two segments of the isotherm in coordinates of quantity of substance versus pressure.

Without going into a detailed analysis of data obtained by Furukawa and McCoskey [54], Kuenen and Clark [78], and Michels et al. [84], we note that those in the last study are the most accurate and reliable. The authors [84] estimate that their data were obtained with a temperature error of 0.05 K near the critical point and 0.01 K far from it, which corresponds to an error of 0.2 to 0.05% in the values of pressure. Data obtained by Blanke [34] are very important for reliable description of thermodynamic properties of air. His measure-

Table 1.5 Studies of pressure at boiling and condensation curves

Year	Author	Reference	Temperature range ΔT, K	Number of points	Pressure branch**
1917	Kuenen and Clark	[78]	123.0–132.5	14*	p'
1953	Furukawa and McCoskey	[54]	60.0–85.0	8	p''
1954	Michels et al.	[84]	118.2–131.9	10	p'
1966	Walker et al.	[108]	60.5–64.1	5	p'
1973	Blanke	[34]	60.4–128.7	9	p'
			67.2–132.3	11	p''

*Including the critical point at $T = 132.46$ K.
**p', Boiling line; p'', dew line.

ments were performed with extreme care over a wide temperature range. He obtained data at $T < 118$ K for the first time, and as of now they are the only ones available. The value of experimental data in [34, 84] on properties at phase equilibrium also consists of the fact that they were brought into agreement with data for the single-phase region.

Kuenen and Clark [78] obtained detailed experimental data for the transcritical region, which allowed their determining the parameters of critical points. The technique of measurements in the immediate vicinity of the critical point they used contained a number of original features which allowed them to follow the phase transition in extensive detail. They observed retrograde condensation in a very narrow temperature range. The data in [78] exhibit a large scatter. However, this is natural for the transcritical region, but at the same time it reduces the practical value of data in [78].

The values of pressure on the boiling curve obtained by Walker and his coworkers [108] are also not too accurate. In the first place, this was due to the use of an ionization vacuum gauge for measuring the pressure. In addition, the mixture they used contained 22.6% O_2, 76.7% N_2, 0.4% Ar, and 0.39% CO_2, i.e., its composition differed from that of ordinary air. The measurements of pressure at low temperature may be particularly affected by the presence of CO_2.

Many investigators calculated the vapor pressure curves of air on the basis of the behavior of the N_2–O_2 system. In particular, Ishkin and Kaganer [13] and Baehr and Schwier [30] used data of Dodge and Dunbar [50], whereas Vasserman et al. [7] additionally used data of Armstrong and his coworkers [29] as well as Cockett [38] and Din [47]. Attainment of satisfactory agreement required making a number of corrections. Nevertheless, the scatter of calculated values of p' and p'' for air obtained from data of various investigators on the

N_2–O_2 mixture was as high as 5% at low temperatures. Vasserman et al. [7] used data on the properties of mixtures because no experimental data were available for p' and p'' of air over certain ranges of variables. After their book was published, Narinskiy [16] performed a thorough study of thermodynamic properties of binary and ternary mixtures containing nitrogen, oxygen, and argon. These data are in much better mutual agreement than the results of many preceding studies. However, we did not use information on properties of mixtures since experimental data obtained directly for air [34] have become available.

In the case of mixtures, as a result of the complex nature of relationships at phase equilibrium, one must consider a special situation different than for pure fluids. The mixture critical point also called the plait point or first critical point, corresponds to the *T, P* coordinates at the mating of the incipient boiling (bubble point) and incipient condensation (dew point) curves. The point on the phase envelope which corresponds to the maximum temperature is located on the dew point curve and is called the point of contact (second critical point) or cricondentherm. The difference between the temperatures of the mixture critical and cricondentherm points for air is approximately 0.1 K, with the latter point the higher temperature. There also exists a maximum pressure point on the phase envelope called the cricondenbar, which in principle should not be identical to the pressure at the mixture critical point but be slightly higher [11, 14]. However, certain investigators [30, 46, 84] did not differentiate between the two in the case of air. Due to the limited body of information on the critical parameters of air, Table 1.6 lists experimental values and data from the best-known analytic studies. The table does not present the study by Chashkin with his coworkers

Table 1.6 Parameters of critical points of air

			Critical mixture point		Cricondentherm point	
Year	Author	Reference	ΔT, K	T, K	Δp, MPa	T, K
			Experiment			
1917	Kuenen and Clark	[78]	3.774	132.41	3.766	132.51
1954	Michels et al.	[84]	—	—	3.766	132.55
			Analytic			
1956	Din	[46]	3.774	132.43	3.766	132.52
1961	Baehr and Schwier	[30]	—	—	3.766	132.52
1966	Vasserman et al.	[7]	3.774	132.45	3.766	132.55

[24], which is erroneously mentioned by Hall [57] among experimental studies of critical parameters. He found the temperature T'_{cr} = 131.41 K, corresponding to maximum specific heat c_v of air containing 1.2% impurities, at the transcritical isochore.

1.3.3 Density at Phase Equilibrium

Very few measurements of the density of air at phase equilibrium have been performed, which is seen from Table 1.7. There is definitely no point in currently using values of the density of liquid air at normal boiling temperature obtained at the end of the nineteenth century [45, 80].

Data obtained by Michel with his coworkers [84] are basically fairly reliable, but the values of ρ' at temperatures below 124 K are rather on the low side, as noted by Vasserman et al. [7]. For this reason one may find it useful, in determining the density of liquid air at phase equilibrium, to employ the results of Blagoy and Rudenko [3], who obtained experimental data on the density of nitrogen–oxygen and argon–oxygen mixtures of different compositions over the range of 65 to 80 K. This study is not mentioned in Table 1.7, since the test substance was not air.

Results obtained by Blanke [34] at low temperature, where no experimental data were available previously, are of great interest. Unfortunately, the number of experimental points in [34] is small given the wide temperature range explored.

1.3.4 Calorific Properties: Heat of Vaporization

The specific heat c_s over the range of 80–120 K at liquid–vapor phase equilibrium was measured only by Eucken and Hauck [51], who presented five smoothed values obtained from workup of test points.

The heat of vaporization of air depends on the manner in which it is vapor-

Table 1.7 Density of saturated air

Year	Author	Reference	Temperature range ΔT, K	Number of points	Phase measured*
1917	Kuenen and Clark	[78]	123.6–132.5	11	ρ'
			129.8–132.5	8	ρ''
1954	Michels et al.	[84]	118.2–131.9	10	ρ'
			118.2–132.6	12	ρ''
1973	Blanke	[34]	60.4–128.7	9	ρ'
			67.2–132.3	11	ρ''

*ρ', Liquid at the bubble point line; ρ'', vapor at the dew point line.

ized and is defined by the majority of investigators as the quantity of heat needed for converting 1 kg of liquid air into vapor at constant pressure. This definition corresponds to the integral heat of vaporization r of a mixture [14]. The heat of vaporization was measured by Behn [32], Shearer [101], Fenner and Richtmyer [53], Witt [112], and Dana [43]. As a rule, the result of each such study is the average value of the heat of vaporization at atmospheric pressure, obtained on the basis of a number of measurements. In all of them the heat of vaporization was obtained at atmospheric pressure. Note that these studies were performed long ago and are not too accurate. The most detailed results are presented by Fenner and Richtmyer [53], who gave values of r as a function of the oxygen concentration in the mixture. They arrived at the erroneous conclusion that r depends little on the oxygen content and list the mean of all the measurements as the heat of vaporization of air. Their data also exhibit a significant scatter.

The scarcity of data on the heat of vaporization of air makes it necessary to use Dana's results [43] obtained in investigating nitrogen–oxygen mixtures at atmospheric pressure and then represented as values of r versus the mixture composition. According to data in [43], the heat of vaporization rises with the oxygen content from 199.7 kJ/kg without oxygen to 214.3 kJ/kg at an 80–85% oxygen content and then changes little. This behavior of the heat of vaporization of a N_2–O_2 mixture as a function of composition appears more reliable than the virtual absence of such a relationship claimed by Fenner and Richtmyer [53].

The various computational procedures for determining the heat of vaporization, for example, from the enthalpy–composition diagram, curve of $r/(p_{cr} v_{cr})$ versus τ_s and the modified Clausius–Clapeyron equation, are analyzed in sufficient detail in the book by Vasserman et al. [7].

CHAPTER

TWO

PROCEDURE FOR COMPILING TABLES OF THERMODYNAMIC PROPERTIES OF AIR

The experimental pvT data for air analyzed in Chapter 1 can be used for compiling a thermal equation of state suitable for calculating the thermodynamic properties for the range of parameters of practical importance. The procedure for compiling the equation of state of a real gas on the basis of experimental data is described by the authors in detail in a previous book [21]. For this reason we present in this chapter only the basic tenets of the procedure, with particular attention paid only to specifics of the behavior of air as a mixture of gases in compiling the equation of state and in calculating the tables of thermodynamic properties.

2.1 METHOD OF COMPILING THE EQUATION OF STATE

As in the case of nitrogen [21] we compiled a unified equation of state for gaseous and liquid air, since the advantages of such an equation over local equations of state are quite significant. The experimental data were described by an equation having the form

$$Z = 1 + \sum_{i=1}^{r} \sum_{j=0}^{s_i} b_{ij}\omega^i/\tau^j, \tag{2.1}$$

where $Z = pv/RT$; $\tau = T/T_{cr}$ is the reduced temperature and $\omega = \rho/\rho_{cr}$ is the reduced pressure.

Equation (2.1) can be treated as a part of a virial expansion. This form has not been rigorously validated for use in the unified equation, but it is successfully used in practice.

Determination of constants of the empirical equation of state from experimental data on thermodynamic properties of a real gas reduces to using the generalized method of least squares. The functional being minimized, in the case when the data being worked up are diverse (thermal, calorific, acoustic) and in the presence of additional conditions (critical conditions, the Planck–Gibbs rule), is

$$S = \sum_{q=1}^{Q} \sum_{k_q=1}^{n_q} w_{k_q} (x_{qk_q} - x_q(\omega_{k_q}, \tau_{k_q}, \vec{b}))^2 + \sum_{k_p=1}^{n_p} w_{k_p} (p(\omega''_{k_p}, \tau_{s_{k_p}}, \vec{b}) - p(\omega'_{k_p}, \tau_{s_{k_p}}, \vec{b}))^2 + \sum_{k_\Phi=1}^{n_\Phi} w_{k_\Phi} (\Phi(\omega''_{k_\Phi}, \tau_{s_{k_\Phi}}, \vec{b}) - \Phi(\omega'_{k_\Phi}, \tau_{s_{k_\Phi}}, \vec{b}))^2 + \sum_{l=1}^{m} \lambda_l \varphi_l (\omega_l, \tau_l, \vec{b}) \tag{2.2}$$

where $X_q \in \{Z, h, c_v, c_p, r, \delta, \mu, w, B_1, B_2, \ldots\}$, w_{k_q}, w_{k_p}, and w_{k_Φ} are the weights of the pertinent data and λ_l are Lagrange multipliers.

Minimization of function (2.2) yields a complex system of nonlinear equations, for which reason we have previously [21] used a functional which, of all the calorific variables, contains only the values of enthalpy, specific heat at constant volume, and heat of vaporization. This made it possible to define the coefficients of equation of state (2.1) by means of a system of linear equations. For air the functional being minimized can be simplified even more since information on enthalpy c_v and heat of vaporization of air are highly limited. In addition, as shown by a large number of calculations, the enthalpy is determined reliably by means of an equation compiled on the basis of pvT data.

The thermodynamic properties of saturated air, treated in the first approximation as a binary gas mixture, differ from those of a pure substance. For this reason, when compiling the equation of state of air from experimental data, it is not necessary to ensure equality of the saturated-vapor and boiling-liquid pressures along the isotherms, and the term responsible for satisfaction of the Maxwell rule (i.e., equality of the isobaric–isothermal potentials of the coexisting phases), should be a reduction of information.

Subcritical isotherms, calculated from empirical equations of state do not conform with real isotherms for the range in which the specific volume changes from v' to v'' (Fig. 3). For this reason in order to satisfy the equality

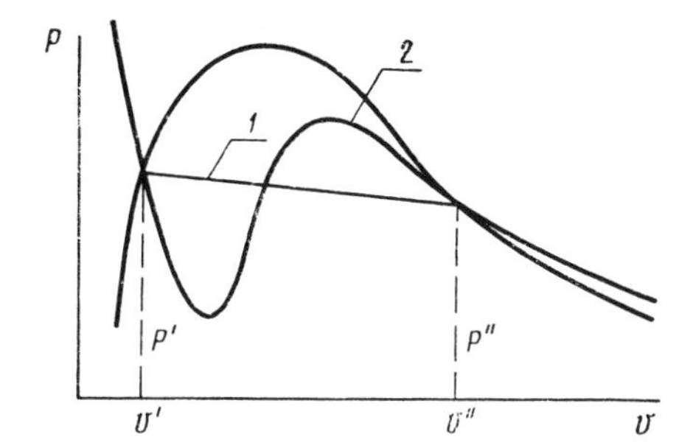

Figure 3 Experimental (1) and analytic (2) subcritical isotherms of a binary mixture.

$$F' = F'' - \int_{v''}^{v'} p dv \tag{2.3}$$

along isotherms in compiling the unified equation of state for a mixture, one must satisfy the condition

$$\int_{v'}^{v''} p_{\text{calc}}\, dv = \int_{v'}^{v''} p_{\text{exp}}\, dv \tag{2.4}$$

analogous to the Maxwell rule for pure substances. Satisfaction of Eq. (2.4) will ensure thermodynamic agreement between the predicted values of calorific quantities of the liquid and gaseous phases.

The curve shape of real isotherms of mixtures in the two-phase region and scarcity of *pvT* data in this range make it difficult to employ Maxwell's rule in compiling equations of state. However, for many mixtures whose components have rather close thermodynamic properties, for example, for the nitrogen–oxygen system [79] and for air [84], the experimental isotherms in the two-phase region are virtually rectilinear in *pv* coordinates. In this case, condition (2.4) can be written as

$$\int_{v'}^{v''} p dv = \frac{p'' + p'}{2}(v'' - v') \tag{2.5}$$

Consequently, when compiling the unified equation of state for air one must include, in the functional being minimized, the term

$$\sum_{k_\Phi=1}^{n_\Phi} w_{k_\Phi} \left[\frac{p'' + p'}{2} \left(\frac{Z''}{p''} - \frac{Z'}{p'} \right) - \int_{\omega''_{k_\Phi}}^{\omega'_{k_\Phi}} \frac{Z}{\omega} d\omega \right]^2 \tag{2.6}$$

In determining the quantities contained in Eq. (2.6), it is desirable to select the same temperatures at which one uses the *pvT* data at the saturation curve for forming the first term of functional (2.2). This aids in better satisfaction of the Maxwell rule and more exact description of the saturation curve.

The terms of functional (2.2) ensuring satisfaction of critical conditions must also be modified when compiling the equation of state of a binary mixture. The critical isotherm of a binary mixture is tangent to the saturation curve at the critical contact point [14] and has an inflection, but the tangent to the isotherm at that point in pv coordinates is not horizontal, i.e., for a mixture

$$\left(\frac{\partial p}{\partial v}\right)^{cr} < 0; \quad \left(\frac{\partial^2 p}{\partial v^2}\right)_T^{cr} = 0 \tag{2.7}$$

In the reduced coordinates Z, ω, ϑ and $\pi = p/p_{cr}$ we obtain the following expressions for the partial derivatives of the compressibility coefficient with respect to density

$$\left(\frac{\partial Z}{\partial \omega}\right)_\vartheta = -\frac{Z}{\omega}\left[1 - \frac{\omega}{\pi}\left(\frac{\partial \pi}{\partial \omega}\right)_\vartheta\right]; \quad \left(\frac{\partial^2 Z}{\partial \omega^2}\right)_\vartheta = -\frac{2}{\omega}\left(\frac{\partial Z}{\partial \omega}\right)_\vartheta + \frac{Z}{\pi}\left(\frac{\partial^2 \pi}{\partial \omega^2}\right)_\vartheta \tag{2.8}$$

When the parameters of the critical point of contact (cricondentherm) are selected as the normalizing parameters, we find

$$\left(\frac{\partial Z}{\partial \omega}\right)_\vartheta^{cr} = -Z_{\text{кр}}\left[1 - \left(\frac{\partial \pi}{\partial \omega}\right)_\vartheta^{cr}\right]$$
$$\left(\frac{\partial^2 Z}{\partial \omega^2}\right)_\vartheta^{cr} = -2\left(\frac{\partial Z}{\partial \omega}\right)_\vartheta^{cr} \tag{2.9}$$

We thus see that in compiling the equation of state for a mixture the pertinent components of function (2.2) are obtained from Eqs. (2.9).

The derivative $(\partial\pi/\partial\omega)_\vartheta^{cr}$ can be estimated on the basis of experimental data for the thermal properties of the mixture at supercritical temperatures or from data at saturation. The inevitable extrapolation of $(\partial\pi/\partial\omega)_\vartheta^{\omega=1} = F(T)$ or $(\Delta\pi/\Delta\omega)_\vartheta^S = f(T)$ to $T = T_{cr}$ reduces the accuracy of definition of derivative $(\partial\pi/\partial\omega)_\vartheta^{cr}$. For this reason the value of the derivative thus found must be corrected in calculations, since it has a significant effect on the accuracy of approximation of data in the critical region.

Since a mixture does not obey the Planck–Gibbs rule, the corresponding terms in functional (2.2) are dropped. In analyzing the question of including data on the heat of vaporization of mixtures as we compile the equation of state, we note the thermal and calorific quantities in phase transitions in binary systems are related by general equations of phase transfer [14]. Here the Clausius–

Clapeyron relation for a pure substance is a particular case. But since the general equations contain differential heats of phase transition at constant p and T, and the experimental data were obtained primarily on the integral heat of vaporization of mixtures r_p at constant pressure, it is best to establish a relationship between r_p and the thermal properties.

The values of T' and T'' obtained at the same pressure do not differ significantly in mixtures made up of constituents with similar properties, and the isobars in the T, s coordinates are close to straight lines (Fig. 4). Then the integral heat of vaporization r_p can be defined as

$$r_p = \frac{T'' + T'}{2}(s'' - s')_p \tag{2.10}$$

Since the absolute values of derivatives ds''/dT and ds'/dT along an isobar are fairly close (not at very low pressures), we replace difference $(s'' - s')_p$ by its analogous difference along the isotherm $T_m = (T'' + T')/2$ and using the thermodynamic relationship $\partial s/\partial v_T = (\partial p/\partial T)_v$, we find

$$r_p = T_m (s'' - s')_{T_m} = T_m \int_{v'}^{v''} \left(\frac{\partial p}{\partial T}\right)_v dv \tag{2.11}$$

According to Eq. (2.11), one may use experimental data on heat of vaporization r_p in determining the coefficients of the equation of state. When such data are not available, we write, for the mixtures under study on the basis of Eq. (2.5), the expression

$$\int_{v'}^{v''} \left(\frac{\partial p}{\partial T}\right)_v dv = \frac{1}{2}\left(\frac{dp''}{dT} + \frac{dp'}{dT}\right)(v'' - v') + \frac{p' - p''}{2}\left(\frac{dv''}{dT} + \frac{dv'}{dT}\right) \tag{2.12}$$

This makes it possible, in principle, to use thermal data at saturation along isotherm T_m instead of the value of r_p. However, the accuracy in determining dv''/dT and dv'/dT is poor and the second term in the righthand side of Eq.

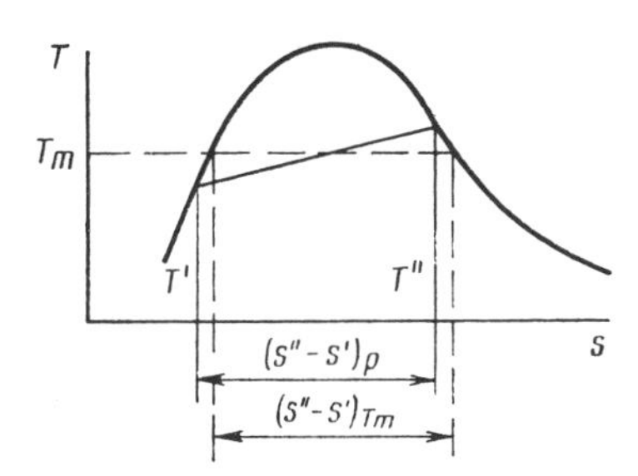

Figure 4 Isobar of a binary mixture made of components with similar properties.

(2.12) at low temperatures is comparable to the first. Hence it is best to calculate the integral heat of vaporization r_p from an equation analogous to the Clausius–Clapeyron relation, on the basis of average values of temperature and of the derivative dp_s/dT along the isobar

$$r_p = \frac{1}{2} T_m \left(\frac{dp''}{dT} + \frac{dp'}{dT}\right)_p (v'' - v')_p \tag{2.13}$$

As shown previously [7], the use of average quantities in Eq. (2.13) is permissible, since the values not only of T'' and T' but also of dp''/dT and dp'/dT along isobars are fairly close. This means that, lacking experimental data on the heat of vaporization, one can include in functional (2.2) the term

$$\sum_{l=1}^{L} w_l \left[\frac{v_{\text{cr}}}{R}\left(\frac{dp''}{dT} + \frac{dp'}{dT}\right)\left(\frac{1}{\omega_l''} - \frac{1}{\omega_l'}\right) - \int_{\bar{\omega}_l''}^{\bar{\omega}_l'} \left(Z - v\left(\frac{dZ}{\partial v}\right)_\omega\right)\frac{d\omega}{\omega}\right]^2 \tag{2.14}$$

in which the integral is taken along isotherm T_m.

These transformations of terms of the functional being minimized are of fundamental importance for making allowance for specifics in the thermodynamic behavior of mixtures in compiling the equations of state. The need for and the feasibility of incorporation of a given term of the functional is governed by the availability of pertinent data and their accuracy. In particular, for air we made no attempt to satisfy the critical conditions, since data on the parameters of critical points are insufficiently reliable, and an insignificant change in T_{cr} brings about a significant change in ρ_{cr} [2]. Most likely, due to the poor agreement between T_{cr} and ρ_{cr}, the satisfaction of the critical point and critical conditions by means of Lagrange multipliers, as noted by many investigators, reduces the accuracy of analytic description of pvT data. Due to lack of experimental data on the heat of vaporization at temperatures differing from the normal boiling temperature, and the poor accuracy of data on the pressure of condensation and boiling of air, the terms of Eqs. (2.11) or (2.14) were not included in the functional. At the same time, in compiling the equation of state for air we ensured satisfaction of condition (2.4), since it is of importance in calculations from the unified equation of state for the gas and liquid.

Transforming the functional being minimized in an appropriate manner, one can compile the unified equation of state of gaseous and liquid air by a procedure developed for pure substances [21]. As previously, the compilation of the equation of state is made up of three stages:

1. Obtaining an approximate equation of state and using it for calculating the weights of experimental data;

2. compiling an auxiliary equation with consideration of the weights and correction of the latter;
3. compiling the main equation of state and comparing the analytic and experimental values.

As before, the weights of experimental values of the compressibility coefficient are represented by quantities $1/\Delta Z^2$ where

$$\Delta Z^2 = [Z(\delta p + \delta\rho + \delta T)]^2 + \left[T\delta T\left(\frac{\partial Z}{\partial T}\right)_\rho\right]^2 + \left[\rho\delta\rho\left(\frac{\partial Z}{\partial \rho}\right)_T\right]^2 \quad (2.15)$$

δp, $\delta\rho$, and δT are the relative errors of the thermal quantities. The partial derivatives of the compressibility coefficient with respect to temperature and density are calculated from the approximate equation of state.

Rigorously speaking, in this case there are errors both in function $Z(\rho, T)$ being approximated and in the independent variables ρ and T. The determination of coefficients of the regression equation for such a case is a difficult problem which has not been satisfactorily solved up to now [23]. Our artificial scheme for referring all the errors to function Z using Eq. (2.15) makes it possible to simplify the problem while still retaining an acceptable accuracy of the approximation.

The values of weights for a component of the functional which ensures the satisfaction of condition (2.6) were calculated from the expression

$$w_{k_\Phi} = \frac{1}{\left[\frac{p'' + p'}{2}\left(\frac{Z''}{\rho''} - \frac{Z'}{\rho'}\right)\right]^2 (\delta p_s)^2} \quad (2.16)$$

where δp_s is the specified value of the relative error of saturation pressure. After the equation of state is compiled, the values of $\int_{v'}^{v''} p dv$ and $(p'' + p')/2 \times (v'' - v')$ are compared on the basis of values of v' and v'' calculated from this equation for experimental values of p' and p''.

This technique for incorporating features of thermodynamic behavior of binary mixtures has been checked out by Vasserman [5] in the case of air for an equation of state represented by elementary functions. The fact that the approach to incorporation of these features is independent of the form of the equation makes it possible to use the above schemes also in the present case. The method of compilation of the equation of state in virial form is suitable for many individual substances. This means that the entire set of algorithms and programs used for implementing the above principal methodological tenets has been verified and can be used for compiling a unified equation of state of gaseous and liquid air.

2.2 METHOD OF CALCULATION OF TABLES OF THERMODYNAMIC PROPERTIES

The thermodynamic properties of air, which is a constant-composition mixture, were calculated in accordance with a method developed by the present authors and described in detail in their previous monograph [21]. The procedure of computations for single-phase air does not differ from the analogous procedure for individual substances. The calculations are based on the thermal equation of state, averaged over an ensemble of equations, which are equivalent from the point of view of accuracy of approximation of the starting *pvT* data

$$Z = 1 + \sum_{i=1}^{r} \sum_{j=0}^{S_i} b_{ij} \frac{\omega^i}{\tau^j}$$

For convenience of programming we, as before [21], introduce similarly structured quantities A_0, A_1, A_2, A_3, A_4, and A_5:

$$\left.\begin{aligned}
A_0 &= \sum_{i=1}^{r} \sum_{j=0}^{S_i} b_{ij}\omega^i/\tau^j \\
A_1 &= \sum_{i=1}^{r} \sum_{j=0}^{S_i} (i+1)\, b_{ij}\, \omega^i/\tau^j \\
A_2 &= -\sum_{i=1}^{r} \sum_{j=0}^{S_i} (j-1)\, b_{ij}\omega^i/\tau^j \\
A_3 &= \sum_{i=1}^{r} \sum_{j=0}^{S_i} \frac{(i+j)}{i}\, b_{ij}\omega^i/\tau^j \\
A_4 &= \sum_{i=1}^{r} \sum_{j=0}^{S_i} \frac{(j-1)}{i}\, b_{ij}\omega^i/\tau^j \\
A_5 &= -\sum_{i=1}^{r} \sum_{j=0}^{S_i} \frac{j\,(j-1)}{i}\, b_{ij}\omega^i/\tau^j
\end{aligned}\right\} \tag{2.17}$$

Using the above quantities, the computational equations will have the following form:

compressibility

$$Z = 1 + A_0 \tag{2.18}$$

enthalpy

$$h/RT = h_0/RT + A_3 \tag{2.19}$$

entropy

$$s/R = s_0/R - \ln(\omega/\omega_0) + A_4 \tag{2.20}$$

internal energy

$$u/RT = h/RT - Z \tag{2.21}$$

Helmholtz function

$$F/RT = u/RT - s/R \tag{2.22}$$

Gibbs function

$$\Phi/RT = h/RT - s/R \tag{2.23}$$

specific heat at constant volume (isochoric)

$$c_v/R = c_{v_0}/R + A_5 \tag{2.24}$$

specific heat at constant pressure (isobaric)

$$c_p/R = c_v/R + (1 + A_2)^2/(1 + A_1) \tag{2.25}$$

speed of sound

$$w/w_0 = \sqrt{1 + A_1} \tag{2.26}$$

isothermal Joule–Thomson effect

$$\delta/\delta_0 = (A_2 - A_1)/(1 + A_1) \tag{2.27}$$

adiabatic Joule–Thomson effect

$$\mu/\mu_0 = (A_2 - A_1)/(1 + A_1) \tag{2.28}$$

volumetric expansion coefficient

$$\alpha/\alpha_0 = (1 + A_2)/(1 + A_1) \tag{2.29}$$

isothermal compression coefficient

$$\beta/\beta_0 = (1 + A_0)/(1 + A_1) \tag{2.30}$$

thermal pressure coefficient

$$\gamma/\gamma_0 = (1 + A_2)/(1 + A_0) \tag{2.31}$$

adiabatic index

$$k/k_0 = (1 + A_1)/(1 + A_0) \tag{2.32}$$

volatility (fugacity)

$$f/f_0 = \exp\,(A_3 - A_4) \tag{2.33}$$

Here h_0/RT, s_0/RT, c_{v0}/R are the reduced enthalpy, entropy, and isochoric specific heat in the ideal gas, respectively; R is the gas constant, T is the temperature; ω_0, w_0, δ_0, μ_0, α_0, β_0, γ_0, k_0, and f_0 are thermodynamic normalizing functions, calculated from the expressions

$$\left.\begin{aligned} \omega_0 &= p_{st}/(\rho_{cr}RT) \\ w_0 &= \sqrt{RTc_p/c_v} \\ \delta_0 &= 1/\rho \\ \mu_0 &= 1/\rho c_p \\ \alpha_0 &= 1/T \\ \beta_0 &= 1/p \\ \gamma_0 &= 1/T \\ k_0 &= c_p/c_v \\ f_0 &= \rho RT \end{aligned}\right\} \tag{2.34}$$

at $p_{st} = 0.101325$ MPa.

The algorithm for calculating the thermodynamic functions involves initial determination of density ω at known π and τ from the equation

$$\pi - \frac{\omega\tau}{Z_{\text{кр}}}\left(1 + \sum_{i=1}^{r}\sum_{j=0}^{S_i} b_{ij}\omega^i/\tau^j\right) = 0 \tag{2.35}$$

The "step-by-step" bipartition method is used for finding the roots. The calculated values of density ω together with the specified temperature τ are used for calculating A_0, A_1, A_2, A_3, A_4, and A_5. This is followed by calculating all the necessary thermodynamic functions using the formulas presented above.

The calculation of thermodynamic functions of air along the boiling and condensation curves differs fundamentally from analogous calculation of properties for individual substances.

In the case of individual substances the location of the saturation curve in the pvT space is determined in accordance with the method we developed from the thermal equation of state upon imposition of the phase-transition conditions ($T' = T''$, $p' = p''$, $\Phi' = \Phi''$). However, the implementation of such a procedure for air, which is a constant-composition mixture, is much more difficult. For this reason we used a scheme for calculating the thermodynamic properties along the boiling and condensation curves which includes, in addition to the thermal equation of state, two equations for the vapor pressure curves. In the present study these equations were used in the form

$$\lg \pi_s' = \frac{a_{-1}}{\tau_s} + a_0 + a_1\tau_s + a_2\tau_s^2 + a_3\tau_s^3 \tag{2.36}$$

$$\lg \pi''_s = \frac{b_{-1}}{\tau_s} + b_0 + b_1\tau_s + b_2\tau_s^2 + b_3\tau_s^3 \tag{2.37}$$

The numerical values of coefficients in Eqs. (2.36) and (2.37) were found by independent workup of experimental data by the method of least squares.

With consideration of the above, the algorithm for calculating the properties of air at the saturation curve is composed of two stages. At the first stage the set of values of $\{p'\}$, $\{p''\}$ or $\{T'\}$, $\{T''\}$ is calculated for the specified values of temperature or pressure from Eqs. (2.36) or (2.37) respectively. The second stage of the computational procedure reduces to formal application of Eqs. (2.17)–(2.34) according to the scheme used for the single-phase region.

Quantities r, c_s', c_s'', $d\pi_s'/d\tau_s$, $d\pi_s''/d\tau_s$, $d^2\pi_s'/d\tau_s^2$, and $d^2\pi_s''/d\tau_s^2$ were calculated from the equations that follow:

heat of vaporization

$$r = h'' - h'; \tag{2.38}$$

the derivatives

$$\frac{d\pi'_s}{d\tau_s} = \pi'_s\left(-\frac{a_{-1}}{\tau_s^2} + a_1 + 2a_2\tau_s + 3a_3\tau_s^2\right) \cdot \ln 10 \tag{2.39}$$

$$\frac{d\pi''_s}{d\tau_s} = \pi''_s\left(-\frac{b_{-1}}{\tau_s^2} + b_1 + 2b_2\tau_s + 3b_3\tau_s^2\right) \cdot \ln 10 \tag{2.40}$$

$$\frac{d^2\pi'_s}{d\tau^2_s} = \frac{d}{d\tau_s}\left(\frac{d\pi''_s}{d\tau_s}\right) \tag{2.41}$$

$$\frac{d^2\pi''_s}{d\tau^2_s} = \frac{d}{d\tau_s}\left(\frac{d\pi''_s}{d\tau_s}\right) \tag{2.42}$$

specific heat of boiling liquid

$$\frac{c'_s}{R} = \frac{c'_p}{R} - \frac{\alpha'}{\alpha_0}\frac{Z_{cr}}{\omega'}\frac{d\pi'_s}{d\tau_s} \tag{2.43}$$

specific heat of saturated vapor

$$\frac{c''_s}{R} = \frac{c''_p}{R} - \frac{\alpha''}{\alpha_0}\frac{Z_{cr}}{\omega''}\frac{d\pi''_s}{d\tau_s} \tag{2.44}$$

CHAPTER

THREE

THE EQUATIONS OF STATE AND THERMODYNAMIC TABLES OF AIR

The analytic tools for calculating the thermal, calorific, and acoustic properties of single-phase air and of air at phase-equilibrium curves include the thermal equation of state, analytic temperature dependence of the isobaric ideal gas specific heat, and two independent equations for the vapor pressure curves. The methodological problems of construction of the thermal equation of state from experimental data and the scheme for calculating the thermodynamic properties were analyzed in Chapter 2. In this chapter we present a quantitative description of the pertinent equations, list numerical values of the coefficients of approximations, and analyze the results of comparison of the calculated values of thermodynamic properties with experimental data. In addition, we present information which validates the selection of tolerances for the tabulated values of thermodynamic quantities. This information allows determination of the reliability of the tabulated data. In the last section we present a comparative analysis of previously published tables of thermodynamic properties of air.

3.1 THERMODYNAMIC FUNCTIONS IN THE IDEAL GAS STATE

The thermodynamic functions of air in the ideal gas state can be calculated on the basis of the corresponding data for individual components [22, 104, 105]. In this case, air is treated as a constant-composition mixture of ideal gases. However, the composition of air differs, depending on the data of investigators. The degree of deviation of compositions according to different publications can be determined from Table 3.1 for consistency in this study we assumed that the composition of air is that given in [7], where it is assumed that the air does not contain CO_2 and consists of 78.11% N_2, 20.96% O_2, and 0.93% Ar by volume.

Table 3.2 lists values of the ideal gas isobaric specific heat as a function of temperature. They were calculated for the composition assumed in this study and compared to values taken from the publication by Hilsenrath et al. [104]. The differences between the two groups of values of isobaric specific heat can be regarded as fully acceptable. The data in [104] are in satisfactory agreement with the results in the book by Glushko [22], but unlike them, they pertain to the low temperature range. For this reason they are used in the present study. As noted in [21], the data in [105] are less reliable.

For convenience of computer calculation, the tabulated values of c_p^0/R extracted from [104], were approximated by a generalized power-law polynomial in temperature

$$\frac{c^0{}_p}{R} = \sum_{j=0}^{6} \alpha_j \tau^j + \sum_{j=1}^{6} \beta_j \tau^{-j} \tag{3.1}$$

The standard deviation of the error in reproducing the original data on $c_p{}^0/R$ at temperatures from 50 to 2000 K is 0.009%, the maximum being ~0.02%. The numerical values of the coefficients of generalized polynomial (3.1) are the following:

$$\alpha_0 = 0{,}661738 \cdot 10^1$$

$$\alpha_1 = -0{,}105885 \cdot 10^1 \qquad \beta_1 = -0{,}549169 \cdot 10^1$$

$$\alpha_2 = 0{,}201650 \cdot 10^0 \qquad \beta_2 = 0{,}585171 \cdot 10^1$$

$$\alpha_3 = -0{,}196930 \cdot 10^{-1} \qquad \beta_3 = -0{,}372865 \cdot 10^1$$

$$\alpha_4 = 0{,}106460 \cdot 10^{-2} \qquad \beta_4 = 0{,}133981 \cdot 10^1$$

$$\alpha_5 = -0{,}303284 \cdot 10^{-4} \qquad \beta_5 = -0{,}233758 \cdot 10^0$$

$$\alpha_6 = 0{,}355861 \cdot 10^{-6} \qquad \beta_6 = 0{,}125718 \cdot 10^{-1}$$

The enthalpy in the ideal gas state was calculated from the expression

$$h_0 = \int_{T_0}^{T} c^0{}_p dT + h_{00} + h_0{}^0$$

where h_{00} is the enthalpy at temperature T_0; h_0^0 is the heat of sublimation at T = 0 K; also reduced enthalpy in terms of the isobaric specific heat constants is:

Table 3.1 Volumetric composition of air according to different investigators

Substance	Molecular weight	Composition according to [104], % by volume	Composition according to [15] % by volume	Composition according to [7] % by volume	Heat of sublimation according to [81], kJ/kg
N_2	28.0134	78.09	78.14	78.11	247.56
O_2	31.9988	20.95	20.90	20.96	275.54
Ar	39.948	0.93	0.93	0.93	195.55
CO_2	44.0100	0.03	0.03	—	596.00

$$\frac{h_0}{RT} = \sum_{j=0}^{m} \frac{\alpha_j}{j+1} \tau^j - \sum_{j=2}^{n} \frac{\beta_j}{j-1} \tau^{-j} + \frac{1}{\tau}(\beta_1 \ln \tau + \Delta_1 + \widetilde{h}_{00}) + \frac{h_0{}^0}{RT}$$

$$\Delta_1 = \sum_{j=2}^{n} \frac{\beta_j}{j-1} - \sum_{j=0}^{m} \frac{\alpha_j}{j+1} \tag{3.2}$$

$$\widetilde{h}_{00} = h_{00}/RT_0$$

The entropy in the ideal gas was calculated from the expression:

$$s_0 = \int_0^{T_0} \frac{c_p{}^0}{T} dT + s_{00} + s_0{}^0 \tag{3.2a}$$

where s_{00} is the entropy at temperature T_0; s_0^0 is some reference point constant (in this study $s_0^0 = 0$; also reduced entropy can be expressed in terms of the isobaric specific heat constants by

$$\frac{s_0}{R} = \sum_{j=1}^{m} \frac{\alpha_j}{j} \tau^j - \sum_{j=1}^{n} \frac{\beta_j}{j} \tau^{-j} + \alpha_0 \ln \tau + \Delta_2 + \widetilde{s}_{00} + \frac{s_0{}^0}{R}$$

$$\Delta_2 = \sum_{j=1}^{n} \frac{\beta_j}{j} - \sum_{j=1}^{m} \frac{\alpha_j}{j}, \quad \widetilde{s}_{00} = s_{00}/R \tag{3.3}$$

The value of heat of sublimation h_0^0 for air was calculated on the basis of data in Table 3.1 and is 253.4 kJ/kg.

The reference point temperature is $T_0 = 100$ K. The values of reduced

Table 3.2 Reduced ideal gas isobaric specific heat for air

T, K	$\left(\frac{Cp_0}{R}\right)_{N_2}$ [104]	$\left(\frac{Cp_0}{R}\right)_{O_2}$ [104]	$\left(\frac{Cp_0}{R}\right)_{Ar}$ [104]	$\left(\frac{Cp_0}{R}\right)^{*}_{Air}$	$\left(\frac{Cp_0}{R}\right)_{Air}$ [104]	$\left(\frac{Cp_0}{R}\right)^{**}_{Air}$
50	3.5003	3,5029	2.5	3,4915	3,4915	3,4915
60	3,5003	3.5023	2.5	3,4914	3,4914	3.4915
70	3,5003	3.5019	2.5	3.4913	3.4914	3.4914
80	3,5004	3,5016	2.5	3.4913	3,4913	3.4915
90	3,5004	3.5015	2.5	3,4913	3.4913	3,4916
100	3.5004	3,5014	2.5	3,4913	3,4913	3.4915
150	3,5006	3.5013	2,5	3.4914	3,4915	3,4913
200	3.5008	3,5032	2,5	3,4920	3.4922	3,4924
250	3.5013	3,5122	2,5	3,4943	3.4945	3,4945
300	3.5030	3.5344	2.5	3,5003	3,5005	3,5002
350	3,5078	3,5717	2.5	3,5118	3,5122	3,5119
400	3.5179	3.6212	2.5	3.5301	3,5305	3.5305
450	3.5344	3.6787	2.5	3.5550	3,5555	3,5558
500	3,5578	3.7396	2.5	3.5861	3,5865	3,5869
600	3.6214	3.8599	2.5	3,6610	3.6615	3.6615
700	3.6990	3.9672	2,5	3,7441	3.7447	3,7444
800	3.7806	4,0577	2.5	3.8268	3.8275	3,8274
900	3.8596	4.1327	2,5	3,9042	3.9049	3.9054
1000	3,9326	4,1948	2,5	3.9742	3,9750	3,9756
1100	3.9982	4,2469	2.5	4,0364	4,0371	4.0375
1200	4.0562	4,2912	2.5	4,0910	4,0917	4,0917
1300	4.1072	4.3300	2.5	4,1389	4,1398	4.1392
1400	4,1518	4,3651	2,5	4.1811	4,1820	4.1814
1500	4.1909	4.3975	2.5	4.2185	4.2193	4,2192

*Calculated for the composition assumed in this study.

**Approximated by polynomial (3.1).

enthalpy and entropy at this temperature are respectively [104] $h_{00} = h_{00}/RT = 3.48115$ and $s_{00} = s_{00}/R = 20.0824$.

3.2 EQUATIONS FOR CALCULATING THE THERMODYNAMIC PROPERTIES OF AIR

It was noted in Chapter 1 that reliable experimental pvT data are available for air over a wide temperature and pressure range whereas information on calorific properties is limited. For this reason the unified equation of state for gaseous and liquid air was compiled on the basis of experimental pvT data to conform to the requirement that $\int_{v'}^{v} p_{exp}\, dv = \int_{v'}^{v} p_{calc}\, dv$.

To obtain an average equation of state which was then used for calculating tables of thermodynamic functions, we compiled a system of 53 statistically equivalent equations on the basis of data listed in Table 3.3. The ranges of parameters for which data were obtained by various investigators are given in Fig. 5. Table 3.3 also lists values of errors $\delta\rho$, specified in calculating the weights of test points in six series of calculations and the standard deviations $\delta\rho_m$ of experimental values of density from those calculated from the averaged equation of state.

The data of the majority of investigators, including those used solely for comparison with calculated values of density (Table 3.4), were corrected so as to bring them into conformance with the IPTS-68 temperature scale. The temperatures in the recent studies [6, 34, 96] are already given for the new scale whereas the data of Kuenen and Clark [78] and Witkowski [110] cannot be corrected due to lack of information on calibration of the thermometers they used.

Unlike pvT data for the single-phase region, those on the thermal properties of saturated air are less reliable. This is apparently because air at saturation starts behaving as a mixture of gases, and changes in the liquid and vapor compositions may result in fluctuations in the values of density. For this reason, data on the properties of saturated air were brought into mutual agreement on coordinates ρ', T, and τ lg v'', τ. Figures 6 and 7 show, in these sets of

Table 3.3 Experimental data used in compiling equations of state of air, and standard deviations $\delta\rho_m$ for the averaged equation of state

Year	Author	Reference	Number of points	Error weights $\delta\rho$, %	$\delta\rho_m$, %
1915	Holborn and Schultze	[67]	42	0.10	0.03
1953	Michels et al.	[85]	157	0.05	0.02
1954	Michels et al.	[84]	197	0.05	0.14
1968	Vukalovich et al.	[10]	349	0.10	0.04
1971	Romberg	[96]	124	0.10	0.05
1973	Blanke	[34}	111	0.05–0.10	0.18
1976	Vasserman et al.	[6]	109	0.05–0.10	0.09
	Experimental values of [34, 84] and smoothed data on the density of saturated vapor		40	0.10	0.33
	Experimental values of [34, 84] and smoothed data on the density of boiling liquid		40	0.05	0.18
	For data in the single-phase region		1089		0.09
	For the entire body of data		1169		0.11

coordinates, experimental values of ρ' and v'', the analytic results of certain investigators, and the assumed reference curves, which are in agreement with the bulk of experimental data of Blanke and with data of Michels et al. [84] for saturated vapor. It is seen from Fig. 6 that the data in [84] on the density of boiling liquid at temperatures below 125 K deviate considerably on the low side, which was previously noted by Vasserman et al. [7]. This figure also illustrates the poor reliability of analytic values of Baehr and Schwier [30] on density ρ' and the high accuracy of analogous data by Vasserman and Rabinovich [9], which were obtained before the publication of Blanke's experimental data [34].

The data for the segment of the saturation curve adjoining the critical region were additionally fit to the set of coordinates lg v, $\sqrt[3]{1 - \tau}$ (Fig. 8). In these coordinates the segment of the saturation curve under study has a small

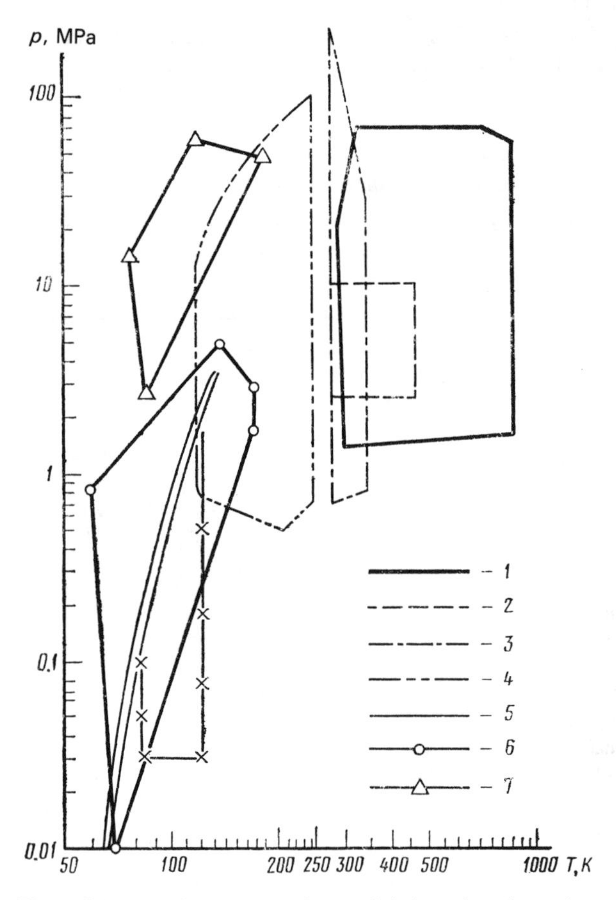

Figure 5 Ranges of parameters of state of air investigated experimentally by (1) Kozlov [15]; (2) Holborn and Schultze [67]; (3) Michels et al. [85]; (4) Michels et al. [84]; (5) Romberg [96]; (6) Blanke [34]; (7) Vasserman et al. [6].

Table 3.4 Values of $\delta\rho_m$ for data used for comparison with the analytically obtained values of density

Year	Author	Reference	Number of points	$\delta\rho_m$, %
		Single-phase region (experimental data)		
1893	Amagat	[27]	152	0.32
1896	Witkowski	[110]	169	0.63
1908	Koch	[77]	45	0.41
1923	Penning	[91]	60	0.18
		Single-phase region (smoothed data)		
1915	Holborn and Schultze	[67]	32	0.06
1924	Holborn and Otto	[65]	48	0.02
1925	Holborn and Otto	[66]	45	0.03
1953	Michels et al.	[85]	43	0.03
1954	Michels et al.	[84]	117	0.09
1968	Vasserman and Rabinovich*	[9]	157	0.31
		Saturation curve (experimental data)		
1917	Kuenen and Clark	[78]	19	6.9
1954	Michels et al.	[84]	18	2.6
1973	Blanke	[34]	19	0.58

*Analytic data on the density of liquid at T = 75–135 K and p = 2–50 MPa.

curvature which facilitates the correlation of data. By postulating a smooth matching of curves of $\lg v' = f(\sqrt[3]{1-\tau})$ and $\lg v'' = F(\sqrt[3]{1-\tau})$, it is possible not only to select the most reliable experimental data but also to refine the value of v_{cr} corresponding to the selected critical temperature. Figure 8 points to scatter of experimental data in the critical region, in particular, data of Kuenen and Clark [78].

Finally, the body of data used for compiling the equations of state also contained 40 values each of v'' and v' at temperatures between 65 and 131.12 K (p'' = 0.0077 to 3.505 MPa, p' = 0.0159 to 3.598 MPa). This was done using a part of the experimental data in [34, 84], and a part of values of v'' and v' were taken from reference curves (see Figs. 6 and 7), which provide optimum correlation of experimental data. Due to insufficient correlation of experimental data on boiling and condensation pressures obtained by different investigators, the values of p' and p'' corresponding to selected values of specific volumes were calculated from the equations we compiled, which are analyzed below.

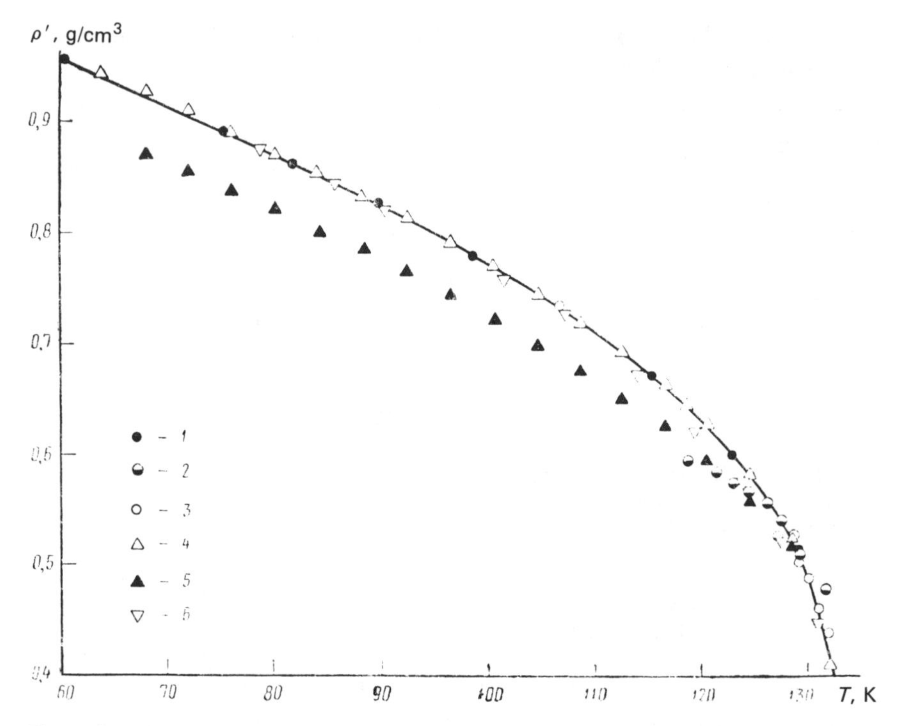

Figure 6 Reference values of density of boiling liquid (solid curve). Experimental data of (1) Blanke [34]; (2) Michels et al. [84]; (3) Kuenen and Clark [78]. Analytic values of (4) Vasserman and Rabinovich [9]; (5) Baehr and Schwier [30]; (6) Din [46].

To satisfy the fundamental condition $\int_{v'}^{v} p_{\text{exp}}\, dv = \int_{v'}^{v} p_{\text{calc}}\, dv$, we included in our data values of p', p'', v' and v'' at 25 temperatures over the range from 65 to 131.12 K. We have selected values of temperature at approximately the same intervals from one another, basically corresponding to experimental values of v' or v''.

The values of $\delta\rho$ in calculating the weights of data points were selected from an estimate of errors of experimental data given by experimenters and in certain analytic studies. Since the experimental data on the thermal properties of air obtained by various investigators and used for compiling the equations are in quite satisfactory agreement with one another, the value of $\delta\rho$ in different series of calculations did not change significantly. The specified values of the permissible error of saturation pressure also ranged within narrow limits (0.4 to 0.5%).

In each series of calculations, after obtaining the approximate equation, using it for calculating weights of data points, and assigning zero values to the points that dropped out, we compiled a series of equations with a different number of coefficients. Large deviations were observed in all series of computations for two points from [84] and seven points from [34]. These were automatically assigned zero weight and were not used in determining the coefficients of equations. These points are not listed in Table 3.3 and are not represented on the graphs of deviations.

The number of equations in different series ranged from 8 to 14, and the number of coefficients ranged from 44 to 48. In addition, the distribution of coefficients among the temperature functions changes. All these equations were for density to the $i = 1\text{–}8$ power with respect to frequency. This can be seen in the averaged equation of state based on Z computation. The power was determined by statistical analysis of data on isotherms extending over the greatest density range by means of Fisher's test.

The values of rms (standard) deviations $\delta\rho_m$ of experimental data obtained by various investigators from the analytic for the 53 equations we compiled are stable. This fact can be treated as evidence of statistical equivalence of the equations. For example, the values of $\delta\rho_m$ for data in [84, 85] at $T \geq 0\,°\text{C}$ lie between 0.02 and 0.04%, and at $T < 0\,°\text{C}$ lie between 0.11 and 0.17%; for

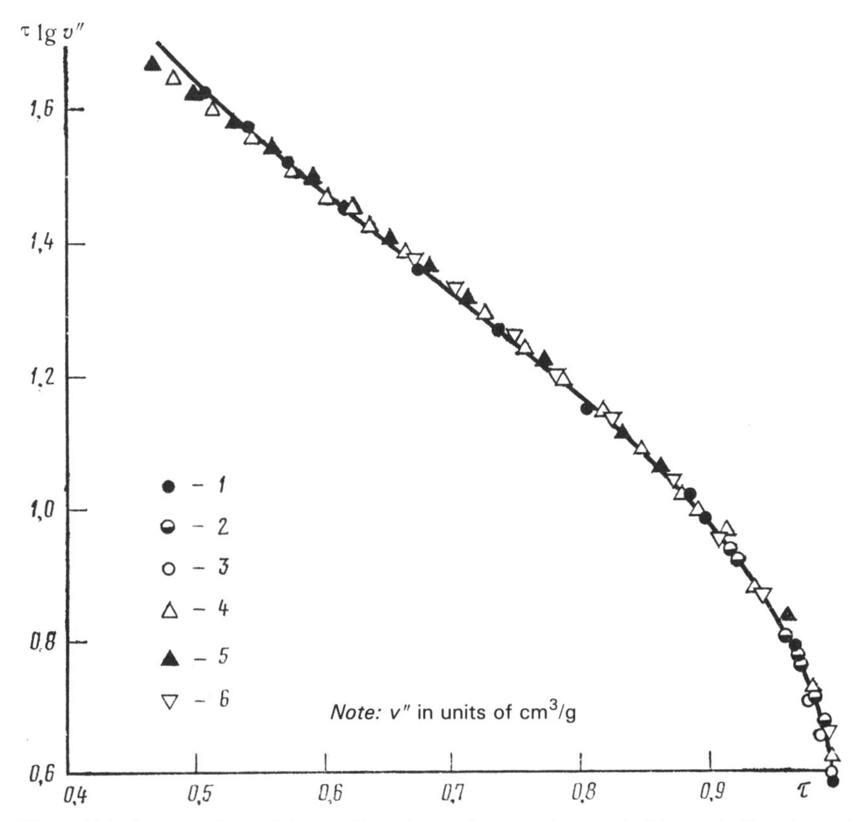

Figure 7 Reference values of the specific volume of saturated vapor (solid curve). Experimental data of (1) Blanke [34]; (2) Michels et al. [84]; (3) Kuenen and Clark [78]. Analytic values of (4) Vasserman et al. [7]; (5) Baehr and Schwier [30]; (6) Din [46].

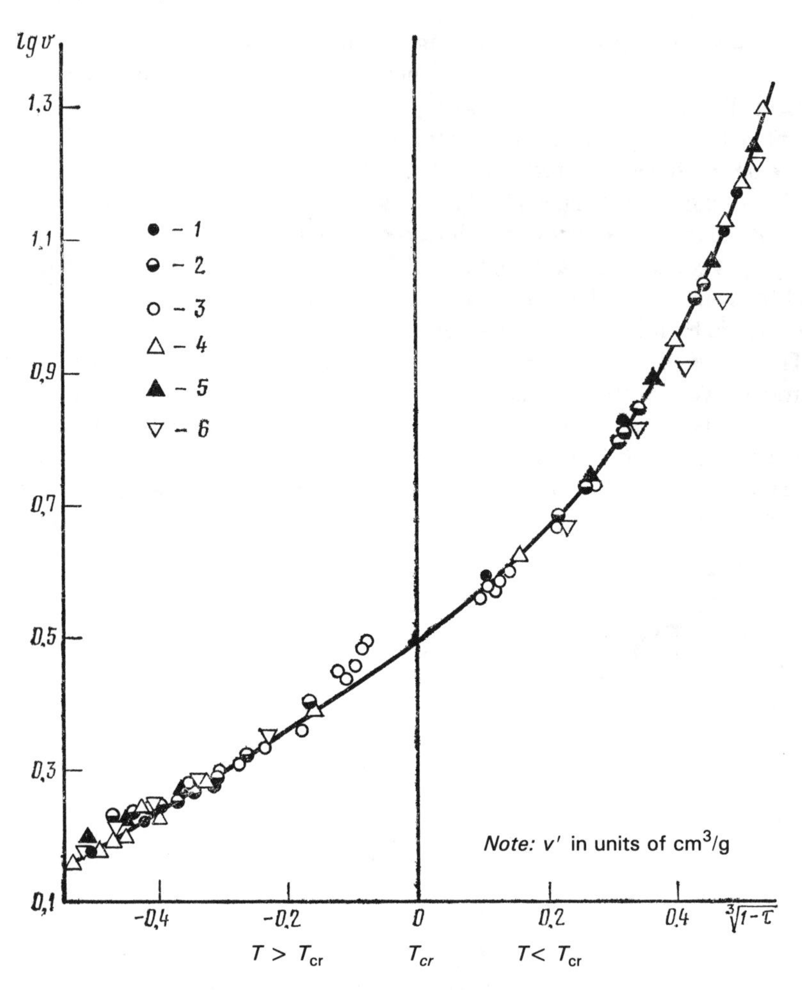

Figure 8 Coordination of data in the transcritical region at the saturation curve from studies of (1) Blanke [34]; (2) Michels et al. [84]; (3) Kuenen and Clark [78]; (4) Vasserman et al. [7], Vasserman and Rabinovich [9]; (5) Baehr and Schwier [30]; (6) Din [46].

data of Vukalovich et al. [10] they lie between 0.04 and 0.06%; and for data of Vasserman et al. [6] they lie between 0.08 and 0.12%. The rms deviations for the 1169 points listed in Table 3.3 amounted to 0.10–0.14%, the rms deviations of values of $\int_{v'}^{v} p_{\text{exp}}\, dv$ from $\int_{v'}^{v} p_{\text{exp}}\, dv$ for different equations of state ranged from 0.09 to 0.3%.

On the basis of the system of equations of state we obtained an averaged equation in the form

$$Z = 1 + \sum_{i=1}^{r} \sum_{j=0}^{S_i} b_{ij}\omega^i/\tau^j, \quad \omega = \rho/\rho_{\text{cr}}, \quad \tau = T/T_{\text{cr}}.$$

The coefficients of the averaged equation are as follows:

$b_{10} = 0.366812 \cdot 10^{0}$	$b_{42} = 0{,}634585 \cdot 10^{0}$
$b_{11} = -0.252712 \cdot 10^{0}$	$b_{43} = -0{,}162912 \cdot 10^{0}$
$b_{12} = -0.284986 \cdot 10^{1}$	$b_{44} = -0{,}217973 \cdot 10^{0}$
$b_{13} = 0{,}360179 \cdot 10^{1}$	$b_{45} = 0.925251 \cdot 10^{-1}$
$b_{14} = -0{,}318665 \cdot 10^{1}$	$b_{46} = 0.893863 \cdot 10^{-3}$
$b_{15} = 0.154029 \cdot 10^{1}$	$b_{50} = -0.444978 \cdot 10^{0}$
$b_{16} = -0{,}260953 \cdot 10^{0}$	$b_{51} = -0{,}734544 \cdot 10^{0}$
$b_{17} = -0.391073 \cdot 10^{-1}$	$b_{52} = 0{,}199522 \cdot 10^{-1}$
$b_{20} = 0.140979 \cdot 10^{0}$	$b_{53} = -0{,}176007 \cdot 10^{0}$
$b_{21} = -0.724337 \cdot 10^{-1}$	$b_{54} = -0{,}998455 \cdot 10^{-1}$
$b_{22} = 0.780803 \cdot 10^{0}$	$b_{55} = -0.620965 \cdot 10^{-1}$
$b_{23} = -0{,}143512 \cdot 10^{0}$	$b_{60} = 0.285780 \cdot 10^{0}$
$b_{24} = 0.633134 \cdot 10^{0}$	$b_{61} = 0{,}258413 \cdot 10^{0}$
$b_{25} = -0.891012 \cdot 10^{0}$	$b_{62} = 0{,}749790 \cdot 10^{-1}$
$b_{26} = 0{,}582531 \cdot 10^{-1}$	$b_{63} = 0{,}859487 \cdot 10^{-1}$
$b_{27} = 0.172908 \cdot 10^{-1}$	$b_{64} = -0.884071 \cdot 10^{-3}$
$b_{30} = -0{,}790202 \cdot 10^{-1}$	$b_{70} = -0.636588 \cdot 10^{-1}$
$b_{31} = -0.213427 \cdot 10^{0}$	$b_{71} = -0{,}105811 \cdot 10^{0}$
$b_{32} = -0.125167 \cdot 10^{1}$	$b_{72} = -0{,}345172 \cdot 10^{-1}$
$b_{33} = -0.164970 \cdot 10^{0}$	$b_{73} = 0{,}429817 \cdot 10^{-1}$
$b_{34} = 0{,}684822 \cdot 10^{0}$	$b_{74} = 0{,}631385 \cdot 10^{-2}$
$b_{35} = 0.221185 \cdot 10^{0}$	$b_{80} = 0{,}116375 \cdot 10^{-3}$
$b_{36} = 0.634056 \cdot 10^{-1}$	$b_{81} = 0{,}361900 \cdot 10^{-1}$
$b_{40} = 0.313247 \cdot 10^{0}$	$b_{82} = -0{,}195095 \cdot 10^{-1}$
$b_{41} = 0{,}885714 \cdot 10^{0}$	$b_{83} = -0{,}379583 \cdot 10^{-2}$

It was determined that rounding off the coefficients to six significant figures does not reduce the accuracy of the calculated values of thermodynamic properties.

The calculations were based on the following values of critical parameters and the gas constant: $T_{cr} = 132.5$ K; $v_{cr} = 0.00316$ m^3/kg; and $R = 287.1$ (m$^3 \cdot$MPa)/(kg$\cdot$K).

The averaged equation of state described satisfactorily all the reliable experimental and smoothed data. This is indicated by values of $\delta\rho_m$ listed in Tables 3.3 and 3.4. The high accuracy of approximation of experimental data is illustrated by the histogram of deviations (Fig. 9) constructed on the basis of values

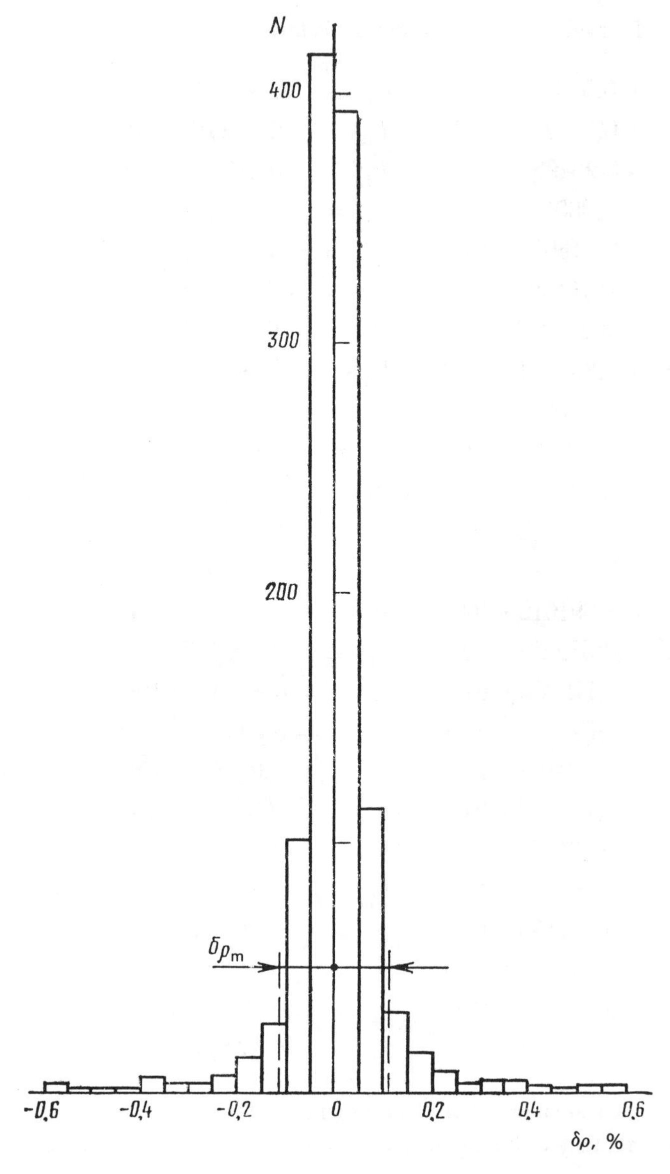

Figure 9 Histogram of deviations of calculated values of the density of air from the measured.

of $\delta\rho$ for 1169 points which are included in Table 3.3. The histogram does not represent nine points in which the absolute values of deviations lie between 0.63 and 0.85%, but the value of $\delta\rho_m = 0.11$ for the 1169 points was calculated with these points included. The averaged equation of state satisfies the condition $\int_{v'}^{v} p_{exp}\, dv = \int_{v'}^{v} p_{calc}\, dv$ with an rms deviation of 0.12%. A detailed assessment of the accuracy of analytic description of data obtained by different investigators can be obtained using graphs of deviations presented in the following section. Comparison of the second and third virial coefficients calculated from the averaged equation of state with experimental data points shows quite satisfactory agreement (Figs. 10 and 11). The deviations of experimental values of B_1 and B_2 from the analytic do not go past the limits of tolerances given in these figures. Figure 12 shows the temperature dependence of all virial coefficients B_2–B_9 of the averaged equation of state.

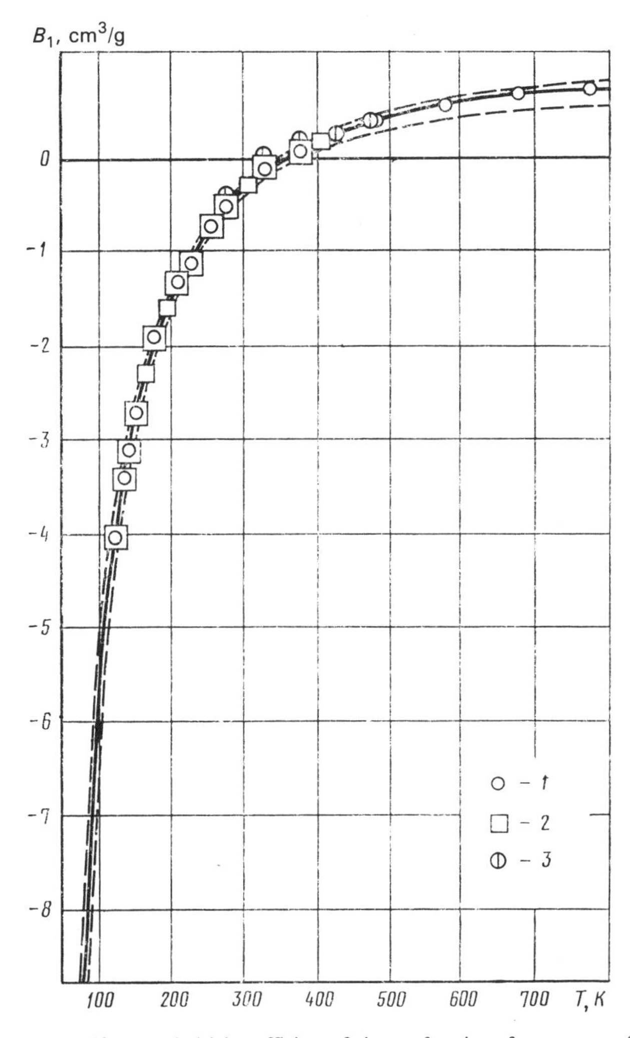

Figure 10 Second virial coefficient of air as a function of temperature from data of (1) Vasserman et al. (in *Teplofiz. Temp.*, vol. 9, no. 5, pp. 915–919, 1971); (2) Michels et al. [84, 85]; (3) Holborn and Otto [65, 66]; the solid curve was calculated from the averaged equation; the dashed curves are tolerance limits for the system of equivalent equations.

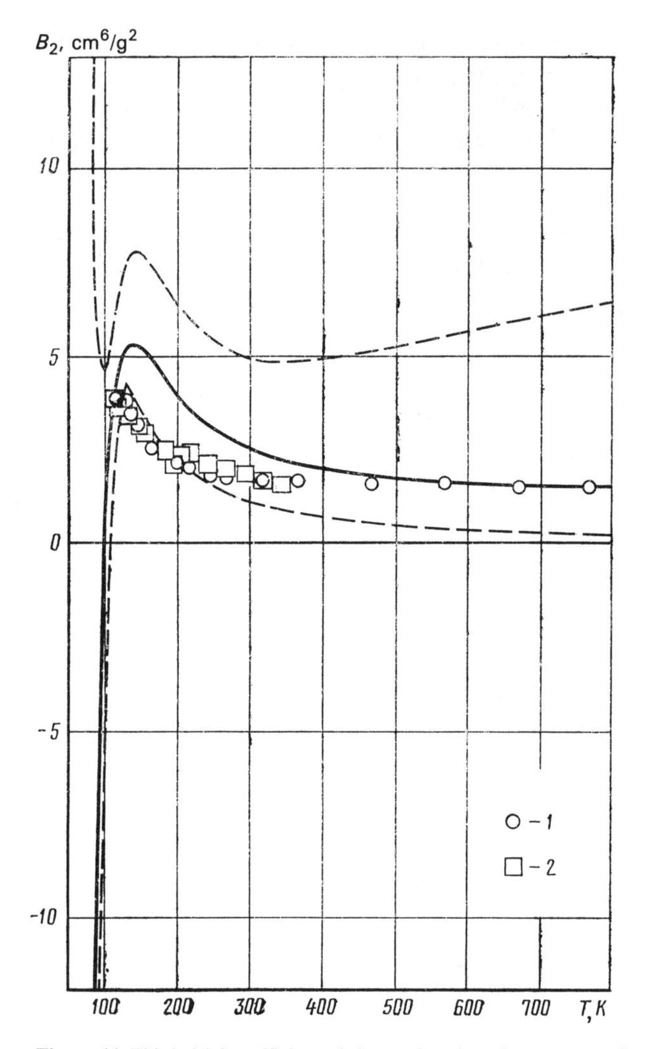

Figure 11 Third virial coefficient of air as a function of temperature from data of (1) Vasserman et al. (see legend for Fig. 10); (2) Michels et al. [84, 85]; the solid curve was calculated from the averaged equation; the dashed curves are tolerance limits for the system of equivalent equations.

The system of equations of state we compiled were based on new experimental data in [6, 34] which pertain to the low-temperature range including the liquid phase. The data in these references significantly improved the reliability of the averaged equation of state as compared with previously obtained equations. It should be emphasized that, given the unavailability of pertinent experimental *pvT* data, Vasserman et al. [7] and Baehr and Schwier [30] did not attempt to provide an analytic description of the properties of liquid air. Vasserman and Rabinovich [9] obtained their reference data on the density of liquid air by a graphical method. Their rms deviation from the values calculated from the equation of state is 0.31% (Table 3.4), which is quite satisfactory for the analytic results in [9], but exceeds the error of modern experiments.

Table 3.4 lists experimental *pvT* data not used in compiling the equations of state either due to poor accuracy, or because they pertain to a range of parameter which were investigated in detail in subsequent studies. The value of $\delta\rho_m$ = 0.32% for Amagat's data can be regarded as satisfactory if it is remembered that these data were obtained as far back as the nineteenth century and represent a wide temperature and pressure range. In comparing with data in [91, 110], we

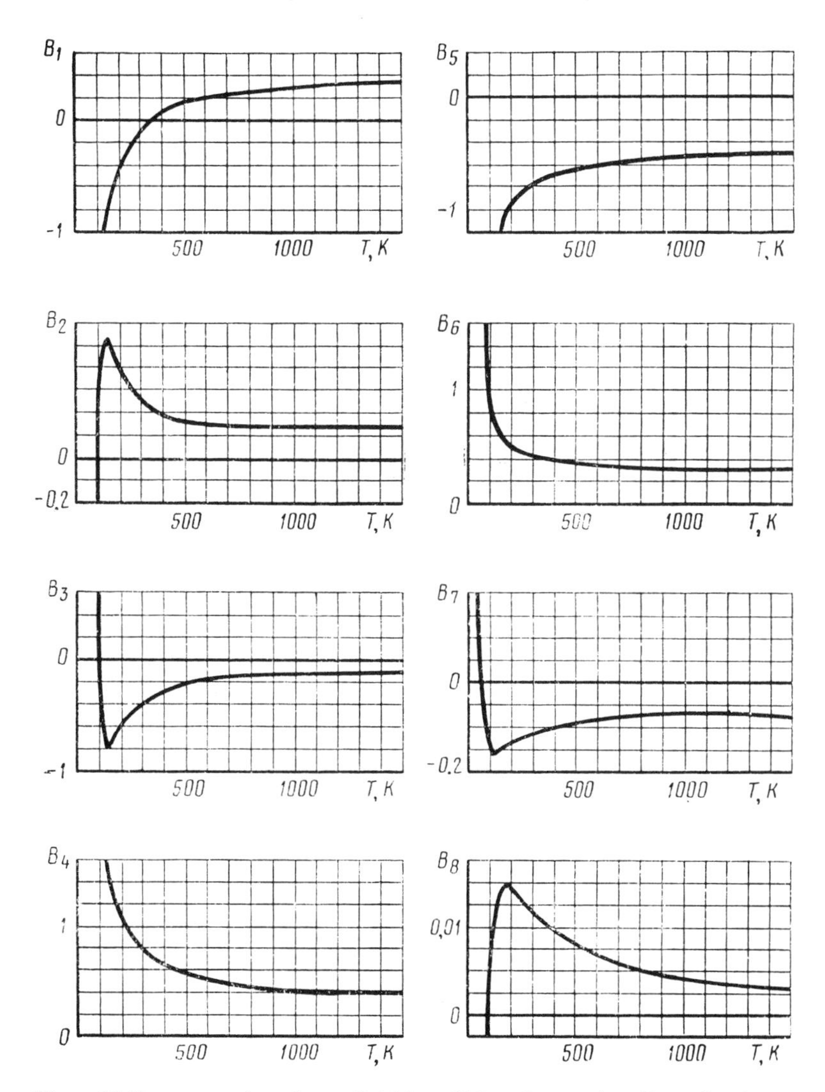

Figure 12 Temperature dependence of virial coefficients for equation of state for air.

excluded several points close to the critical region in which significant deviations occurred for the values of density, but the pressure deviations are small. The relatively high values of $\delta\rho_m$ at saturation for data in [84] are basically due to poor reliability of values of ρ' used there. For data in [34] at saturation, with the exception of a single value of ρ'' at 132.34 K, which is close to critical, the value of $\delta\rho_m$ decreases to 0.22%. The experimental data in [34, 78, 84] for the saturation curve were compared with values of ρ' and ρ'' listed in experimental studies.

The values of pressure, in MPa, for all the points at saturation were calculated from our equations:

$$\lg \pi_s' = a_{-1}/\tau_s + a_0 + a_1\tau_s + a_2\tau_s^2 + a_3\tau_s^3 \tag{3.4}$$

$$\lg \pi_s'' = b_{-1}/\tau_s + b_0 + b_1\tau_s + b_2\tau_s^2 + b_3\tau_s^3 \tag{3.5}$$

where

$$a_{-1} = -0{,}221789 \cdot 10^1; \quad b_{-1} = -0{,}398400 \cdot 10^1$$
$$a_0 = 0{,}191330 \cdot 10^1; \quad b_0 = 0{,}926210 \cdot 10^1$$
$$a_1 = 0{,}828443 \cdot 10^0; \quad b_1 = -0{,}112591 \cdot 10^2$$
$$a_2 = -0{,}915594 \cdot 10^0; \quad b_2 = 0{,}803356 \cdot 10^1$$
$$a_3 = 0{,}396525 \cdot 10^0; \quad b_3 = -0{,}205615 \cdot 10^1$$
$$\pi_s' = p_s'/p_{cr}; \quad \pi_s'' = p_s''/p_{cr}; \quad \tau_s = T_s/T_{cr}.$$

Equations (3.4) and (3.5) describe the bulk of experimental data in [34, 84] used in compiling them, with an error of less than 0.3% for the boiling curve and less than 0.5% for the condensation curve (Figs. 13 and 14). The figures do not show three experimental points obtained by Kuenen and Clark [78] for the critical range (δp_s = 2.0–1.8%), one experimental point of Blanke [34] at 71.7 K (δp_s = −1.9%), and also a number of analytic points [7, 30, 46] at temperatures below 95 K (δp_s = −(1.6–4.2)%). Unlike the previous points [7, 30, 46], Eqs. (3.4) and (3.5) at temperatures below 118 K are based on experimental data of Blanke [34], which puts more confidence in their reliability.

We did not compile an equation of the melting curve for reasons analyzed in detail in Sec. 1.3.1. We simply defined arbitrarily the location of the crystal-to liquid-phase equilibrium curve and limited the data along the 70 and 75 K isotherms to pressures of 45 and 75 MPa respectively.

3.3 ESTIMATION OF THE RELIABILITY OF THE TABLES

The accuracy of tabulated values of thermodynamic properties of air can be determined by analysis of the results of their comparison with experimental data and also by calculating the rms errors (standard deviations) of the tabulated quantities using a system of equivalent equations.

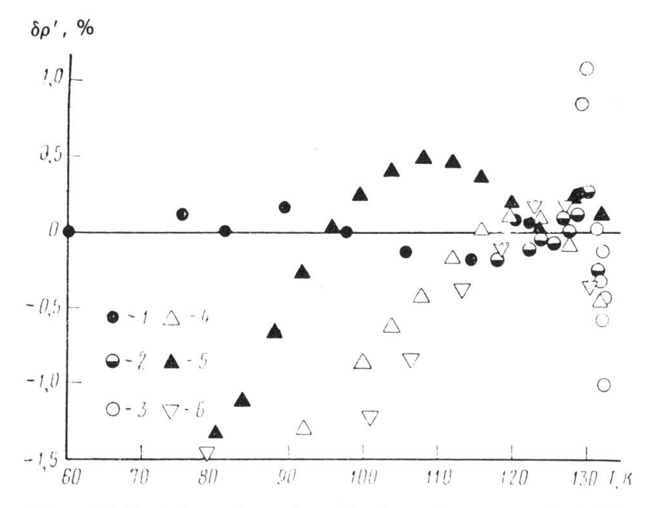

Figure 13 Deviations of experimental values of pressure at the boiling curve obtained by various investigators from those calculated from Eq. (3.4). (1) Blanke [34]; (2) Michels et al. [84]; (3) Kuenen and Clark [78]; (4) Vasserman et al. [7], Vasserman and Rabinovich [9]; (5) Baehr and Schwier [30]; (6) Din [46].

The results of comparison of the calculated values of density with experimental data obtained by various investigators are plotted in Figs. 15 through 22. It is seen from these figures that on the majority of isotherms the value of $\delta\rho$ for gas and liquid do not exceed 0.1%, with deviations in many points lying within the limits of $\pm 0.05\%$. Only for data by Vasserman et al. [6] and for data at the

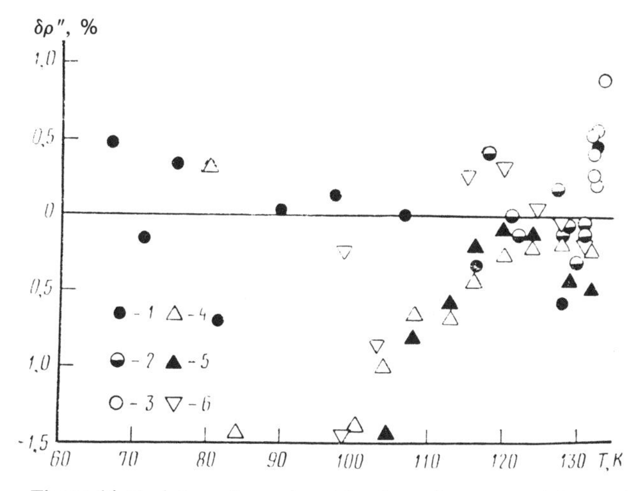

Figure 14 Deviation of experimental values of pressure at the condensation curve obtained by various investigators from those calculated from Eq. (3.5). (1) Blanke [34]; (2) Michels et al. [84]; (3) Kuenen and Clark [78]; (4) Vasserman et al. [7], Vasserman and Rabinovich [9]; (5) Baehr and Schwier [30]; (6) Din [46].

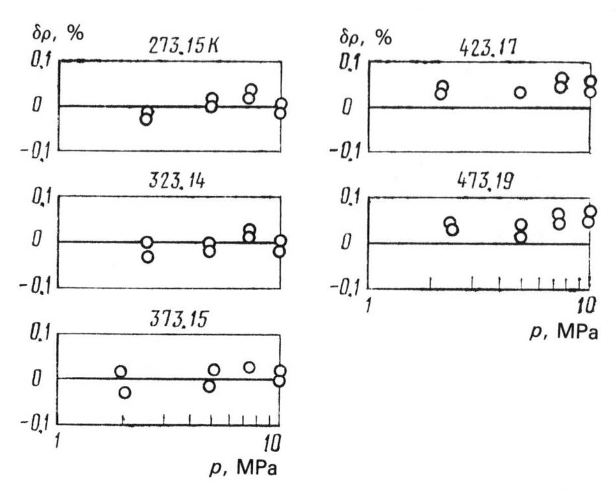

Figure 15 Deviations of experimental data of Holborn and Schultze [67] on the density of air from the analytic.

saturation curve (Fig. 23) do the deviations in a number of points exceed 0.2–0.4%, which has necessitated widening the limits of $\delta\rho$ for the corresponding graphs.

The graphs do not show deviations for certain points in [34, 84] which were assigned zero weight in the course of calculations. These points are listed

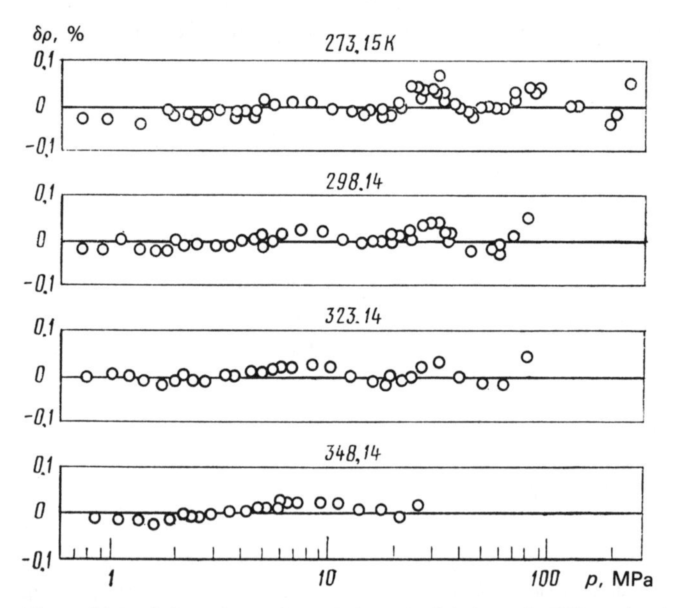

Figure 16 Deviation of experimental data of Michels et al. [85] on the density of air from the analytic.

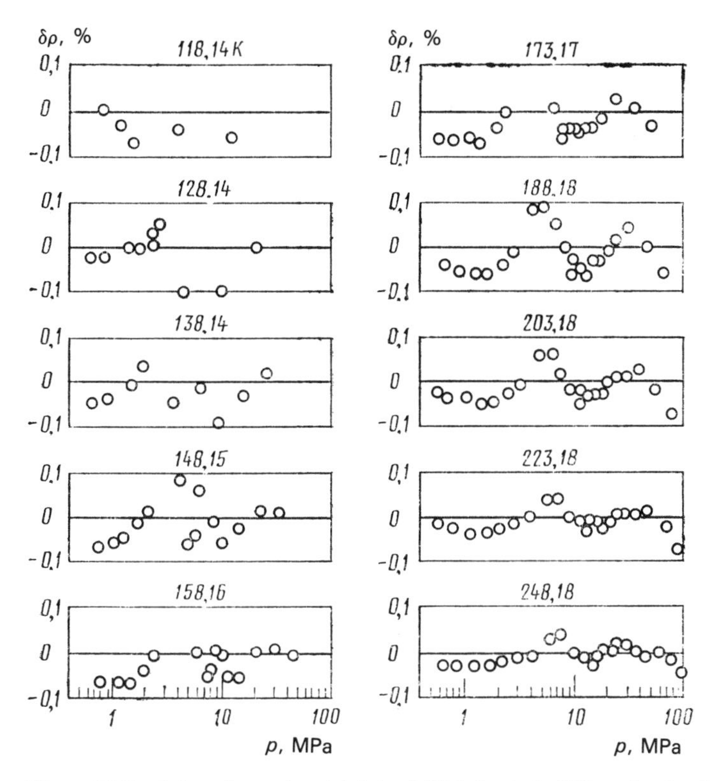

Figure 17 Deviation of experimental data of Michels et al. [84] on the density of air from the analytic.

in Table 3.5, which also shows deviations $\delta\rho$ and δZ. Tables 3.6 and 3.7 list points used in compiling the equations but not plotted because the values of $\delta\rho$ for them exceed the maximum values selected for the figures, or because no plots of deviations were constructed for the corresponding isotherms due to the small number of experimental points on them. Of the 68 points listed in Tables 3.6 and 3.7, approximately one half have values of $\delta\rho$ smaller than 0.15%.

The plots of deviations and the tables do not contain 30 points from the paper by Vukalovich et al. [10], obtained at individual values of temperature and density over the range of T = 287.84–295.87 K and p = 1.38–59.91 MPa. The values of $\delta\rho$ for these points range from −0.06 to +0.08.

We also compared the analytic results with the majority of experimental data on calorific and acoustic properties of air.

The analytic values of enthalpy were compared with data points of Dawe and Snowdon [44] only in those points in which the values of h were determined by the authors themselves. As seen from Table 3.8, the analytic values of enthalpy are in excellent agreement with the experimental over the entire exper-

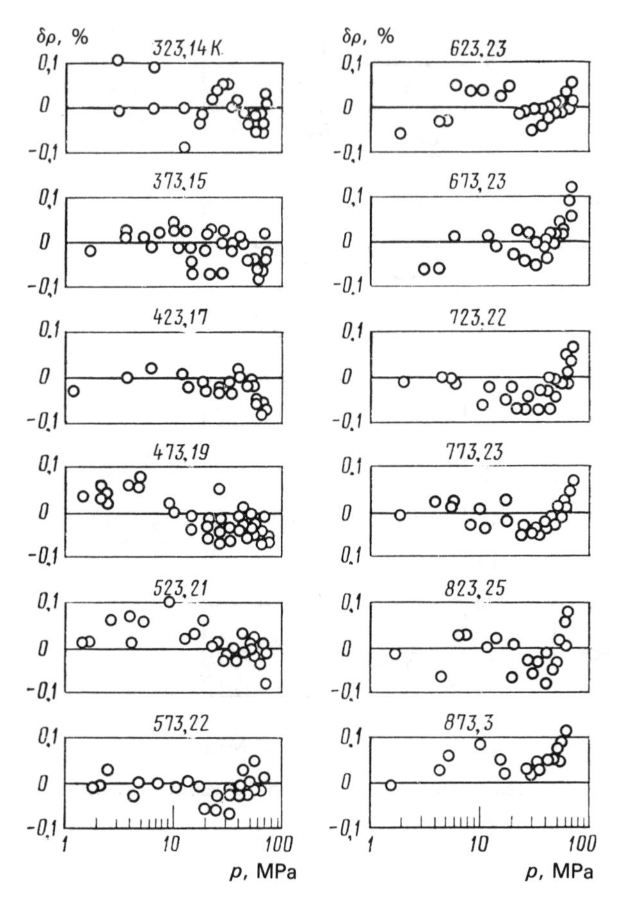

Figure 18 Deviations of experimental data of Vukalovich et al. [10] on the density of air from the analytic.

imentally investigated range of parameters, and the differences do not exceed the value of 0.3 kJ/kg.

Unfortunately, no new experimental data are available on c_p. For this reason the values we calculated had to be compared with data of Jakob [72] and Roebuck [94], which were measured rather long ago. The analytic values of c_p are in agreement with the overwhelming majority of data of Jakob (Table 3.9) within less than 1%, with the differences exceeding 1% in only 7 out of 56 points. The greatest differences are observed at the minimum temperature (193.85 K) at virtually all the pressures. These differences increase with pressure, becoming as high as 16% at p = 19.613 MPa. However, in 1943, in discussing thermodynamic properties of air at low temperatures [109], Jakob communicated that his previously published data [72] on c_p at 193.85 K and pressures of 14.7 and 19.6 MPa should be corrected. The deviations of the corrected values at these points are —5.6 and 1.8%, respectively (compared to 9.2 and 16.3 in Table 3.9). The analytic values are in somewhat poorer agreement with data of Roebuck (Table 3.10). The standard deviation for 84 points amounts to 1.54% with the deviation exceeding 1% in 33 points. At the same time the maximum divergence, unlike that in the paper by Jakob [72], is —3.7%.

The analytic values of isochoric specific heat are higher (Table 3.11) than those of Eucken and Hauck [51] by 4.4 to 8.5%. Apparently, the values in [51] were obtained with a systematic error. Eucken and Hauck list smoothed values of c_s' at temperatures of 80 to 120 K obtained on the basis of their experimental data. The disagreement with these values is not systematic and ranges from 9.9% at 90 K to —4.1% at 120 K. The deviation at 80 K is as high as 32%. On the whole, such an accuracy of values of c_v and c_s' obtained more than 60 years ago at relatively low temperatures can be regarded as satisfactory. The only exception is the value of c_s' at T = 80 K.

As mentioned previously, measurements of the Joule–Thomson effect are among the most numerous of investigations of calorific properties of air. The analytic values of this effect μ (adiabatic J–T coefficient) are compared with

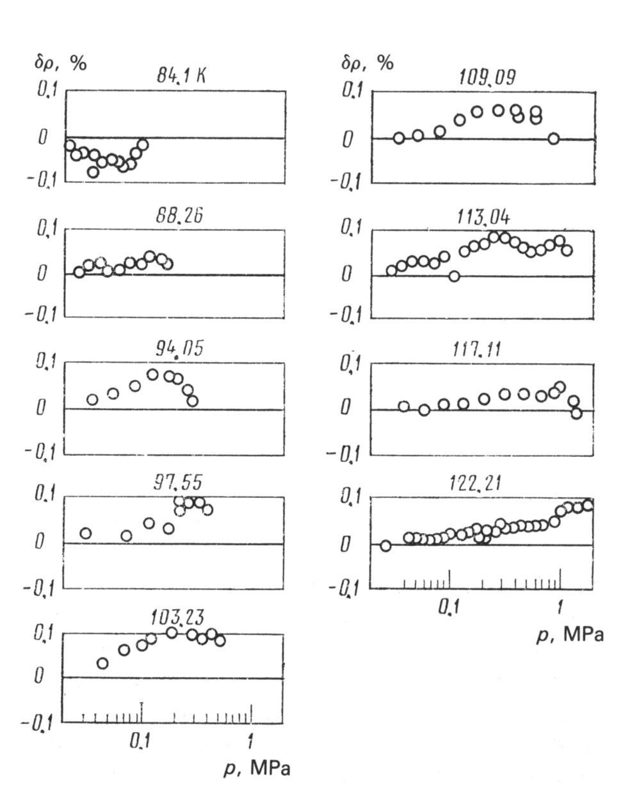

Figure 19 Deviations of experimental data of Romberg [96] on the density of air from the analytic.

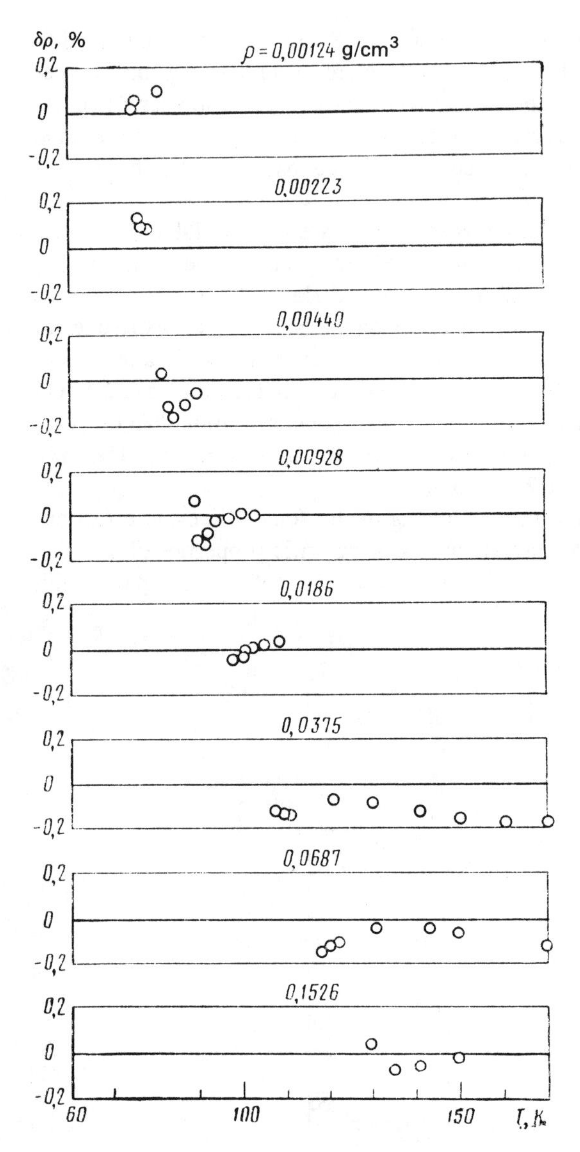

Figure 20 Deviations of experimental data of Blanke [34] on the density of air along isochores, from the analytic at $\rho < \rho_{cr}$.

experimental data of Roebuck [93] and Hausen [59] in Tables 3.12 and 3.13 respectively. These tables list absolute values of deviations $\Delta\mu$ of analytic from the experimental data. The deviations for experimental data in [93] range from —0.79 to 0.34 K/MPa. However, these numbers do not fully reflect the nature of the differences. For 117 of 145 values of μ the deviations do not exceed ± 0.15 K/MPa. The greatest deviations are observed within the same range of parameters, i.e., at low temperatures (153 to 248 K) and relatively low pressures (0.1 to 4 MPa). It was pointed out previously (Sec. 1.2) that the data in

[93] should be corrected for the error detected by Roebuck and due to calibration of the manometer. When this is done, the deviations become significantly smaller. The data of Hausen [59] deviate typically between —0.62 and 0.83 K/MPa. However, here also the deviations in 52 out of 87 points do not exceed ±0.15 K/MPa.

The extensive information on the isothermal Joule–Thomson effect δ obtained by Ishkin and Kaganer [12] contains a systematic error, i.e., it is known that the data in [12] are 5% on the high side. We calculated the values of δ for all the experimental temperatures and pressures from [12]. However, the final tables of thermodynamic properties of air in this book do not list the isothermal Joule–Thomson effect. In fact, the overwhelming majority of values in [12] are higher than these calculated by the analytical equation. However, this excess frequently exceeds the previously mentioned systematic error and becomes as high as 21% at 138 K and 3.98 MPa. Six values of δ are 1 to 11% higher than the analytic. As a whole the data in [12] were obtained with an error (standard deviation) of 7.8%.

The analytical values of the speed of sound w in gaseous air are in agreement with experimental data of van Itterbeek and de Rop within the limits of

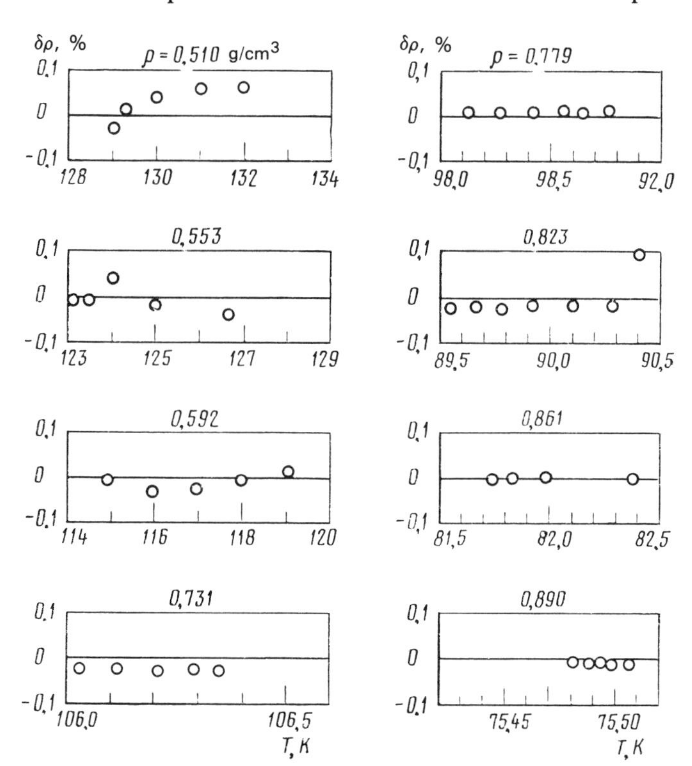

Figure 21 Deviations of experimental data of Blanke [34] of the density of air along isochores at $\rho > \rho_{cr}$ from the analytic.

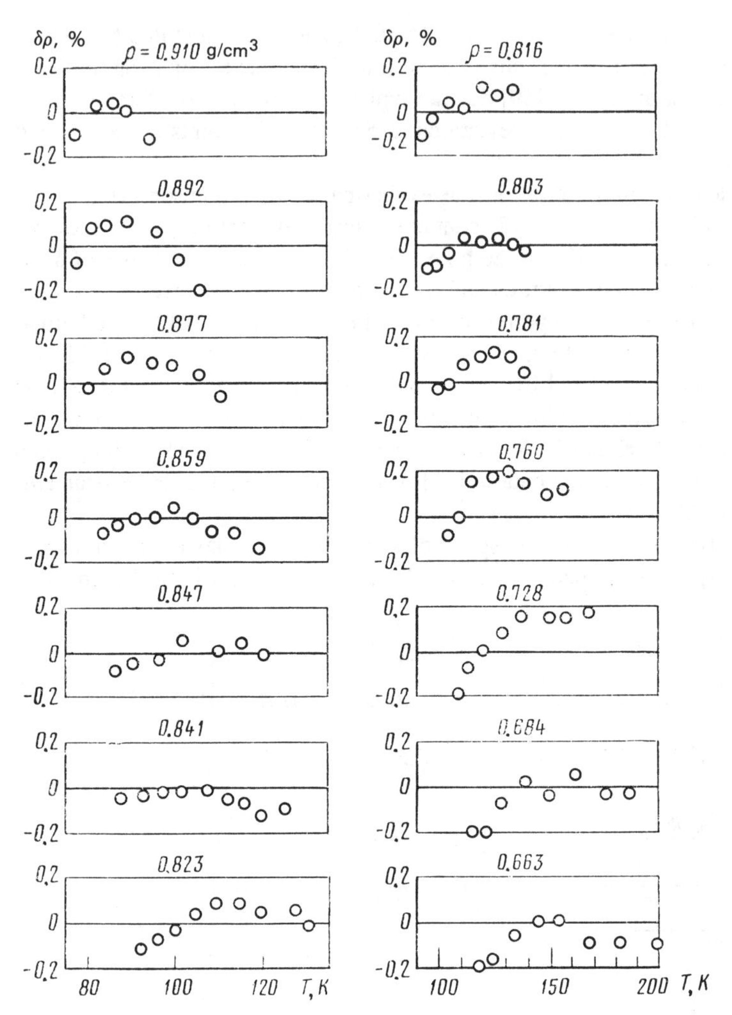

Figure 22 Deviation of experimental values of density of air obtained by Vasserman et al. [6] from the analytic.

—0.16 to +0.35% (Table 3.14). The analytic values of w in liquid air are in satisfactory agreement with data of van Itterbeek and van Dael [70] at 80.98 K (Table 3.15). The deviation of data in [70] along the 77.20 K isotherm is between —10.0 and —20.4% and increases with pressure. The analytic values of w for liquid air at this temperature are apparently unreliable.

Figure 24 shows Boyle, ideal gas, and Joule–Thomson inversion curves, calculated from the averaged equation of state. The figure shows that these are in satisfactory agreement with the results of most reliable experiments.

Tables 3.16 through 3.23 list the standard deviations of the analytic values

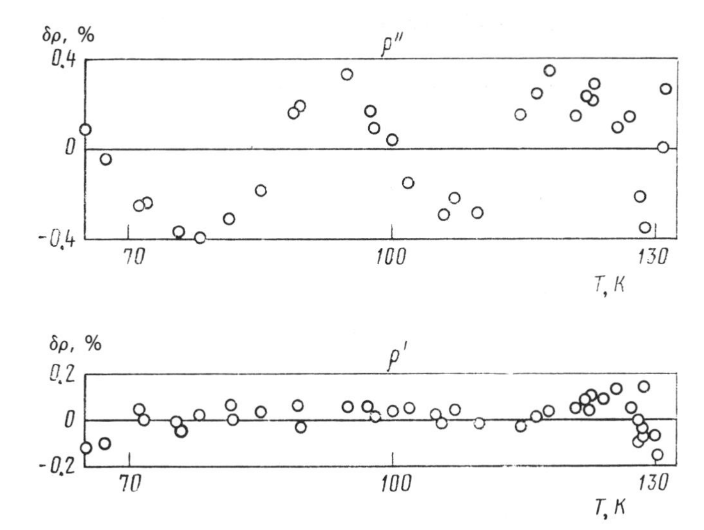

Figure 23 Deviations of experimental values of the density of saturated air obtained by various investigators from the analytic.

of thermodynamic functions. These deviations were calculated from the expression

$$\sigma_x = \sqrt{\sum_{k=1}^{N} (\bar{x} - x_k)^2/(N-1)}$$

and describe the error of an individual equation from the system of equivalent equations. Quantities $\sigma_{\bar{x}}$ and $3\sigma_{\bar{x}}$ can be calculated by the technique described in our previous book [21] on the basis of data given in Table 3.16–3.23.

Figures 25 through 34 give the scattering ellipses for values of thermodynamic

Table 3.5 Experimental data not used in compiling the equation of state

T, K	p, MPA	δρ, %	δp, %	Reference
60.53	0.0964	43.9	—89.2	[34]
60.73	0.3018	—21.3	—57.4	[34]
60.82	0.4728	—21.2	—26.7	[34]
132,14	3,755	—3,54	0.42	[34]
132.96	3,879	—1.44	0.07	[34]
133.02	3,893	—2.44	0.14	[34]
133,05	3,815	1,64	—0.17	[34]
133,94	3,930	0.78	—0,11	[84]
134.14	4,001	0,87	—0.10	[84]

Table 3.6 Experimental data along isotherms not represented on plots of deviation

T, K	p, MPA	δρ, %	Number of figure and reference
273.15	38.46	0.11	Fig. 13 [85]
298.14	117.7	0.12	
122.64	2.255	—0.04	Fig. 14 [84]
124.14	2.314	—0.07	
125.64	2.372	—0.03	
127.14	2.429	0	
130.14	3.208	—0.14	
	3.710	—0.02	
132.14	3.348	—0.18	
	3.603	0.34	
	4.390	0.05	
134.14	3.801	—0.16	
	4.166	0.59	
135.14	3.899	—0.39	
	4.142	0.35	
135.14	4.379	0.81	Fig. 14 [84]
	5.424	0.03	
138.14	2.770	0.14	
	3.261	0.12	
	4.106	—0.23	
	4.343	—0.47	
	4.400	—0.24	
	4.566	0.11	
	4.711	0.46	
	4.861	0.77	
	5.091	0.65	
	5.564	0.23	
148.15	3.109	0.15	
	3.727	0.16	
	5.416	—0.21	
	5.540	—0.13	
	6.778	0.18	
	7.384	0.17	
158.16	3.439	0.14	
	4.179	0.16	
	5.043	0.11	
	6.453	—0.12	
323.14	2.953	0.11	Fig. 15 [10]
673.23	1.945	—0.11	
	67.94	0.12	
873.30	59.24	0.11	
103.23	0.161	0.11	Fig. 16 [96]
	0.243	0.11	
122.21	0.777	0.14	

Table 3.7 Experimental data along isochores, not represented on plots of deviations

ρ, cm³	T, K	p, MPa	δρ, %	Number of figure and reference
0,00062	67,77	0,0119	0,63	Fig. 17 [34]
	69,03	0,0122	0,55	
0,00121	71,81	0,0244	0,43	
	74,80	0,0255	0,21	
	77,70	0,0265	0,23	
	80,60	0,0280	0,14	
	80,80	0,0276	0,28	
	86,87	0,0297	0,36	
0,258	133,95	3,934	0,33	
	133,96	3,934	0,35	
		3,935	0,37	
	136,81	4,291	0,05	
	140,01	4,687	0	
0.349	133.88	4.043	0.24	Fig. 18 [34]
	135.56	4.344	0.70	
		4.345	0.65	
	138.66	4.908	0.51	
0.509	128.96	3.303	—0.12	
0.861	82.18	0.533	0.11	
0.954	60.92	0.678	0.20	
	61.03	0.842	0.59	
0.910	99.50	57.96	—0.24	Fig. 19 [6]

Table 3.8 Comparison of experimental values of enthalpy [44] (line 1) with analytic values (line 2). Line 3 is the difference Δh, kJ/kg

p, MPA	Values of specific heat at T, K				
	222.00	273.15	298.15	333.15	366.45
0.1	475.3	526.7	551.9	587.1	620.7
	475.2	526.7	551.8	587.1	620.7
	0.1	0	0.1	0	0
2.5	465.4	520.2	546.5	582.9	617.5
	465.2	520.0	546.3	582.8	617.3
	0.2	0.2	0.2	0.1	0.2
5.0	455.1	513.6	541.1	578.8	614.3
	455.0	513.5	540.9	578.6	614.0
	0.1	0.1	0.2	0.2	0.3
7.5	445.1	507.4	536.0	574.9	611.2
	444.9	507.3	535.3	574.8	611.1
	0.2	0.1	0.1	0.1	0.1
10.0	435.7	501.6	531.3	571.3	608.4
	435.4	501.5	531.2	571.2	608.3
	0.3	0.1	0.1	0.1	0.1

namic functions obtained from the system of equivalent equations compared to the value from the averaged equation of state identified as ⊕ in each figure. The values of σ = standard deviation, from sections of Tables 3.16, 3.19, and 3.20, are also shown in Figs. 25–34.

3.4 COMPARISON OF PREVIOUSLY PUBLISHED TABLES

Tables by Baehr and Schwier [30], Vasserman et al. [7, 9], and Sychev et al. [20] are the most recent and detailed of the numerous published tables of thermodynamic properties. For this reason the values of density, enthalpy, entropy, specific heats c_v and c_p, and speed of sound were compared with data in these tables. The results of this comparison are listed in Tables 3.24 through 3.29. No comparison was performed with data of Din [46] since this was done in [7, 9]. In comparisons over the range of parameters where data in [7, 9] overlapped, preference was given to data in [9] as being more recent.

For the extrapolated range (T from 873 to 1500 K) the tables of this study were compared with tables calculated by V. N. Zubarev (Investigation of Thermodynamic Properties of Water Vapor and Air. Author's abstract of doctoral thesis. Moscow, 1975) to 2000 K on the basis of an analytically validated equation of state which hence ensured more reliable extrapolation. The comparison of tables over the extrapolation range showed that they are in satisfactory agreement. Deviations with respect to density do not exceed 0.5% with respect to c_v at 1% and to c_p at 0.3%.

Table 3.9 Comparison of values of specific heat c_p obtained by Jakob [72] (line 1) with the analytic results of the present study (line 2). Line 3 is the deviation δc_p, %

p, MPA	Values of specific heat at T, K; c_p, kJ/(kg·K)				
	193.85	223.15	273.15	323.15	332.15
0	1.009	1.009	1.009	1.009	1.009
	1.003	1.003	1.004	1.006	1.007
	0.59	0.59	0.50	0.30	0.20
0.098	1.019	1.013	1.009	1.009	1.011
	1.007	1.006	1.006	1.008	1.008
	1.19	0.69	0.30	0.10	0.30
4.903	1.327	1.185	1.110	1.078	1.074
	1.309	1.187	1.106	1.072	1.069
	1.38	—0.17	0.36	0.56	0.46
9.807	1.742	1.367	1.202	1.139	1.130
	1.729	1.398	1.206	1.133	1.125
	0.75	—2.20	—0.33	0.53	0.44
14.710	2.077	1.507	1.277	1.191	1.183
	1.903	1.552	1.291	1.186	1.175
	9.15	—2.90	—1.09	0.42	0.68

Table 3.9 (*Continued*)

p, MPA	Values of specific heat at T, K; c_p, kJ/(kg·K)				
	193.85	223.15	273.15	323.15	332.15
19.613	2.156 1.854 16.3	1.591 1.612 —1.30	1.340 1.350 —0.75	1.237 1.228 0.73	1.225 1.214 0.90
24.517	—	—	—	—	1.252 1.244 0.64
29.420	—	—	—	—	1.269 1.265 0.32

Table 3.9 (*Continued*)

p, MPA	Values of specific heat at T, K; c_p, kJ/(kg·K)			
	373.15	423.15	473.15	523.15
0	1.013 1.010 0.30	1.017 1.017 0	1.022 1.025 —0.30	1.026 1.034 —0.78
0.098	1.013 1.011 0.20	1.017 1.017 0	1.022 1.025 —0.29	1.026 1.035 —0.87
4.903	1.059 1.057 0.19	1.049 1.051 —0.19	1.045 1.051 —0.57	1.045 1.055 —0.95
9.807	1.103 1.099 0.36	1.080 1.082 —0.19	1.070 1.074 —0.37	1.063 1.073 —0.93
14.710	1.141 1.135 0.52	1.110 1.109 0.09	1.091 1.095 —0.37	1.080 1.090 —0.91
19.613	1.174 1.166 0.68	1.133 1.132 0.09	1.110 1.113 —0.27	1.097 1.105 —0.72

Table 3.10 Comparison of values of specific heat c_p obtained by Roebuck [94] (line 1) with the analytic results of the present study (line 2). Line 3 is the deviation δc_p, %

T, K	Values of specific heat at p, MPa; c_p, kJ/(kg·K)			
	0.101	2.027	6.080	10.13
173.15	0.999 1.009 —0.99	1.154 1.154 0	— — —	— — —

Table 3.10 (*Continued*)

T, K	Values of specific heat at p, MPa; c_p, kJ/(kg·K)			
	0.101	2.027	6.080	10.13
198.15	1.001	1.101	1.333	1.650
	1.007	1.101	1.372	1.677
	−0.60	0	−2.85	−1.61
223.15	1.002	1.070	1.207	1.367
	1.006	1.073	1.238	1.411
	−0.40	−0.28	−2.51	−3.12
248.15	1.004	1.052	1.149	1.247
	1.006	1.057	1.170	1.284
	−0.20	−0.48	−1.80	−2.88
273.15	1.007	1.043	1.112	1.188
	1.006	1.046	1.130	1.212
	0.10	−0.29	−1.59	−1.98
298.15	1.009	1.041	1.100	1.156
	1.007	1.039	1.105	1.167
	0.20	0.19	−0.45	−0.94
323.15	1.011	1.038	1.090	1.138
	1.008	1.034	1.087	1.137
	0.30	0.39	0.28	0.09
348.15	1.013	1.036	1.081	1.122
	1.009	1.031	1.076	1.116
	0.39	0.48	0.46	0.53

Table 3.10 (*Continued*)

T, K	Values of specific heat at p, MPa; c_p, kJ/(kg·K)		
	14.19	18.24	22.29
198.15	1.853	—	—
	1.832	—	—
	1.14	—	—
223.15	1.491	1.569	1.583
	1.540	1.603	1.617
	−3.18	−2.12	−2.10
248.15	1.328	1.389	1.421
	1.378	1.441	1.475
	−3.63	−3.60	−3.66
273.15	1.250	1.295	1.333
	1.283	1.336	1.372
	−2.57	−3.07	−2.85
298.15	1.203	1.231	1.264
	1.222	1.266	1.299
	−1.56	−2.12	−2.70

Table 3.10 (*Continued*)

T, K	Values of specific heat at p, MPa; c_p, kJ/(kg·K)		
	14.19	18.24	22.29
323.15	1.179 1.181 −0.17	1.213 1.218 −0.41	1.238 1.247 −0.72
348.15	1.158 1.153 0.43	1.189 1.183 0.52	1.211 1.208 0.25

Table 3.10 (*Continued*)

T, K	Values of specific heat at p, MPa; c_p, kJ/(kg·K)			
	0.101	2.027	6.080	10.13
373.15	1.015 1.012 0.30	1.034 1.030 0.39	1.073 1.067 0.56	1.110 1.101 0.82
423.15	1.019 1.017 0.20	1.032 1.031 0.10	1.060 1.059 0.09	1.089 1.084 0.46
473.15	1.023 1.025 −0.20	1.031 1.036 −0.48	1.052 1.057 −0.48	1.074 1.076 −0.19
523.15	1.027 1.035 −0.77	1.033 1.043 −0.96	1.047 1.060 −1.22	1.062 1.075 −1.20
553.15	1.029 1.041 −1.16	1.035 1.048 −1.24	1.043 1.063 −1.88	1.055 1.076 −1.95

Table 3.10 (*Continued*)

T, K	Values of specific heat at p, MPa; c_p, kJ/(kg·K)		
	14.19	18.24	22.29
373.15	1.141 1.132 0.80	1.168 1.158 0.87	1.188 1.180 0.67
423.15	1.113 1.106 0.63	1.133 1.126 0.62	1.151 1.143 0.70
473.15	1.091 1.093 −0.18	1.107 1.109 −0.18	1.121 1.122 −0.09

Table 3.10 (*Continued*)

T, K	Values of Joule–Thomson effect at p, MPa		
	14,19	18,24	22,29
523,15	1,074 1,088 —1,28	1,087 1.101 —1.27	1.098 1,112 —1.26
553,15	1,065 1.088 —2.11	1.076 1,099 —2,09	1,086 1.109 —2.08

Table 3.11 Comparison of values of c_v obtained by Eucken and Hauch [51] with analytic values

T, K	p, MPA	c_v[51]	c_v	δc_v, %
		kJ/(kg·K)		
148.0	11.15	0.823	0.892	—7.8
152.0	12.16	0.814	0.872	—6.6
156.2	13.17	0.807	0.862	—6.4
160.4	14.19	0.802	0.854	—6.1
163.0	15.20	0.799	0.848	—5.7
137.6	5.472	0.904	0.988	—8.5
151.6	11.15	0.804	0.878	—8.4
153.55	12.16	0.801	0.870	—7.9
159.0	13.17	0.797	0.858	—7.2
162.0	14.19	0 792	0.851	—6.9
164.5	15.20	0.789	0.846	—6.7
137.6	6.080	0.904	0.958	—5.7
147.0	11.15	0.836	0.883	—5.3
151.0	12.16	0.818	0.873	—6.3
159.5	14.19	0.812	0.855	—5.0
164.6	16.21	0.808	0.845	—4.4

Table 3.12 Comparison of Roebuck's [93] experimental data on the Joule–Thomson effect (line 1) with the analytic values (line 2). Line 3 represents the absolute values of deviations $\Delta\mu = \mu_{an}$, K/MPa

T, K	Values of Joule–Thomson effect on p, MPa					
	0.101	2.026	4.053	6.079	8.106	10.13
123.15	—	—	0.61 0.54 0.07	0.43 0.28 0.15	0.31 0.13 0.18	0.19 0.03 0.16

Table 3.12 (*Continued*)

T, K	Values of Joule–Thomson effect on p, MPa					
	0.101	2.026	4.053	6.079	8.106	10.13
133.15	9.58	9.62	5.63	1.09	0.67	0.43
	9.77	10.15	3.48	1.02	0.55	0.31
	—0.19	—0.53	2.15	0.07	0.12	0.12
153.15	6.98	6.96	6.63	5.15	2.90	1.63
	7.77	7.62	7.01	5.16	2.75	1.53
	—0.79	—0.66	—0.38	—0.01	0.15	0.10
173.15	5.67	5.46	5.14	4.64	3.82	2.81
	6.36	6.04	5.60	4.89	3.84	2.76
	—0.69	—0.58	—0.46	—0.25	—0.02	0.05
198.15	4.56	4.32	4.05	3.72	3.29	2.83
	5.08	4.73	4.37	3.96	3.45	2.88
	—0.52	—0.41	—0.32	—0.24	—0.16	—0.05
223.15	3.71	3.52	3.28	3.04	2.75	2.44
	4.14	3.82	3.52	3.22	2.89	2.54
	—0.43	—0.30	—0.24	—0.18	—0.14	—0.10
248.15	3.12	2.93	2.71	2.51	2.30	2.12
	3.41	3.14	2.88	2.65	2.41	2.16
	—0.29	—0.21	—0.17	—0.14	—0.11	—0.04
273.15	2.63	2.46	—	2.11	—	1.76
	2.84	2.60		2.20		1.82
	—0.21	—0.14		—0.09		—0.06
298.15	2.24	2.09	—	1.79	—	1.50
	2.37	2.17		1.84		1.53
	—0.13	—0.08		—0.05		—0.03

Table 3.12 (*Continued*)

T, K	Values of Joule–Thomson effect on p, MPa					
	12.16	14.19	16.21	18.24	20.26	22.29
123.15	0.07	—0.03	—0.12	—0.21	—0.28	—0.38
	—0.05	—0.10	—0.15	—0.19	—0.22	—0.24
	0.12	—0.07	—0.03	—0.02	—0.06	—0.14
133.15	0.28	0.15	0.02	—0.08	—0.17	—0.28
	0.16	0.06	—0.02	—0.08	—0.13	—0.17
	0.12	0.09	0.04	0	—0.04	—0.11
153.15	1.00	0.83	0.45	0.28	0.13	—0.15
	0.94	0.61	0.40	0.25	0.14	0.05
	0.06	0.22	0.05	0.03	—0.01	—0.20
173.15	1.96	1.40	1.02	0.74	0.48	0.31
	1.93	1.35	0.96	0.69	0.48	0.33
	0.03	0.05	0.06	0.05	0	—0.02
198.15	2.67	1.93	1.46	1.15	0.91	0.68
	2.33	1.84	1.44	1.12	0.87	0.66
	0.34	0.09	0.02	0.03	0.04	0.02

Table 3.12 (*Continued*)

T, K	Values of Joule–Thomson effect on p, MPa					
	12.16	14.19	16.21	18.24	20.26	22.29
223.15	2.11 2.18 —0.07	1.79 1.84 —0.05	1.53 1.53 0	1.29 1.26 0.03	1.09 1.03 0.06	0.92 0.84 0.08
248.15	1.84 1.91 —0.07	1.62 1.66 —0.04	1.42 1.43 —0.01	1.23 1.22 0.01	1.06 1.03 0.03	0.96 0.87 0.09
273.15	—	1.43 1.45 —0.02	—	1.11 1.11 0	—	0.80 0.82 —0.02
298.15	—	1.22 1.24 —0.02	—	0.96 0.97 —0.01	—	0.71 0.73 —0.02

Table 3.12 (*Continued*)

T, K	Values of Joule–Thomson effect on p, MPa			
	0.101	2.027	6.080	10.13
323.15	1.86 1.99 —0.13	1.75 1.82 —0.07	1.51 1.54 —0.03	1.27 1.29 —0.02
348.15	1.56 1.67 —0.11	1.47 1.53 —0.06	1.26 1.28 —0.02	1.06 1.08 —0.02
373.15	1.31 1.40 —0.09	1.23 1.28 —0.05	1.04 1.07 —0.03	0.88 0.89 —0.01
423.15	0.91 0.97 —0.06	0.84 0.88 —0.04	0.70 0.72 —0.02	0.58 0.59 —0.01
473.15	0.62 0.65 —0.03	0.56 0.58 —0.02	0.44 0.46 —0.02	0.34 0.36 —0.02
523.15	0.40 0.40 0	0.34 0.34 0	0.25 0.25 0	0.16 0.17 —0.01
553.15	0.29 0.28 0.01	0.24 0.23 0.01	0.16 0.15 0.01	0.08 0.08 0

Table 3.12 (*Continued*)

T, K	Values of Joule–Thomson effect on p, MPa		
	14.19	18.24	22.29
323.15	1.03 1.05 —0.02	0.82 0.83 —0.01	0.62 0 63 —0 01
348.15	0.86 0.88 —0.02	0.79 0.69 0.10	0.53 0.53 0
373.15	0.71 0.73 —0.02	0.57 0.57 0	0.45 0.43 0.02
423.15	0.46 0.47 —0.01	0.36 0.36 0	0.28 0.25 0.03
473.15	0.25 0,27 —0,02	0.18 0.18 0	0.12 0.10 0.02
523.15	0.09 0.10 —0.01	0,03 0,03 0	—0,02 —0,03 0,01
553.15	0.01 0.02 —0.01	—0,05 —0.04 —0.01	—0,11 —0.10 —0.01

Table 3.13 Comparison of experimental data of Hausen [59] on the Joule–Thomson effect with analytic values

T, K	p, MPa	μ [59]	μ	$\Delta\mu$
		K/MPa		
97.65	4.903	—0.10	—0.25	0,15
97.65	7.443	—0.11	—0.28	0.17
98.45	2.530	0.03	—0.20	0.23
98.85	0.990	0.06	—0.16	0.22
100.75	12.33	—0.08	—0.30	0.22
102.85	14.87	—0.28	—0.30	0.02
107.25	18.59	—0.34	—0.30	—0.04
109.55	0.451	13.16	14.40	—1.24*
116,35	9.934	—0.01	—0.09	0.08
116,55	12,31	—0.04	—0.15	0.11
116.55	14.53	—0.15	—0.19	0.04
117.15	2,452	0.39	0.36	0,03
118.95	1.510	12.99	12.97	0.02
120.55	2.991	0.38	0.53	—0.15
121.15	0.745	12.45	12.00	0.45
122.85	4.943	0.13	0.39	—0,26

*At saturation.

Table 3.13 (*Continued*)

T, K	p, MPa	μ [59]	μ	Δμ
			K/MPa	
126.25	7.541	0.04	0.28	—0.24
127.55	2.981	7.37	1.96	5.41*
128.35	0.726	11.00	10.55	0.45
128.45	1.481	11.36	10.92	0.44
129.15	14.76	—0.17	—0.03	—0.14
130.05	12.33	—0.04	0.08	—0.12
130.65	1.549	11.00	10.55	0.45
131.05	2.962	10.03	10.17	—0.14
131.45	4.894	1.20	1.30	—0.10
136.15	4.913	2.68	2.84	—0.16
137.35	4.001	8.38	8.01	0.37
137.45	1.265	9.86	9.48	0.38
137.45	2.520	9.80	9.40	0.40
137.55	12.35	0.06	0.27	—0.21
142.55	4.903	6.19	6.23	—0.04
142.95	14.78	0.11	0.25	—0.14
143.25	5.011	6.31	6.08	0.23
144.25	7.414	1.80	1.97	—0.17
145.75	9.954	1.07	1.01	0.06
146.55	14.84	0.11	0.34	—0.23
146.75	12.35	0.47	0.61	—0.14
150.35	0.755	8.41	8.00	0.41
151.25	4.050	7.15	7.16	—0.01
152.85	4.913	6.32	6.44	—0.12
155.45	19.32	—0.03	0.22	—0.25
156.05	14.85	0.45	0.63	—0.18
158.45	9.885	2.05	2.08	—0.03
159.55	2.511	7.50	6.95	0.55
159.85	11.64	1.42	1.46	—0.04
160.05	7.394	4.04	4.08	—0.04
160.65	12.69	1.17	1.20	—0.03
161.65	14.70	0.80	0.84	—0.04
164.65	7.414	4.10	4.21	—0.11
164.65	12.55	1.41	1.44	—0.03
171.45	12.33	1.75	1.81	—0.06
171.55	0.667	6.60	6.37	0.23
172.35	9.875	2.77	2.86	—0.09
172.45	14.48	1.20	1.27	—0.07
173.75	4.903	5.86	5.31	0.55
174.65	14.44	1.37	1.34	0.03
175.85	2.373	6.44	5.81	0.63
184.15	4.965	5.40	4.78	—0.62
184.55	9.905	3.08	3.03	0.05
186.15	9.993	3.07	2.99	0.08
187.85	12.39	2.14	2.20	—0.06
188.95	9.728	3.10	3.08	0.02
189.85	20.12	0.45	0.78	—0.33
193.95	14.81	1.72	1.67	0.05
194.25	7.551	4.09	3.70	0.39
194.45	9.875	3.01	2.99	0.02
195.35	4.962	4.80	4.30	0.50

*At saturation.

Table 3.13 (*Continued*)

T, K	p, MPa	μ [59]	μ	Δμ
		K/MPa		
198.35	0.530	5.10	4.99	0.11
199.05	2.520	5,53	4,61	0,42
212.35	7.482	4,08	3,25	0.83
212,55	9.973	2,91	2,73	0.18
244,75	9 875	2.22	2,24	—0,02
246.35	9,875	2,19	2.22	—0.03
247 75	12.35	2.01	1.89	0.12
269,85	9.895	1.89	1.89	0
278.25	11.48	1,66	1,64	0.02
280.75	1.589	2,52	2.51	0,01
280.75	1,736	2,49	2.49	0
284,25	19.43	0.97	0.96	0,01
284,45	7,551	1.90	1,90	0
284.65	9,954	1.65	1,70	—0.05
285.25	12.71	1.49	1.46	—0,03
285.35	2,638	2,30	2.32	—0,02
285.35	5.031	2,11	2.10	0,01
286.25	2,187	2,26	2,35	—0.09
286,35	14.46	1.33	1,32	0.01
286,65	2,050	2.34	2.36	—0,02

Figure 35 shows the range of parameters included in tables in [7, 9, 20, 30] and in the present book. It is seen that the tables published here, unlike those that appeared previously, are validated experimentally over almost the entire range of parameters.

Deviations $\delta\rho = (\rho_{tab} - \rho/\rho) \times 100$ of the previously published tabulated data from the calculated values of density in this book basically do not exceed 0.2%. In many cases the deviations lie within the limits of ±1% (see Table 3.24). At temperatures below 150 K the deviations in a number of points amount to 0.4–0.8%, which is due to the use in this study of new experimental data [6, 34, 96]. At temperatures of 600 to 1000 K rather satisfactory agreement is observed with data in [20] which were obtained using experimental results in [10]. Significant differences exist with respect to values of density from tables in [7], obtained by extrapolating the equation of state.

It is seen from Table 3.25 that the values of enthalpy we calculated agree with the majority of data in [20] within ±0.5 kJ/kg (with the exception of several points at $T = 150$ K and $p \geq 60$ MPa). A somewhat poorer agreement (within the limits of —3.2 to +0.4 kJ/kg) is observed with data of [7, 9, 30] at 80 to 400 K. At higher temperatures the disagreement with data in [7] increases, amounting to 6–8 kJ/kg at $p = 100$ MPa. This is due to the same reason as the increase in deviations of the values of density.

The calculated values of the entropy are also compared with the tabulated data in [7, 9, 20, 30]. The deviations $\Delta s = s_{tab} - s$ are given in Table 3.26.

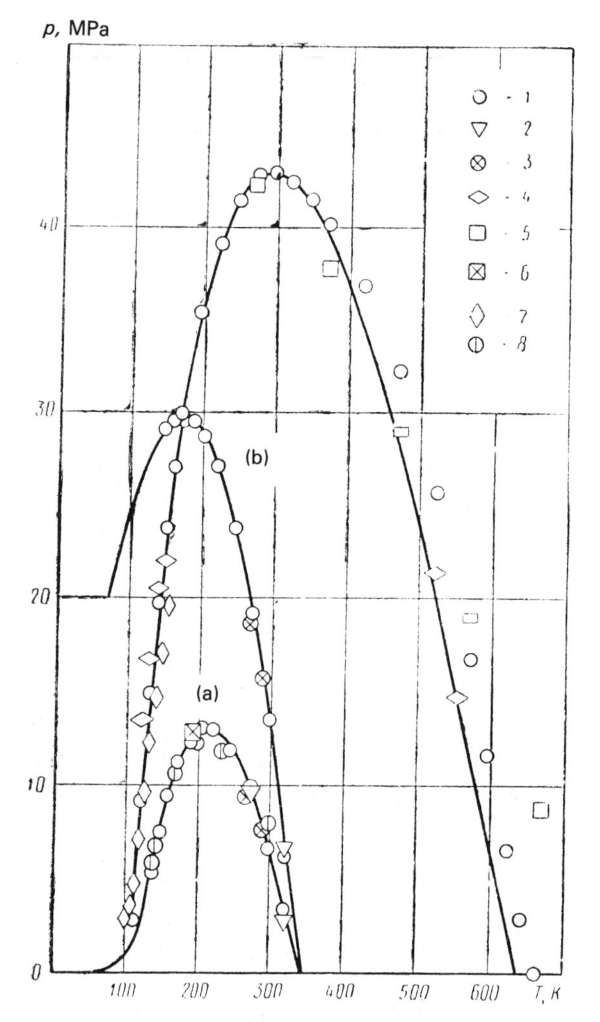

Figure 24 (a) Boyle, (b) ideal gas, and (c) Joule–Thomson inversion curves for air in pT coordinates from data of different investigators. (1) Michels et al. [84, 85]; (2) Holborn and Schultze [67]; 93) Amagat [27]; (4) Roebuck and Murrel [95]; (5) Jakob [72]; (6) Koch [76]; (7) Hausen [59]; (8) Witkowski [110]; the solid curve was plotted from the averaged equation of state.

The values we calculated agree with the majority of data in [20] with —4 to +1 J/(kg·K). The deviations go past these limits only along the 150 K isotherm at high pressures. The agreement with data of others at 150 to 400 K is within —10 to +2 kJ/(kg·K). Outside this interval, the deviations for several points are greater, amounting to 20 kJ/(kg·K).

Table 3.27 lists the deviations of data in [20, 30] from our analytic values of the isochoric specific heat. Since the data in [20, 30] were obtained only for gas phase, in the majority of points the differences are quite acceptable (±3%). At temperatures of 120 to 140 K the differences from the results in [30] become as high as 4–8%. The tables in the present study for this range of parameters were constructed using additional experimental data and are more reliable.

The tabulated data in [20] on c_p agree with those we calculated within ±0.7% (Table 3.28). The deviations at $p \geq 60$ MPa along the 150 and 200 K isotherms become as high as several percentage points since the data in [20] over this range of variables were obtained by extrapolation. The analytic values in [7, 30] at $T \geq 300$ K are mostly on the low side as compared with those we obtained (to 2.5%). At temperatures of 80 to 140 K the deviations from data in [30] and [7] lie within the limits of ±7 and ±6% respectively. Since we utilized new experimental data from [6] on the density of air at temperatures below 200 K and pressures to 60 MPa, it can be assumed that the data on the thermody-

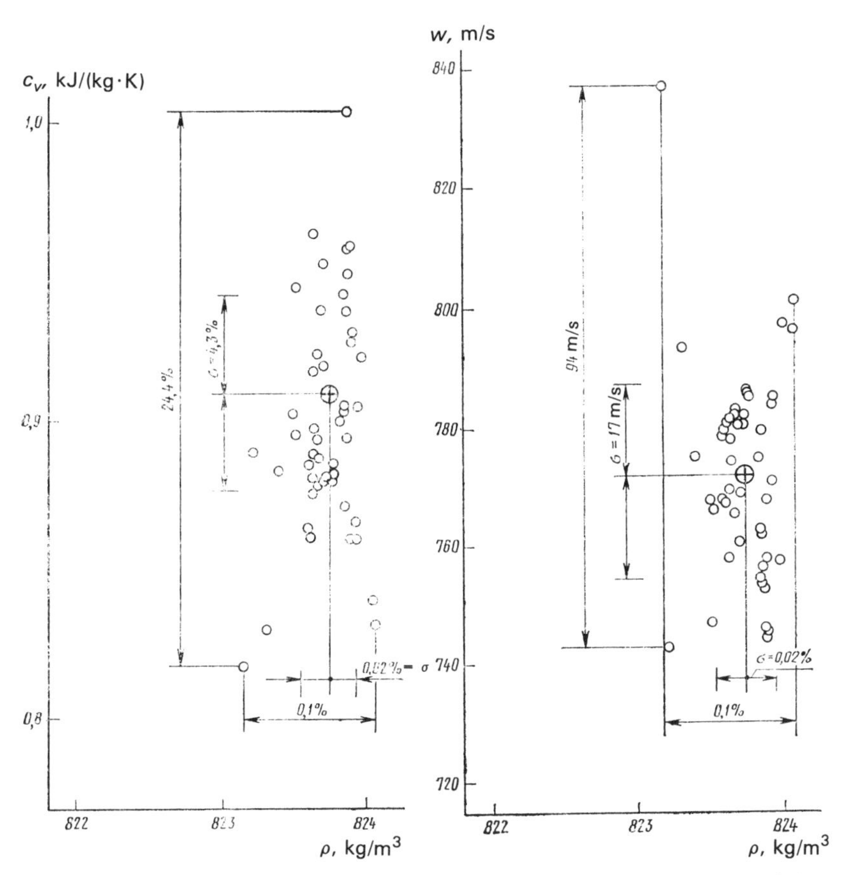

Figure 25 Scattering ellipse for analytic values of isochoric specific heat of air at $T = 90$ K and $p = 1$ MPa.

Figure 26 Scattering ellipse for analytic values of the speed of sound in air at $T = 90$ K and $p = 1$ MPa.

Table 3.14 Comparison of experimental data on van Itterbeek and de Rop [71] on the speed of sound with analytic values

T, K	p, MPa	w [71] (m/s)	w (m/s)	δw, %
229.3	0.101	303.1	303.6	−0.16
	0.437	303.1	303.4	−0.10
	0.747	303.8	303.3	0.16
	1.149	304.0	303.2	0.26
	1.331	304.1	303.2	0.29
249.25	0.101	316.3	316.6	−0.10
	0.236	316.3	316.6	−0.08
	0.336	316.4	316.6	−0.05
	0.436	316.5	316.7	−0.06
	0.518	316.7	316.7	−0.01
	0.758	316.8	316.7	0.04
	0.882	317.0	316.8	0.05
274.2	0.102	331.7	332.1	−0.13
	0.198	331.8	332.2	−0.13
	0.285	331.8	332.2	−0.11
	0.396	332.0	332.3	−0.10
	0.477	332.0	332.4	−0.13
	0.610	332.1	332.5	−0.11
	0.862	332.4	332.7	−0.08
293.1	0.163	343.7	343.4	0.09
	0.260	343.9	343.5	0.13
	0.352	343.8	343.6	0.07
	0.670	344.2	344.0	0.05
	0.881	344.4	344.2	0.06
300.1	0.102	348.5	347.4	0.32
	0.163	348.6	347.5	0.32
	0.258	348.7	347.6	0.32
	0.375	348.9	347.7	0.33
	0.525	349.1	347.9	0.34
	0.664	349.2	348.1	0.32
	0.745	349.4	348.2	0.35
	0.858	349.5	348.4	0.31
300.1	0.102	348.4	347.4	0.29
	0.228	348.5	347.6	0.26
	0.362	348.7	347.7	0.29
	0.476	348.7	347.9	0.24
	0.589	349.0	348.0	0.29
	0.763	349.1	348.2	0.27
	0.871	349.3	348.4	0.27
313.0	0.259	355.2	355.0	0.06
	0.578	356.0	355.4	0.17
	0.682	355.0	355.6	−0.16
	0.788	356.2	355.7	0.13
	0.853	356.3	355.8	0.13

Table 3.15 Comparison of experimental data on van Itterbeek and van Dael [70] on the speed of sound with analytic values

T, K	p, MPa	w [70], m/s	w, m/s	δw, %
89.98	0.363	761.1	760.9	0.03
	0.843	763.7	769.3	—0.73
	1.344	768.9	777.7	—1.13
	1.853	774.4	786.0	—1.48
	2.354	780.7	793.8	—1.65
	3.217	788.4	806.6	—2.26
	3.874	794.5	815.8	—2.61
	4.011	797.5	817.7	—2.47
	4.158	798.8	819.7	—2.54
	5.325	809.0	834.8	—3.09
	6.315	817.1	846.7	—3.49
	7.286	825.3	857.6	—3.76

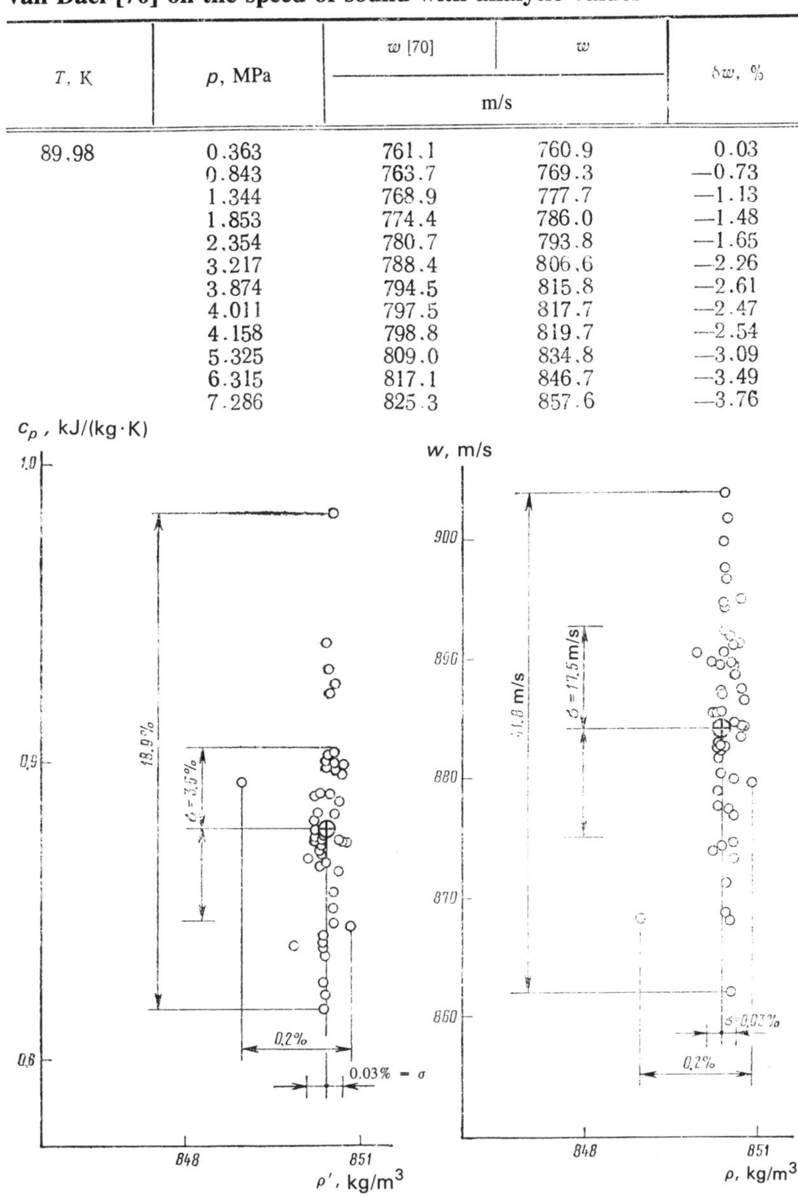

Figure 27 Scattering ellipse for analytic values of isobaric specific heat of air at T = 90 K and p = 10 MPa.

Figure 28 Scattering ellipse for analytic values of the speed of sound in air at T = 90 K and p = 10 MPa.

Table 3.16 Standard deviations of analytic values of the density of air

T, K	Deviations, % at p, MPA, of									
	0.1	1.0	2.0	3.0	4.0	5.0	6.0	7.0	8.0	9.0
70	0.06	0.14	0.22	0.28	0.34	0.38	0.42	0.45	0.48	0.51
80	0.02	0.02	0.02	0.03	0.04	0.05	0.05	0.05	0.06	0.06
90	0.01	0.02	0.02	0.03	0.03	0.03	0.03	0.03	0.03	0.03
100	0.01	0.02	0.02	0.02	0.02	0.02	0.02	0.02	0.02	0.02
110	0.01	0.02	0.02	0.02	0.02	0.02	0.02	0.03	0.03	0.03
120	0.01	0.04	0.07	0.01	0.02	0.02	0.03	0.03	0.03	0.03
130	0.01	0.03	0.02	0.03	0.02	0.02	0.02	0.02	0.02	0.02
140	0.01	0.03	0.02	0.03	0.04	0.13	0.04	0.03	0.02	0.02
150	0.01	0.02	0.01	0.03	0.06	0.05	0.04	0.06	0.03	0.03
200	0.01	0.02	0.03	0.03	0.03	0.04	0.03	0.02	0.01	0.02
250	0.01	0.02	0.02	0.02	0.02	0.02	0.02	0.03	0.03	0.03
300	0.01	0.01	0.02	0.02	0.01	0.01	0.02	0.02	0.02	0.02
400	0.01	0.01	0.02	0.03	0.03	0.03	0.03	0.03	0.03	0.03
500	0.01	0.01	0.02	0.03	0.03	0.03	0.03	0.03	0.03	0.03
600	0.01	0.01	0.02	0.03	0.03	0.03	0.03	0.03	0.03	0.03
700	0.01	0.01	0.02	0.03	0.03	0.03	0.03	0.04	0.04	0.03
800	0.01	0.01	0.02	0.03	0.04	0.04	0.04	0.05	0.05	0.05
900	0.01	0.01	0.02	0.03	0.04	0.05	0.05	0.06	0.06	0.07
1000	0.01	0.01	0.03	0.04	0.05	0.05	0.06	0.07	0.07	0.08
1100	0.01	0.02	0.03	0.04	0.05	0.06	0.07	0.08	0.08	0.09
1200	0.01	0.02	0.03	0.04	0.05	0.06	0.07	0.08	0.09	0.10
1300	0.01	0.02	0.03	0.04	0.06	0.07	0.08	0.09	0.09	0.10
1400	0.01	0.02	0.03	0.04	0.06	0.07	0.08	0.09	0.10	0.11
1500	0.01	0.02	0.03	0.04	0.06	0.07	0.08	0.09	0.10	0.11

Table 3.16 (*Continued*)

T, K	Deviations, % at p, MPa, of									
	10	20	30	40	50	60	70	80	90	100
70	0.54	0.70	0.80	0.97						
80	0.06	0.06	1.25	0.73						
90	0.03	0.02	0.05	0.12	0.24	0.73				
100	0.02	0.02	0.02	0.03	0.06	0.12	0.21	0.36	0.66	
110	0.03	0.02	0.02	0.02	0.05	0.09	0.14	0.22	0.32	0.45
120	0.03	0.03	0.03	0.02	0.01	0.03	0.06	0.09	0.13	0.18
130	0.02	0.03	0.03	0.04	0.04	0.05	0.06	0.08	0.11	0.14
140	0.02	0.02	0.03	0.04	0.05	0.08	0.11	0.15	0.19	0.25
150	0.03	0.03	0.03	0.03	0.05	0.08	0.13	0.18	0.25	0.32
200	0.03	0.02	0.01	0.02	0.03	0.02	0.03	0.07	0.13	0.20
250	0.03	0.01	0.02	0.02	0.02	0.02	0.02	0.02	0.02	0.03
300	0.02	0.02	0.01	0.01	0.02	0.01	0.02	0.04	0.06	0.07
400	0.03	0.02	0.02	0.02	0.02	0.04	0.05	0.06	0.09	0.14
500	0.03	0.03	0.03	0.03	0.02	0.02	0.04	0.07	0.09	0.13
600	0.02	0.02	0.02	0.03	0.03	0.03	0.03	0.05	0.08	0.12

Table 3.16 (*Continued*)

T, K	Deviation, %, at p, MPa, of									
	10	20	30	40	50	60	70	80	90	100
700	0.03	0.03	0.03	0.02	0.02	0.04	0.04	0.05	0.07	0.10
800	0.05	0.05	0.05	0.03	0.02	0.03	0.05	0.06	0.08	0.10
900	0.07	0.08	0.08	0.06	0.05	0.04	0.05	0.07	0.08	0.10
1000	0.08	0.10	0.10	0.09	0.08	0.06	0.06	0.07	0.09	0.10
1100	0.09	0.12	0.13	0.12	0.11	0.09	0.08	0.08	0.09	0.11
1200	0.10	0.14	0.15	0.15	0.14	0.12	0.10	0.10	0.10	0.11
1300	0.11	0.15	0.17	0.17	0.16	0.15	0.13	0.12	0.12	0.12
1400	0.11	0.17	0.19	0.19	0.19	0.17	0.16	0.14	0.14	0.14
1500	0.12	0.18	0.20	0.21	0.21	0.20	0.16	0.17	0.16	0.15

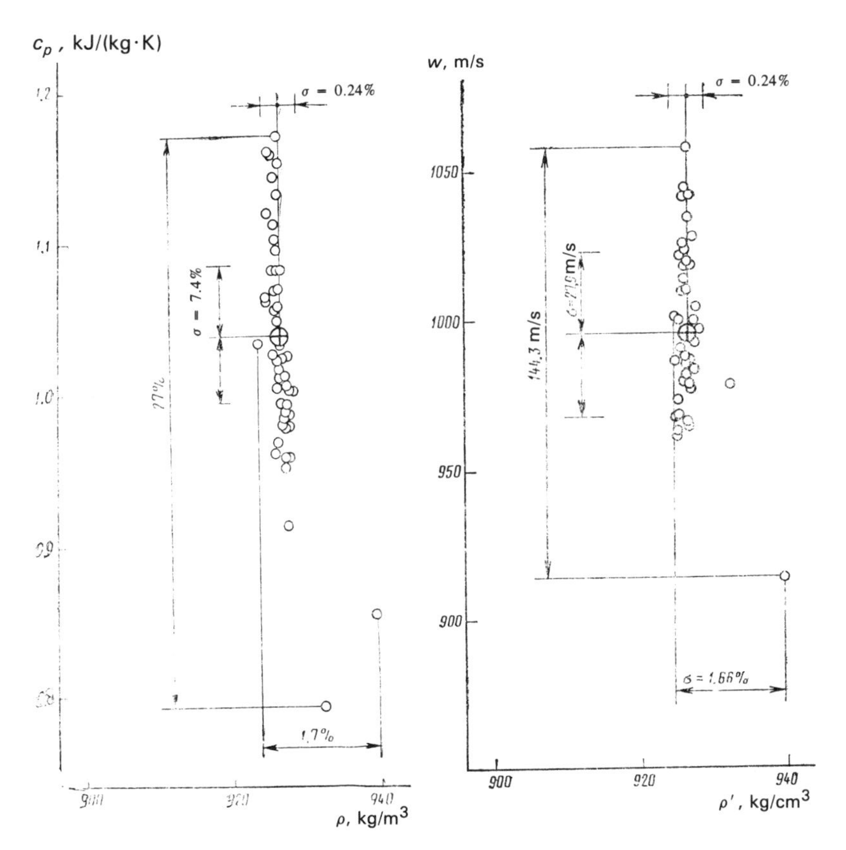

Figure 29 Scattering ellipse for analytic values of the isobaric specific heat of air at T = 90 K and p = 50 MPa.

Figure 30 Scattering ellipse for analytic values of the speed of sound in air at T = 90 K and p = 50 MPa.

Table 3.17 Standard deviations of analytic values of the enthalpy of air

T, K	Deviations, kJ/(kg·K), at p, MPA, of									
	0.1	1.0	2.0	3.0	4.0	5.0	6.0	7.0	8.0	9.0
70	1.8	1.8	1.8	1.8	1.9	1.9	1.9	2.0	2.0	2.1
80	0.4	0.4	0.4	0.4	0.4	0.4	0.4	0.4	0.4	0.4
90	0.1	0.6	0.6	0.6	0.6	0.6	0.6	0.6	0.6	0.6
100	0.1	0.4	0.4	0.4	0.4	0.4	0.4	0.4	0.4	0.4
110	0.1	0.1	0.2	0.2	0.2	0.2	0.2	0.2	0.2	0.2
120	0.1	0.1	0.1	0.2	0.2	0.2	0.2	0.2	0.2	0.2
130	0.1	0.1	0.1	0.1	0.2	0.2	0.2	0.2	0.2	0.2
140	0.1	0.1	0.1	0.2	0.2	0.2	0.2	0.2	0.2	0.2
150	0.1	0.1	0.1	0.1	0.2	0.2	0.2	0.1	0.2	0.2
200	0.1	0.1	0.1	0.1	0.1	0.1	0.1	0.1	0.1	0.1
250	0.1	0.1	0.1	0.1	0.1	0.2	0.2	0.2	0.2	0.2
300	0.1	0.1	0.1	0.1	0.1	0.2	0.2	0.2	0.2	0.2
400	0.1	0.1	0.1	0.1	0.1	0.1	0.1	0.1	0.1	0.1
500	0.1	0.1	0.1	0.1	0.1	0.2	0.2	0.2	0.2	0.2
600	0.1	0.1	0.1	0.1	0.2	0.2	0.2	0.3	0.3	0.3
700	0.1	0.1	0.1	0.2	0.2	0.2	0.3	0.3	0.4	0.4
800	0.1	0.1	0.1	0.2	0.2	0.3	0.3	0.3	0.4	0.4
900	0.1	0.1	0.1	0.2	0.2	0.2	0.3	0.3	0.4	0.4
1000	0.1	0.1	0.1	0.1	0.2	0.2	0.3	0.3	0.4	0.4
1100	0.1	0.1	0.1	0.1	0.2	0.2	0.3	0.3	0.3	0.4
1200	0.1	0.1	0.1	0.1	0.2	0.2	0.2	0.3	0.3	0.3
1300	0.1	0.1	0.1	0.1	0.1	0.2	0.2	0.2	0.3	0.3
1400	0.1	0.1	0.1	0.1	0.1	0.2	0.2	0.2	0.2	0.3
1500	0.1	0.1	0.1	0.1	0.1	0.1	0.2	0.2	0.2	0.2

Table 3.17 (*Continued*)

T, K	Deviations, kJ/(kg·K), at p, MPA, of									
	10	20	30	40	50	60	70	80	90	100
70	2.1	2.9	4.0	5.4						
80	0.4	0.5	1.2	0.9						
90	0.6	0.6	0.6	0.6	0.7	1.4				
100	0.4	0.4	0.4	0.5	0.5	0.5	0.7	1.0	2.0	
110	0.2	0.2	0.2	0.2	0.2	0.3	0.3	0.5	0.7	1.1
120	0.2	0.2	0.2	0.2	0.2	0.3	0.4	0.5	0.8	1.1
130	0.2	0.2	0.2	0.2	0.3	0.3	0.3	0.4	0.6	0.8
140	0.2	0.2	0.2	0.2	0.2	0.2	0.2	0.2	0.3	0.4
150	0 2	0.2	0.2	0.2	0.2	0.2	0.2	0.2	0.2	0.2
200	0.1	0.1	0.1	0.1	0.1	0.1	0.1	0.2	0.3	0.4
250	0.2	0.2	0.1	0.1	0.1	0.2	0.2	0.2	0.2	0.3
300	0.2	0.3	0.3	0.3	0.3	0.2	0.2	0.2	0.2	0.3
400	0.2	0.2	0.2	0.2	0.3	0.3	0.3	0.3	0.3	0.2
500	0.2	0.3	0.3	0.3	0.3	0.2	0.2	0.3	0.3	0.4
600	0.4	0.5	0.6	0.7	0.7	0.7	0.6	0.6	0.6	0.6
700	0.4	0.7	0.9	1.0	1.1	1.1	1.1	1.1	1.0	1.0
800	0.5	0.8	1.0	1.2	1.4	1.4	1.5	1.5	1.5	1.5

Table 3.17 (*Continued*)

T, K	Deviations, kJ/(kg·K), at p, MPA, of									
	10	20	30	40	50	60	70	80	90	100
900	0.5	0.8	1.1	1.3	1.5	1.7	1.8	1.8	1.9	1.9
1000	0.4	0.8	1.1	1.4	1.6	1.8	2.0	2.1	2.1	2.2
1100	0.4	0.8	1.1	1.4	1.6	1.9	2.1	2.2	2.3	2.4
1200	0.4	0.7	1.0	1.3	1.6	1.9	2.1	2.3	2.4	2.6
1300	0.3	0.6	0.9	1.2	1.5	1.8	2.1	2.3	2.5	2.6
1400	0.3	0.5	0.8	1.1	1.4	1.7	2.0	2.2	2.4	2.6
1500	0.2	0.5	0.7	1.0	1.3	1.6	1.9	2.1	2.4	2.6

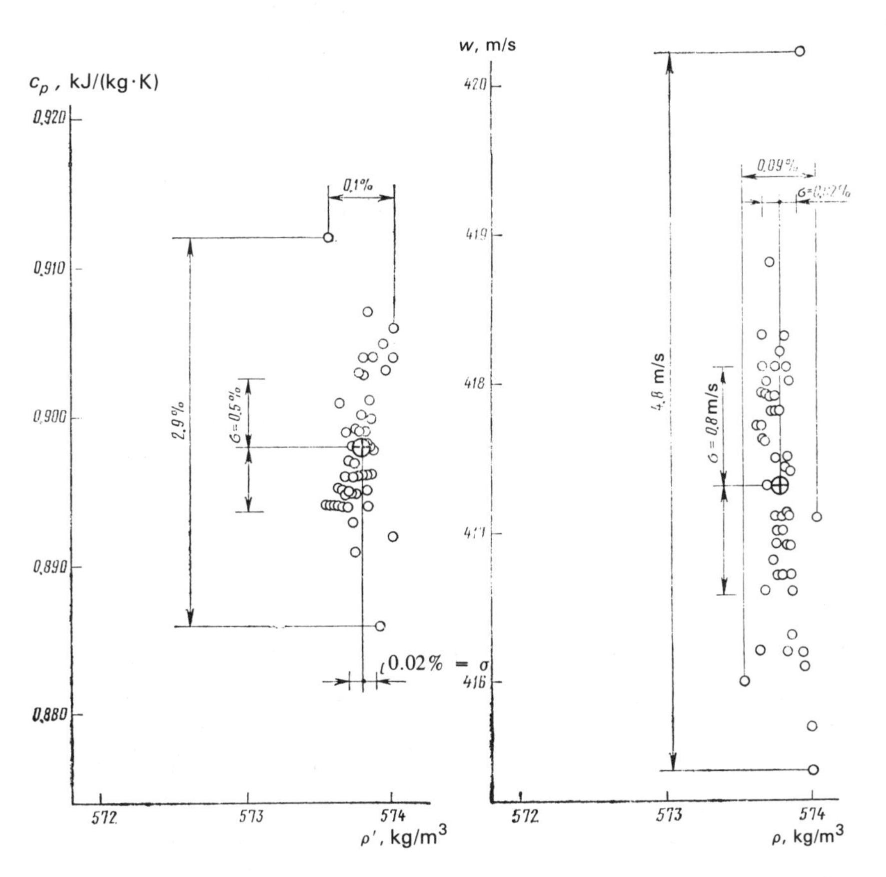

Figure 31 Scattering ellipse for analytic values of the isobaric specific heat of air at T = 90 K and p = 10 MPa.

Figure 32 Scattering ellipse for analytic values of the speed of sound in air at T = 90 K and p = 10 MPa.

Table 3.18 Standard deviations of analytic values of entropy of air

T, K	Deviation, %, at p, MPa, of									
	0.1	1.0	2.0	3.0	4.0	5.0	6.0	7.0	8.0	9.0
70	0.85	0.86	0.87	0.89	0.90	0.92	0.94	0.96	0.99	1.02
80	0.18	0.18	0.18	0.18	0.18	0.18	0.19	0.19	0.19	0.19
90	0.01	0.19	0.19	0.19	0.19	0.19	0.19	0.19	0.19	0.19
100	0.01	0.13	0.13	0.13	0.13	0.13	0.13	0.13	0.13	0.13
110	0.01	0.02	0.05	0.05	0.05	0.05	0.05	0.05	0.05	0.05
120	0.01	0.01	0.02	0.03	0.03	0.03	0.03	0.03	0.03	0.03
130	0.01	0.01	0.02	0.01	0.04	0.03	0.03	0.03	0.03	0.04
140	0.01	0.01	0.02	0.02	0.02	0.03	0.03	0.03	0.03	0.03
150	0.01	0.01	0.02	0.02	0.02	0.02	0.02	0.02	0.02	0.03
200	0.01	0.01	0.01	0.01	0.01	0.01	0.01	0.01	0.01	0.01
250	0.01	0.01	0.01	0.01	0.01	0.01	0.01	0.01	0.01	0.01
300	0.01	0.01	0.01	0.01	0.01	0.01	0.01	0.01	0.01	0.01
400	0.01	0.01	0.01	0.01	0.01	0.01	0.01	0.01	0.01	0.01
500	0.01	0.01	0.01	0.01	0.01	0.01	0.01	0.01	0.01	0.01
600	0.01	0.01	0.01	0.01	0.01	0.01	0.01	0.01	0.01	0.01
700	0.01	0.01	0.01	0.01	0.01	0.01	0.01	0.01	0.01	0.01
800	0.01	0.01	0.01	0.01	0.01	0.01	0.01	0.01	0.01	0.01
900	0.01	0.01	0.01	0.01	0.01	0.01	0.01	0.01	0.01	0.01
1000	0.01	0.01	0.01	0.01	0.01	0.01	0.01	0.01	0.01	0.01
1100	0.01	0.01	0.01	0.01	0.01	0.01	0.01	0.01	0.01	0.01
1200	0.01	0.01	0.01	0.01	0.01	0.01	0.01	0.01	0.01	0.01
1300	0.01	0.01	0.01	0.01	0.01	0.01	0.01	0.01	0.01	0.01
1400	0.01	0.01	0.01	0.01	0.01	0.01	0.01	0.01	0.01	0.01
1500	0.01	0.01	0.01	0.01	0.01	0.01	0.01	0.01	0.01	0.01

Table 3.18 (*Continued*)

T, K	Deviation, %, at p, MPa, of									
	10	20	30	40	50	60	70	80	90	100
70	1.05	1.44	1.99	2.72						
80	0.19	0.23	0.56	0.42						
90	0.20	0.20	0.21	0.22	0.25	0.51				
100	0.13	0.13	0.14	0.14	0.15	0.17	0.22	0.33	0.62	
110	0.05	0.05	0.06	0.06	0.06	0.07	0.08	0.11	0.17	0.27
120	0.03	0.04	0.04	0.04	0.05	0.06	0.09	0.13	0.18	0.25
130	0.04	0.04	0.04	0.05	0.05	0.05	0.07	0.09	0.13	0.18
140	0.03	0.04	0.04	0.04	0.04	0.04	0.04	0.05	0.07	0.10
150	0.03	0.03	0.03	0.03	0.04	0.04	0.04	0.04	0.04	0.04
200	0.01	0.02	0.02	0.01	0.01	0.01	0.02	0.02	0.04	0.05
250	0.01	0.01	0.01	0.01	0.01	0.01	0.02	0.02	0.02	0.02
300	0.02	0.02	0.02	0.02	0.02	0.01	0.01	0.01	0.02	0.02
400	0.01	0.01	0.01	0.01	0.02	0.02	0.02	0.02	0.01	0.01
500	0.01	0.01	0.01	0.01	0.01	0.01	0.01	0.01	0.01	0.01
600	0.01	0.02	0.02	0.02	0.02	0.02	0.02	0.02	0.02	0.02
700	0.01	0.02	0.02	0.03	0.03	0.03	0.03	0.03	0.03	0.03
800	0.01	0.02	0.03	0.03	0.03	0.04	0.04	0.04	0.04	0.04

Table 3.18 ***(Continued)***

T, K	Deviation, %, at p, MPa, of 10	20	30	40	50	60	70	80	90	100
900	0.01	0.02	0.03	0.03	0.04	0.04	0.04	0.04	0.04	0.04
1000	0.01	0.02	0.03	0.03	0.04	0.04	0.04	0.05	0.05	0.05
1100	0.01	0.02	0.03	0.03	0.04	0.04	0.04	0.05	0.05	0.05
1200	0.01	0.02	0.02	0.03	0.04	0.04	0.04	0.05	0.05	0.05
1300	0.01	0.02	0.02	0.03	0.03	0.04	0.04	0.05	0.05	0.05
1400	0.01	0.01	0.02	0.03	0.03	0.04	0.04	0.05	0.05	0.05
1500	0.01	0.01	0.02	0.03	0.03	0.04	0.04	0.04	0.05	0.05

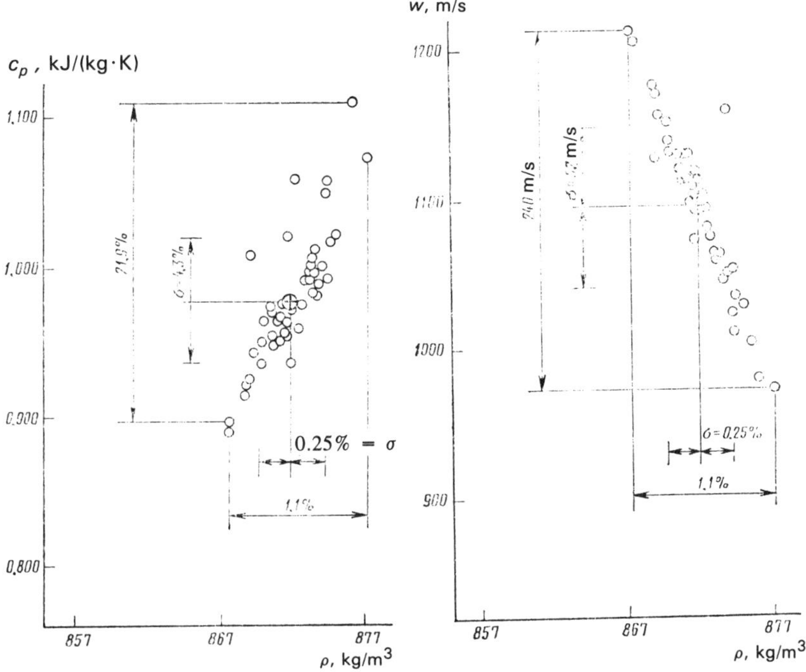

Figure 33 Scattering ellipse for analytic values of the isobaric specific heat of air at T = 140 K and p = 100 MPa.

Figure 34 Scattering ellipse for analytic values of the speed of sound in air at T = 140 K and p = 100 MPa.

namic properties of air at low temperatures obtained in this study are the most dependable.

The values of the speed of sound we calculated are in agreement with the bulk of data tabulated in [20] within the limits of ±0.3% (Table 3.29). Only on the 200 and 150 K isotherms at p = 100 MPa do the differences rise to as high as —0.9 and —2.3% for the previously mentioned reason.

This book presents a more comprehensive set of thermodynamic properties of air than the previously published books [7, 9, 30]. The thermodynamic functions calculated in [20] are the same as in the present study, but only for the gas phase and over a narrower temperature range (150 to 1000 K). The advantages of the present tables from the point of view of their experimental validation, range of values of parameters, and the number of tabulated quantities are illustrated by Fig. 35 and Table 3.30.

The temperature and pressure dependence of the tabulated thermodynamic functions is plotted in Figs. 36 to 46.

Table 3.19 Standard deviations of analytic values of the isochoric specific heat of air

T, K	Deviation, %, at p, MPa, of									
	0.1	1.0	2.0	3.0	4.0	5.0	6.0	7.0	8.0	9.0
70	119.1	63.8	42.5	35.9	34.9	35.8	37.2	38.6	40.0	41.3
80	23.5	23.1	22.9	22.7	22.7	22.7	22.7	22.8	22.9	23.0
90	0.6	4.3	4.1	3.9	3.8	3.7	3.6	3.6	3.6	3.6
100	0.3	3.6	3.6	3.6	3.6	3.6	3.6	3.6	3.6	3.6
110	0.2	2.1	3.1	3.1	3.1	3.1	3.1	3.1	3.1	3.1
120	0.1	1.0	2.5	1.8	1.8	1.8	1.8	1.9	1.9	1.9
130	0.1	0.5	0.9	1.5	0.8	0.8	0.8	0.8	0.8	0.8
140	0.1	0.3	0.5	0.7	0.7	0.5	0.6	0.5	0.5	0.5
150	0.1	0.3	0.5	0.6	0.7	0.7	0.8	0.9	0.8	0.8
200	0.1	0.2	0.3	0.4	0.5	0.6	0.6	0.6	0.6	0.6
250	0.1	0.1	0.2	0.2	0.3	0.3	0.4	0.4	0.4	0.4
300	0.1	0.1	0.1	0.1	0.2	0.2	0.2	0.2	0.2	0.2
400	0.1	0.1	0.1	0.1	0.2	0.2	0.2	0.3	0.3	0.3
500	0.1	0.1	0.1	0.1	0.2	0.2	0.2	0.2	0.3	0.3
600	0.1	0.1	0.1	0.1	0.1	0.2	0.2	0.2	0.2	0.2
700	0.1	0.1	0.1	0.1	0.1	0.1	0.1	0.2	0.2	0.2
800	0.1	0.1	0.1	0.1	0.1	0.1	0.1	0.1	0.1	0.1
900	0.1	0.1	0.1	0.1	0.1	0.1	0.1	0.1	0.1	0.1
1000	0.1	0.1	0.1	0.1	0.1	0.1	0.1	0.1	0.1	0.1
1100	0.1	0.1	0.1	0.1	0.1	0.1	0.1	0.1	0.1	0.1
1200	0.1	0.1	0.1	0.1	0.1	0.1	0.1	0.1	0.1	0.1
1300	0.1	0.1	0.1	0.1	0.1	0.1	0.1	0.1	0.1	0.1
1400	0.1	0.1	0.1	0.1	0.1	0.1	0.1	0.1	0.1	0.1
1500	0.1	0.1	0.1	0.1	0.1	0.1	0.1	0.1	0.1	0.1

Table 3.19 (*Continued*)

T, K	Deviation, %, at p, MPa, of									
	10	20	30	40	50	60	70	80	90	100
70	46.3	64.4	75.0	95.5						
80	10.2	11.3	19.7	21.0						
90	2.0	2.1	2.4	2.7	3.7	83.5				
100	1.8	2.0	2.1	2.4	2.8	3.9	5.9	10.7	49.8	
110	1.4	1.5	1.6	1.7	1.8	2.1	2.7	3.5	4.7	6.2
120	0.8	0.9	0.9	1.0	1.0	1.0	1.0	1.1	1.4	1.9
130	0.4	0.4	0.5	0.6	0.7	0.9	1.1	1.4	1.8	2.3
140	0.2	0.3	0.4	0.4	0.6	0.8	1.1	1.5	2.0	2.5

Table 3.19 (*Continued*)

T, K	Deviation, %, at p, MPa, of									
	10	20	30	40	50	60	70	80	90	100
150	0.2	0.3	0.4	0.5	0.5	0.6	0.9	1.2	1.7	2.2
200	0.4	0.2	0.2	0.2	0.3	0.4	0.5	0.6	0.7	0.8
250	0.3	0.4	0.3	0.3	0.3	0.2	0.2	0.3	0.4	0.5
300	0.2	0.2	0.2	0.2	0.3	0.3	0.3	0.2	0.2	0.3
400	0.2	0.3	0.3	0.3	0.2	0.2	0.2	0.2	0.2	0.3
500	0.2	0.3	0.4	0.4	0.4	0.4	0.4	0.3	0.3	0.3
600	0.1	0.2	0.3	0.3	0.4	0.4	0.5	0.5	0.4	0.4
700	0.1	0.1	0.2	0.2	0.3	0.3	0.4	0.4	0.4	0.4
800	0.1	0.1	0.1	0.2	0.2	0.2	0.3	0.3	0.4	0.4
900	0.1	0.1	0.1	0.1	0.1	0.2	0.2	0.2	0.3	0.3
1000	0.1	0.1	0.1	0.1	0.1	0.1	0.1	0.2	0.2	0.2
1100	0.1	0.1	0.1	0.1	0.1	0.1	0.1	0.1	0.1	0.2
1200	0.1	0.1	0.1	0.1	0.1	0.1	0.1	0.1	0.1	0.1
1300	0.1	0.1	0.1	0.1	0.1	0.1	0.1	0.1	0.1	0.1
1400	0.1	0.1	0.1	0.1	0.1	0.1	0.1	0.1	0.1	0.1
1500	0.1	0.1	0.1	0.1	0.1	0.1	0.1	0.1	0.1	0.1

Table 3.20 Standard deviations of analytic values of isobaric specific heat of air

T, K	Deviation, %, at p, MPa, of									
	0.1	1.0	2.0	3.0	4.0	5.0	6.0	7.0	8.0	9.0
70	26.9	27.1	27.8	29.2	31.3	33.6	36.2	38.8	41.4	43.9
80	9.4	9.5	9.5	9.6	9.7	9.8	9.9	9.9	10.0	10.1
90	0.5	1.9	1.9	1.9	1.9	1.9	1.9	1.9	1.9	2.0
100	0.2	1.6	1.6	1.7	1.7	1.7	1.7	1.8	1.8	1.8
110	0.1	1.7	1.2	1.2	1.2	1.3	1.3	1.3	1.3	1.4
120	0.1	0.7	1.8	0.6	0.6	0.7	0.7	0.7	0.7	0.8
130	0.1	0.3	0.6	0.6	0.3	0.3	0.3	0.3	0.3	0.3
140	0.1	0.3	0.4	0.4	0.4	0.4	0.3	0.2	0.2	0.1
150	0.1	0.2	0.4	0.4	0.4	0.4	0.4	0.3	0.3	0.3
200	0.1	0.1	0.2	0.3	0.3	0.4	0.4	0.4	0.4	0.4
250	0.1	0.1	0.1	0.1	0.2	0.2	0.2	0.2	0.3	0.3
300	0.1	0.1	0.1	0.1	0.1	0.1	0.2	0.2	0.2	0.2
400	0.1	0.1	0.1	0.1	0.1	0.2	0.2	0.2	0.2	0.2
500	0.1	0.1	0.1	0.1	0.1	0.1	0.1	0.1	0.2	0.2
600	0.1	0.1	0.1	0.1	0.1	0.1	0.1	0.1	0.1	0.1
700	0.1	0.1	0.1	0.1	0.1	0.1	0.1	0.1	0.1	0.1
800	0.1	0.1	0.1	0.1	0.1	0.1	0.1	0.1	0.1	0.1
900	0.1	0.1	0.1	0.1	0.1	0.1	0.1	0.1	0.1	0.1
1000	0.1	0.1	0.1	0.1	0.1	0.1	0.1	0.1	0.1	0.1
1100	0.1	0.1	0.1	0.1	0.1	0.1	0.1	0.1	0.1	0.1
1200	0.1	0.1	0.1	0.1	0.1	0.1	0.1	0.1	0.1	0.1
1300	0.1	0.1	0.1	0.1	0.1	0.1	0.1	0.1	0.1	0.1
1400	0.1	0.1	0.1	0.1	0.1	0.1	0.1	0.1	0.1	0.1
1500	0.1	0.1	0.1	0.1	0.1	0.1	0.1	0.1	0.1	0.1

Table 3.20 (*Continued*)

T, K	Deviation, %, at p, MPa, of									
	10	20	30	40	50	60	70	80	90	100
70	42.4	49.8	57.2	74.8						
80	23.1	23.1	22.9	20.8						
90	3.5	3.5	3.9	5.2	7.4	11.7				
100	3.6	3.6	3.5	3.5	3.7	4.1	4.7	6.2	10.2	
110	3.1	3.1	3.0	3.1	3.3	3.9	5.1	7.0	10.2	15.2
120	1.9	1.8	1.8	1.9	2.2	2.8	3.9	5.5	7.6	10.3
130	0.8	0.8	0.8	0.9	1.1	1.7	2.5	3.7	5.0	6.7
140	0.5	0.6	0.7	0.7	0.8	1.2	1.7	2.4	3.3	4.3
150	0.8	0.8	0.9	0.9	1.0	1.1	1.4	1.8	2.3	2.9
200	0.6	0.5	0.5	0.5	0.6	0.6	0.6	0.6	0.6	0.6
250	0.5	0.6	0.5	0.5	0.5	0.6	0.7	0.9	1.0	1.2
300	0.3	0.3	0.3	0.3	0.4	0.4	0.5	0.7	0.9	1.2
400	0.3	0.4	0.4	0.4	0.4	0.4	0.4	0.4	0.5	0.6
500	0.3	0.4	0.5	0.5	0.5	0.6	0.6	0.6	0.6	0.6
600	0.3	0.4	0.5	0.5	0.6	0.6	0.6	0.6	0.6	0.6
700	0.2	0.3	0.4	0.5	0.5	0.5	0.6	0.6	0.6	0.6
800	0.2	0.3	0.3	0.4	0.4	0.5	0.5	0.5	0.5	0.5
900	0.1	0.2	0.3	0.3	0.4	0.4	0.4	0.4	0.5	0.5
1000	0.1	0.2	0.2	0.3	0.3	0.3	0.4	0.4	0.4	0.4
1100	0.1	0.1	0.2	0.2	0.3	0.3	0.3	0.3	0.3	0.3
1200	0.1	0.1	0.2	0.2	0.2	0.2	0.3	0.3	0.3	0.3
1300	0.1	0.1	0.1	0.2	0.2	0.2	0.2	0.2	0.2	0.3
1400	0.1	0.1	0.1	0.1	0.2	0.2	0.2	0.2	0.2	0.2
1500	0.1	0.1	0.1	0.1	0.1	0.2	0.2	0.2	0.2	0.2

Table 3.21 Standard deviations of analytic values of the speed of sound in air

T, K	Deviation, %, at p, MPa, of									
	0.1	1.0	2.0	3.0	4.0	5.0	6.0	7.0	8.0	9.0
70	456.7	535.1	13.9	12.5	11.4	10.5	9.8	9.2	8.7	8.2
80	11.5	10.7	9.9	9.3	8.9	8.5	8.3	8.1	8.0	7.9
90	0.1	2.2	1.8	1.6	1.4	1.2	1.1	1.0	1.0	1.0
100	0.1	1.7	1.6	1.4	1.3	1.3	1.2	1.2	1.2	1.1
110	0.1	0.3	1.3	1.2	1.2	1.1	1.1	1.1	1.0	1.0
120	0.1	0.2	0.5	0.8	0.7	0.7	0.7	0.7	0.6	0.6
130	0.1	0.1	0.2	0.5	0.4	0.3	0.3	0.3	0.3	0.3
140	0.1	0.1	0.1	0.2	0.2	0.2	0.2	0.2	0.2	0.2
150	0.1	0.1	0.1	0.1	0.1	0.2	0.3	0.4	0.4	0.3
200	0.1	0.1	0.1	0.1	0.1	0.1	0.1	0.1	0.1	0.2
250	0.1	0.1	0.1	0.1	0.1	0.1	0.1	0.1	0.1	0.1
300	0.1	0.1	0.1	0.1	0.1	0.1	0.1	0.1	0.1	0.1
400	0.1	0.1	0.1	0.1	0.1	0.1	0.1	0.1	0.1	0.1
500	0.1	0.1	0.1	0.1	0.1	0.1	0.1	0.1	0.1	0.1
600	0.1	0.1	0.1	0.1	0.1	0.1	0.1	0.1	0.1	0.1
700	0.1	0.1	0.1	0.1	0.1	0.1	0.1	0.1	0.1	0.1
800	0.1	0.1	0.1	0.1	0.1	0.1	0.1	0.1	0.1	0.1

Table 3.21 (*Continued*)

T, K	Deviation, %, at p, MPa, of									
	1.0	1.0	2.0	3.0	4.0	5.0	6.0	7.0	8.0	9.0
900	0.1	0.1	0.1	0.1	0.1	0.1	0.1	0.1	0.1	0.1
1000	0.1	0.1	0.1	0.1	0.1	0.1	0.1	0.1	0.1	0.1
1100	0.1	0.1	0.1	0.1	0.1	0.1	0.1	0.1	0.1	0.1
1200	0.1	0.1	0.1	0.1	0.1	0.1	0.1	0.1	0.1	0.1
1300	0.1	0.1	0.1	0.1	0.1	0.1	0.1	0.1	0.1	0.1
1400	0.1	0.1	0.1	0.1	0.1	0.1	0.1	0.1	0.1	0.1
1500	0.1	0.1	0.1	0.1	0.1	0.1	0.1	0.1	0.1	0.1

Table 3.21 (*Continued*)

T, K	Deviation, %, at p, MPa, of									
	10	20	30	40	50	60	70	80	90	100
70	7.9	582.7	457.0	400.1						
80	7.9	8.2	15.6	15.4						
90	1.0	1.2	1.3	1.8	2.7	14.5				
100	1.1	0.9	0.7	1.1	1.9	2.9	4.3	6.3	10.8	
110	1.0	0.8	0.7	0.7	0.9	1.2	1.5	1.9	2.5	3.8
120	0.6	0.6	0.5	0.5	0.7	1.0	1.5	2.3	3.3	4.8
130	0.3	0.3	0.3	0.4	0.7	1.3	2.0	2.8	3.8	5.1
140	0.2	0.2	0.2	0.2	0.7	1.3	2.0	2.8	3.8	4.8
150	0.3	0.3	0.3	0.3	0.6	1.1	1.8	2.6	3.5	4.4
200	0.1	0.2	0.2	0.2	0.2	0.3	0.5	0.8	1.1	1.5
250	0.1	0.2	0.1	0.1	0.1	0.2	0.3	0.3	0.4	0.4
300	0.1	0.1	0.1	0.1	0.1	0.2	0.4	0.5	0.7	0.8
400	0.1	0.1	0.1	0.1	0.2	0.2	0.2	0.4	0.7	0.9
500	0.1	0.1	0.1	0.1	0.1	0.2	0.3	0.3	0.5	0.7
600	0.1	0.1	0.1	0.1	0.1	0.2	0.2	0.3	0.4	0.5
700	0.1	0.1	0.1	0.1	0.1	0.2	0.2	0.3	0.3	0.4
800	0.1	0.1	0.1	0.1	0.1	0.1	0.2	0.2	0.3	0.3
900	0.1	0.2	0.2	0.1	0.1	0.1	0.2	0.2	0.3	0.3
1000	0.1	0.2	0.2	0.1	0.1	0.1	0.2	0.2	0.2	0.3
1100	0.1	0.2	0.2	0.2	0.2	0.2	0.2	0.2	0.2	0.3
1200	0.1	0.2	0.2	0.2	0.2	0.2	0.2	0.2	0.2	0.2
1300	0.1	0.2	0.2	0.2	0.2	0.2	0.2	0.2	0.2	0.2
1400	0.1	0.2	0.2	0.2	0.2	0.2	0.2	0.2	0.2	0.2
1500	0.1	0.2	0.2	0.2	0.2	0.2	0.2	0.2	0.2	0.2

Table 3.22 Standard deviations of analytic values of the adiabatic Joule–Thomson effect in air

T, K	Deviation, %, at p, MPa, of									
	0.1	1.0	2.0	3.0	4.0	5.0	6.0	7.0	8.0	9.0
70	389.7	894.3	282.3	95.8	58.4	47.7	45.1	45.2	46.3	47.6
80	9.6	9.7	9.9	10.1	10.2	10.4	10.6	10.8	11.0	11.1
90	1.5	2.1	2.1	2.1	2.1	2.1	2.1	2.2	2.2	2.2

Table 3.22 (*Continued*)

T, K	Deviation, %, at p, MPa, of									
	0.1	1.0	2.0	3.0	4.0	5.0	6.0	7.0	8.0	9.0
100	1.1	2.5	2.4	2.3	2.3	2.2	2.2	2.2	2.2	2.1
110	0.9	0.5	1.8	6.8	4.2	2.3	1.9	1.8	1.7	1.7
120	1.1	0.5	0.7	0.7	0.8	0.9	1.1	1.5	2.7	22.6
130	1.2	0.6	0.4	0.5	0.2	0.2	0.3	0.4	0.5	0.7
140	1.2	0.6	0.3	0.3	0.2	0.2	0.2	0.2	0.3	0.4
150	1.1	0.5	0.3	0.3	0.3	0.2	0.2	0.2	0.2	0.3
200	0.9	0.7	0.5	0.5	0.4	0.3	0.2	0.2	0.2	0.2
250	1.8	1.4	1.1	0.9	0.7	0.6	0.5	0.4	0.4	0.4
300	2.3	1.9	1.5	1.3	1.1	1.0	0.9	0.8	0.7	0.7
400	3.0	2.4	2.0	1.8	1.6	1.5	1.4	1.3	1.2	1.1
500	7.8	7.3	7.0	6.7	6.6	6.4	6.2	5.9	5.6	5.3
600	41.5	45.9	53.2	64.6	84.2	124.5	257.0	1611.6	178.4	89.4
700	40.1	34.4	29.8	26.3	23.5	21.2	19.2	17.5	16.0	14.6
800	17.4	16.1	15.0	14.0	13.2	12.5	11.8	11.2	10.6	10.1
900	11.8	11.2	10.6	10.1	9.7	9.3	9.0	8.7	8.4	8.1
1000	9.0	8.6	8.2	7.9	7.7	7.4	7.3	7.1	7.0	6.8
1100	7.2	6.9	6.6	6.4	6.2	6.1	6.0	5.9	5.8	5.8
1200	6.0	5.7	5.4	5.2	5.1	5.0	5.0	4.9	4.9	4.9
1300	5.0	4.8	4.5	4.3	4.2	4.1	4.1	4.1	4.1	4.1
1400	4.4	4.1	3.8	3.6	3.5	3.4	3.4	3.4	3.4	3.4
1500	3.9	3.6	3.3	3.1	2.9	2.8	2.8	2.7	2.7	2.8

Table 3.22 (*Continued*)

T, K	Deviation, %, at p, MPa, of									
	10	20	30	40	50	60	70	80	90	100
70	49.0	60.3	69.1	82.6						
80	11.3	13.6	1434.4	333.8						
90	2.2	2.3	2.7	3.7	6.6	31.6				
100	2.1	2.1	2.2	2.7	3.8	5.7	8.9	14.1	24.1	
110	1.6	1.6	1.6	1.6	1.6	1.8	2.4	3.5	5.2	8.4
120	3.6	1.0	1.1	1.3	1.5	1.8	2.3	3.1	4.2	5.9
130	0.9	1.5	0.7	0.9	1.3	1.9	2.6	3.5	4.4	5.6
140	0.4	6.2	1.1	0.6	0.7	1.2	1.9	2.8	3.7	4.6
150	0.3	1.9	1.7	1.0	0.7	0.6	0.9	1.6	2.3	3.2
200	0.2	0.4	1.3	2.6	0.8	0.8	1.3	2.0	2.5	3.0
250	0.4	0.3	0.5	4.1	1.1	0.6	0.6	0.7	0.9	1.3
300	0.7	0.3	0.5	3.0	2.3	1.4	0.9	0.8	1.0	1.3
400	1.0	1.3	1.9	6.1	1.9	0.8	0.7	1.1	1.2	1.1
500	5.0	5.1	2.6	1.4	1.8	1.8	1.5	1.0	1.0	1.2
600	57.2	6.7	2.5	1.0	0.8	1.3	1.6	1.7	1.5	1.4
700	13.4	5.9	3.3	2.0	1.1	0.7	1.1	1.5	1.7	1.7
800	9.6	5.8	3.8	2.7	1.9	1.2	0.8	1.1	1.4	1.7
900	7.9	5.6	4.1	3.2	2.5	1.8	1.3	1.0	1.1	1.4
1000	6.7	5.3	4.3	3.5	2.9	2.4	1.8	1.4	1.1	1.2
1100	5.7	5.0	4.3	3.7	3.2	2.8	2.3	1.8	1.5	1.3
1200	4.9	4.6	4.2	3.7	3.4	3.0	2.6	2.2	1.9	1.6
1300	4.1	4.2	4.0	3.7	3.5	3.2	2.9	2.6	2.2	1.9
1400	3.4	3.7	3.7	3.6	3.5	3.3	3.1	2.7	2.5	2.2
1500	2.8	3.3	3.5	3.5	3.4	3.3	3.2	3.0	2.7	2.5

Table 3.23 Standard deviations of analytic values of the adiabatic index of air

T, K	Deviation, %, at p, MPa, of									
	0.1	1.0	2.0	3.0	4.0	5.0	6.0	7.0	8.0	9.0
70	718.0	957.1	80.1	59.9	56.1	56.2	58.6	63.8	73.8	94.5
80	25.4	23.3	21.5	20.0	18.9	18.0	17.5	17.1	16.8	16.7
90	0.2	4.4	3.7	3.1	2.7	2.4	2.2	2.1	2.0	2.0
100	0.1	3.5	3.2	2.9	2.7	2.6	2.5	2.4	2.3	2.3
110	0.1	0.6	2.7	2.5	2.3	2.2	2.2	2.1	2.1	2.1
120	0.1	0.4	1.0	1.6	1.4	1.4	1.3	1.3	1.3	1.3
130	0.1	0.2	0.5	1.0	0.8	0.6	0.6	0.6	0.6	0.6
140	0.1	0.1	0.2	0.3	0.5	0.4	0.6	0.5	0.4	0.4
150	0.1	0.1	0.2	0.2	0.3	0.4	0.6	0.7	0.7	0.6
200	0.1	0.1	0.1	0.2	0.2	0.2	0.2	0.3	0.3	0.3
250	0.1	0.1	0.1	0.1	0.1	0.1	0.1	0.1	0.2	0.2
300	0.1	0.1	0.1	0.1	0.1	0.1	0.1	0.1	0.1	0.1
400	0.1	0.1	0.1	0.1	0.1	0.1	0.1	0.1	0.1	0.1
500	0.1	0.1	0.1	0.1	0.1	0.1	0.1	0.1	0.1	0.1
600	0.1	0.1	0.1	0.1	0.1	0.1	0.1	0.1	0.1	0.2
700	0.1	0.1	0.1	0.1	0.1	0.1	0.1	0.1	0.1	0.2
800	0.1	0.1	0.1	0.1	0.1	0.1	0.1	0.1	0.2	0.2
900	0.1	0.1	0.1	0.1	0.1	0.1	0.1	0.1	0.2	0.2
1000	0.1	0.1	0.1	0.1	0.1	0.1	0.1	0.1	0.2	0.2
1100	0.1	0.1	0.1	0.1	0.1	0.1	0.1	0.1	0.2	0.2
1200	0.1	0.1	0.1	0.1	0.1	0.1	0.1	0.1	0.2	0.2
1300	0.1	0.1	0.1	0.1	0.1	0.1	0.1	0.1	0.1	0.2
1400	0.1	0.1	0.1	0.1	0.1	0.1	0.1	0.1	0.1	0.2
1500	0.1	0.1	0.1	0.1	0.1	0.1	0.1	0.1	0.1	0.2

Table 3.23 ***(Continued)***

T, K	Deviation, %, at p, MPa, of									
	10	20	30	40	50	60	70	80	90	100
70	147.0	936.3	611.4	766.4						
80	16.6	16.7	16.7	19.0						
90	2.0	2.4	2.5	3.4	5.3	7.6				
100	2.2	1.7	1.5	2.2	3.7	5.7	8.4	12.4	23.3	
110	2.0	1.7	1.4	1.5	1.8	2.3	2.9	3.7	5.0	8.1
120	1.3	1.1	1.0	1.1	1.4	2.0	3.1	4.6	6.8	9.9
130	0.6	0.6	0.5	0.7	1.4	2.5	3.9	5.6	7.6	10.1
140	0.4	0.5	0.4	0.5	1.3	2.5	3.9	5.5	7.4	9.4
150	0.5	0.5	0.6	0.5	1.1	2.2	3.5	5.0	6.7	8.4
200	0.3	0.3	0.3	0.4	0.5	0.7	1.1	1.6	2.1	2.8
250	0.2	0.3	0.2	0.2	0.3	0.4	0.6	0.7	0.7	0.8
300	0.1	0.2	0.1	0.2	0.2	0.4	0.7	1.0	1.3	1.5
400	0.1	0.1	0.1	0.2	0.3	0.3	0.4	0.8	1.3	1.7
500	0.1	0.2	0.1	0.2	0.3	0.4	0.5	0.6	0.8	1.2
600	0.2	0.2	0.2	0.2	0.2	0.3	0.4	0.5	0.6	0.9
700	0.2	0.2	0.2	0.2	0.2	0.3	0.4	0.5	0.6	0.7
800	0.2	0.2	0.2	0.2	0.2	0.3	0.3	0.4	0.5	0.6
900	0.2	0.2	0.2	0.2	0.2	0.3	0.3	0.4	0.4	0.5

Table 3.23 (*Continued*)

T, K	Deviation, %, at p, MPa, of									
	10	20	30	40	50	60	70	80	90	100
1000	0.2	0.2	0.2	0.2	0.2	0.2	0.3	0.3	0.4	0.5
1100	0.2	0.2	0.2	0.2	0.2	0.2	0.3	0.3	0.4	0.4
1200	0.2	0.2	0.3	0.2	0.2	0.2	0.3	0.3	0.4	0.4
1300	0.2	0.2	0.3	0.2	0.2	0.2	0.3	0.3	0.3	0.4
1400	0.2	0.2	0.3	0.2	0.2	0.2	0.2	0.3	0.3	0.4
1500	0.2	0.2	0.3	0.3	0.2	0.2	0.2	0.3	0.3	0.4

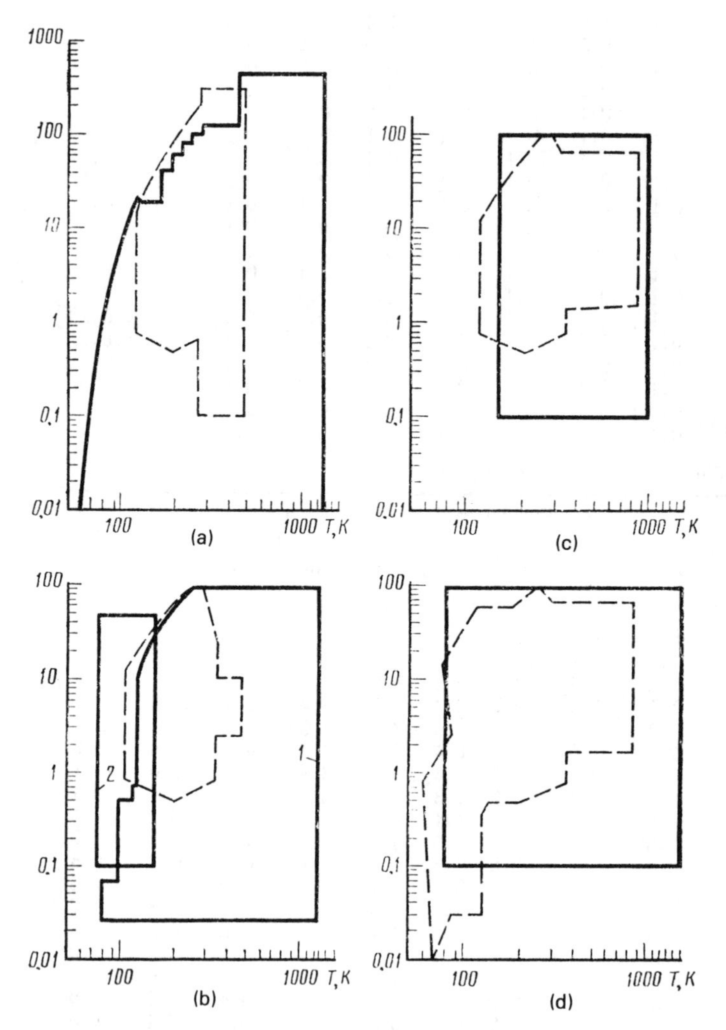

Figure 35 Experimental validation of tables of thermodynamic properties of air. (a) Baehr and Schwier [30]. (b) 1, Vasserman et al. [7]; 2, Vasserman and Rabinovich [9]. (c) Sychev et al. [20]. (d) Present tables. The dashed line designates the boundaries of the experimentally explored region.

Table 3.24 Deviations $\delta\rho$ of tabulated data of [7, 9, 20, 30] from analytic values of density

T, K	Reference	δc_p, %, at p, MPa									
		1	2	3	5	10	20	40	60	80	100
80	[9]	0.43	0.10	0.05	-0.03	-0.17	-0.31				
90	[9]	0.10	0.05	0	-0.08	-0.18	-0.32	-0.47			
100	[9]	0.01	-0.07	-0.15	-0.24	-0.36	-0.42	-0.49			
110	[30]	0.54									
	[7.9]	0.13	-0.08	-0.20	-0.36	-0.48	-0.46	-0.40			
120	[30]	0.42	-0.84								
	[7.9]	0.15	0.37	0.07	-0.25	-0.44	-0.37	-0.22			
130	[30]	0.20	0	-0.57							
	[7.9]	0.07	0.36	0.17	0.09	-0.13	-0.13	0.03			
140	[30]	0.07	0.05	-0.04	0.67	-0.07	-0.03				
	[7]	0	0.25	0.45	0.39	-0.07	-0.03	0.18			
150	[30]	0	0.02	0	-0.07	-0.20	-0.05				
	[7]	0	0.07	0.24	0.04	-0.14	0.06	0.15			
	[20]	-0.04	0.04	0.14	0.07	-0.04	-0.01	0.01	0.15	0.36	0.65
200	[30]	-0.11	-0.08	-0.04	0.04	-0.16	-0.04	-0.02	-0.06		
	[7]	-0.06	-0.05	0.02	0.05	-0.11	0	-0.02	-0.06		
	[20]	0	-0.03	0	0.06	-0.02	-0.04	-0.02	-0.05	-0.12	-0.08
300	[30]	0	-0.04	-0.06	-0.03	0	-0.05	-0.01	-0.01	0	0.01
	[7]	0	0	0	0.05	0	-0.01	0.01	0	0.07	0.16
	[20]	0	0	0	0.03	0.05	0.02	0.03	0	0.03	0.09
400	[30]	-0.02	-0.02	-0.08	-0.09	-0.18	-0.11	-0.21	-0.18	-0.07	0
	[7]	-0.02	0	0	-0.02	-0.11	-0.17	-0.10	-0.09	-0.09	0.02
	[20]	0.02	0.06	0	0.05	0.05	0.02	0.05	-0.02	-0.01	0.06
600	[7]	0.07	0	-0.06	-0.07	-0.27	-0.64	-0.83	-0.69	-0.54	-0.49
	[20]	0.09	0.09	0	0.03	0.05	0.05	0.03	0.05	0.04	0
800	[7]	-0.05	-0.08	-0.08	-0.14	-0.33	-0.79	-1.35	-1.36	-1.23	-1.08
	[20]	-0.02	-0.02	0.05	0.04	0.07	0.05	0.04	0.08	0.12	0.12
1000	[7]	0.03	0	-0.10	-0.17	-0.36	-0.79	-1.56	-1.82	-1.82	-1.66
	[20]	0.09	0.06	0.03	0.03	0.06	0.05	0.06	0.08	0.14	0.16

Table 3.25 Deviations Δh of tabulated data of [7, 9, 20, 30] from analytic values of enthalpy

T, K	Reference	Δh, kJ/kg, at p, MPa									
		1	2	3	5	10	20	40	60	80	100
80	[9]	−1.5	−1.5	−1.5	−1.5	−1.5	−1.3				
90	[9]	0.3	0.4	0.3	0.3	0.2	0.1	0.1			
100	[9]	0.3	0.3	0.2	0.2	0.1	0	0			
110	[30]	−2.0									
	[7.9]	0.1	−0.7	−0.8	−0.8	−0.8	−0.8	−0.6			
120	[30]	−1.6	−0.9								
	[7.9]	−0.3	−0.2	−2.3	−2.1	−2.0	−1.6	−1.1			
130	[30]	−1.2	−1.4	−0.4							
	[7.9]	−0.3	−0.6	−0.3	−2.9	−2.8	−2.1	−1.3			
140	[30]	−0.7	−1.0	−0.9	−1.6	−1.2	−1.2				
	[7.9]	−0.2	−0.5	−0.7	−0.4	−2.6	−1.6	−1.3			
150	[30]	−0.6	−0.9	−0.8	−0.7	−0.8	−0.8				
	[7]	−0.1	−0.4	−0.5	−0.6	−0.6	−0.9	−1.1			
	[20]	−0.1	−0.2	−0.2	−0.2	−0.4	−0.3	−0.3	−0.7	−1.5	−2.8
200	[30]	0.2	0.2	0.2	0.1	0.3	0.2	0.2	0.3		
	[7]	0.1	0.1	0.1	0	0.1	0.1	0.1	0.1		
	[20]	0.1	0.1	0	−0.1	0	0.2	0.1	0.2	0.2	0.1
300	[30]	0.3	0	0.4	0.4	0.2	0.1	0	−0.1	−0.1	−0.1
	[7]	0.1	0.1	0.1	0.1	0	−0.1	−0.2	−0.2	−0.4	−0.4
	[20]	0	0.1	0.2	0.2	0.2	0.2	0.2	0.2	0.1	0.2
400	[30]	0.3	0.1	0.2	0.1	−0.2	−1.1	−1.6	−2.3	−2.7	−3.0
	[7]	0	0	−0.1	−0.2	−0.4	−1.0	−2.1	−2.6	−2.3	−3.2
	[20]	0	0	0	0	0.1	0.1	0.1	0.1	0.2	0
600	[7]	−0.1	−0.1	−0.1	−0.4	−0.6	−1.3	−3.1	−5.0	−6.4	−7.2
	[20]	−0.1	−0.1	−0.1	−0.2	−0.2	−0.1	0	0	0.1	0.4
800	[7]	−0.2	−0.2	−0.2	−0.3	−0.5	−0.3	−2.1	−4.0	−6.1	−7.8
	[20]	−0.2	−0.2	−0.2	−0.2	−0.3	−0.3	−0.2	−0.2	−0.2	0
1000	[7]	−0.1	−0.1	−0.2	−0.2	−0.2	−0.3	−0.7	−2.0	−3.8	−5.9
	[20]	−0.3	−0.3	−0.3	−0.4	−0.4	−0.5	−0.5	−0.5	−0.5	−0.5

Table 3.26 Deviations Δs of tabulated data of [7, 9, 20, 30] from the analytic values of entropy

T, K	Reference	Δs, J/(kg·K)									
		1	2	3	5	10	20	40	60	80	100
80	[9]	−6	−6	−6	−6	−6	−4				
90	[9]	11	12	12	11	11	9	7			
100	[9]	8	8	8	7	6	5	4			
110	[30]	−16									
	[7.9]	2	−4	−5	−5	−5	−6	−5			
120	[30]	−14	−8								
	[7.9]	−1	0	−16	−16	−15	−13	−8			
130	[30]	−10	−11	−4							
	[7.9]	−2	−3	0	−15	−20	−15	−9			
140	[30]	−6	−8	−7	−12	−10	−10				
	[7.9]	−1	−2	−4	−1	−12	−10	−9			
150	[30]	−5	−7	−7	−6	−6	−7				
	[7]	0	−1	−3	−2	−2	−4	−6			
	[20]	−1	−2	−3	−2	−3	−3	−3	−6	−10	−17
200	[30]	−1	−1	−1	−2	−1	−1	−1	−1		
	[7]	1	1	1	0	1	1	1	1		
	[20]	−1	−1	−2	−2	−2	−1	−1	−1	−1	−1
300	[30]	0	0	0	0	−1	−1	−2	−1	−1	−2
	[7]	1	1	2	1	0	1	0	1	0	0
	[20]	−3	−2	−2	−2	−2	−1	−2	−1	−2	−2
400	[30]	−1	0	−1	−1	−2	−4	−6	−8	−9	−10
	[7]	0	1	0	0	−1	−2	−5	−7	−7	−8
	[20]	−1	0	−1	0	0	2	1	0	1	0
600	[7]	1	0	1	0	−1	−3	−8	−12	−15	−17
	[20]	−1	−1	−1	−1	−1	−1	−1	−1	0	−0
800	[7]	0	1	1	1	−1	−2	−6	−10	−14	−17
	[20]	−1	−1	−1	0	−1	0	−1	0	0	−0
1000	[7]	1	0	0	0	0	−2	−5	−8	−12	−16
	[20]	−3	−4	−3	−4	−3	−3	−3	−3	−3	−3

Table 3.37 Deviations δc_v of tabulated data of [20, 30] from the analytic values of isochoric specific heat

T, K	Reference	δc_v, %, at p, MPa									
		1	2	3	5	10	20	40	60	80	100
110	[30]	0.6									
120	[30]	3.9	−2.2								
130	[30]	4.2	3.1	−3.1							

Table 3.27 (*Continued*)

T, K	Reference	δc_v, %, at p, MPa									
		1	2	3	5	10	20	40	60	80	100
140	[30]	3.1	3.4	−1.0	7.7	6.2	7.9				
150	[30]	1.0	2.9	2.8	1.2	1.9	2.8				
	[20]	0.3	0.1	0	−0.8	−1.2	−0.9	−1.7	−1.7	−0.8	0.9
200	[30]	0.9	1.3	1.0	1.5	1.9	2.4	2.3	2.4		
	[20]	0.3	0.4	0.5	0.7	0.8	1.2	1.2	1.4	1.8	2.3
300	[30]	0.1	−0.1	−0.1	−0.4	−1.1	−1.6	−2.3	−2.6	−2.5	−2.4
	[20]	0	0	0	0	−0.1	0	−0.3	−0.3	−0.1	0.1
400	[30]	−0.1	−0.1	−0.3	−0.5	−1.2	−2.0	−3.2	−2.8	−4.3	−4.3
	[20]	0	0	−0.1	0	−0.1	−0.1	0	0	0	0.1
600	[20]	−0.1	0	0	0	−0.1	−0.1	−0.1	0	0.1	0.2
800	[20]	0	0	0	0	−0.1	−0.1	0	0	0.1	0.1
1000	[20]	0	−0.1	−0.1	−0.1	−0.1	−0.1	−0.1	−0.1	−0.1	−0.1

Table 3.28 Deviations δc_p of tabulated data of [7, 9, 20, 30] from the analytic values of isobaric specific heat

T, K	Reference	δc_p, %, at p, MPa									
		1	2	3	5	10	20	40	60	80	100
90	[7]	3.4	3.5	3.4	3.3	3.1	2.6	2.3			
100	[7]	−3.1	−3.0	−3.0	−2.9	−3.2	−3.2	−2.3			
110	[30]	0.4									
	[7]	−4.8	−5.8	−5.7	−5.6	−5.5	−4.9	−3.2			
120	[30]	3.6	−7.5								
	[7]	−0.7	−5.2	−5.8	−6.0	−5.8	−4.2	−2.4			
130	[30]	3.7	0.7	−5.3							
	[7]	0.5	−0.1	−5.4	−3.5	−4.5	−2.7	−1.0			
140	[30]	2.7	−0.4	−1.2	7.1	2.7	3.3				
	[7]	0.8	1.1	0.4	5.8	−0.8	−1.3	0.2			
150	[30]	2.0	2.1	1.2	−0.3	0.8	1.0				
	[7]	0.6	1.2	1.3	−0.9	−0.5	−0.2	1.0			
	[20]	0.2	0.3	0.2	−0.6	−0.4	−0.9	−0.2	1.6	4.2	7.9
200	[30]	0.7	0.9	1.0	0.9	0.9	0.8	1.2	1.2		
	[7]	0.2	0.4	0.5	0.6	0.6	0.9	1.1	1.3		
	[20]	0.2	0.3	0.4	0.5	0.2	0.4	0.6	0.7	1.3	2.2
300	[30]	0	−0.1	−0.1	−0.1	−0.4	−1.0	−1.1	−1.6	−1.6	−1.7
	[7]	0	−0.1	−0.1	−0.2	−0.4	−0.9	−1.3	−1.3	−1.6	−2.1
	[20]	0	−0.1	−0.1	−0.1	−0.1	0	−0.1	−0.1	−0.3	−0.6

Table 3.27 (*Continued*)

T, K	Reference	δc_p, %, at p, MPa									
		1	2	3	5	10	20	40	60	80	100
400	[30]	−0.1	−0.1	−0.2	−0.2	−0.2	−0.8	−1.3	−1.5	−2.1	−2.5
	[7]	−0.1	−0.1	−0.1	−0.2	−0.2	−0.5	−1.3	−1.9	−2.1	−2.2
	[20]	0	−0.1	−0.1	−0.1	−0.1	−0.1	0	0	0.2	0.2
600	[7]	0.1	0	0	0	0	0.1	0.3	−0.2	−0.6	−1.0
	[20]	0	0	0	0	−0.1	−0.1	−0.1	−0.1	−0.1	−0.1
800	[7]	0	0	0	0.1	0.1	0.3	0.6	0.8	0.7	0.4
	[20]	0	0	0	0	0	0	−0.1	−0.1	−0.2	−0.2
1000	[7]	0	0	0	0	0.1	0.3	0.6	0.9	1.1	1.1
	[20]	0	−0.1	−0.1	−0.1	−0.1	−0.1	−0.2	−0.2	−0.2	−0.3

Table 3.29 Deviations δw of tabulated data of [20] from the analytic values of the speed of sound

T, K	δw, %, at p, MPa									
	1	2	3	5	10	20	40	60	80	100
150	0.04	−0.04	−0.04	0.33	0.58	0.02	0.45	0.19	−0.81	−2.31
200	−0.04	−0.11	−0.11	−0.18	−0.20	−0.43	−0.18	0.03	−0.19	−0.86
300	−0.03	−0.03	−0.06	−0.08	0	−0.02	0.15	0	−0.33	−0.45
400	−0.02	−0.05	−0.05	−0.05	0	0.02	0	0.11	−0.11	−0.46
600	−0.02	−0.02	−0.04	−0.04	−0.02	0.02	−0.08	−0.12	−0.04	−0.05
800	−0.02	−0.02	−0.02	−0.02	−0.03	−0.02	−0.06	−0.18	−0.20	−0.12
1000	0	−0.02	−0.02	−0.02	−0.03	−0.02	−0.04	−0.16	−0.24	−0.25

Table 3.30 Comparative description of tables of [7, 9, 20, 30] and the tables in this study

Quantity calculated	[30]	[7]	[9]	[20]	Present tables
Density (specific volume) $\rho(v)$	+	+	+	+	+
Compressibility Z	+	+	+	+	+
Enthalpy h	+	+	+	+	+
Entropy s	+	+	+	+	+
Energy e	+				
Isochoric specific heat c_v				+	
Isobaric specific heat c_p		+	+	+	+
Speed of sound w				+	+
Adiabatic Joule–Thomson effect μ				+	+
Adiabatic index k				+	+
Fugacity f				+	+
Derivatives: $(\partial v/\partial T)_p$				+	+
$(\partial p/\partial T)_v$				+	+
$(\partial v/\partial p)_T$			+	+	

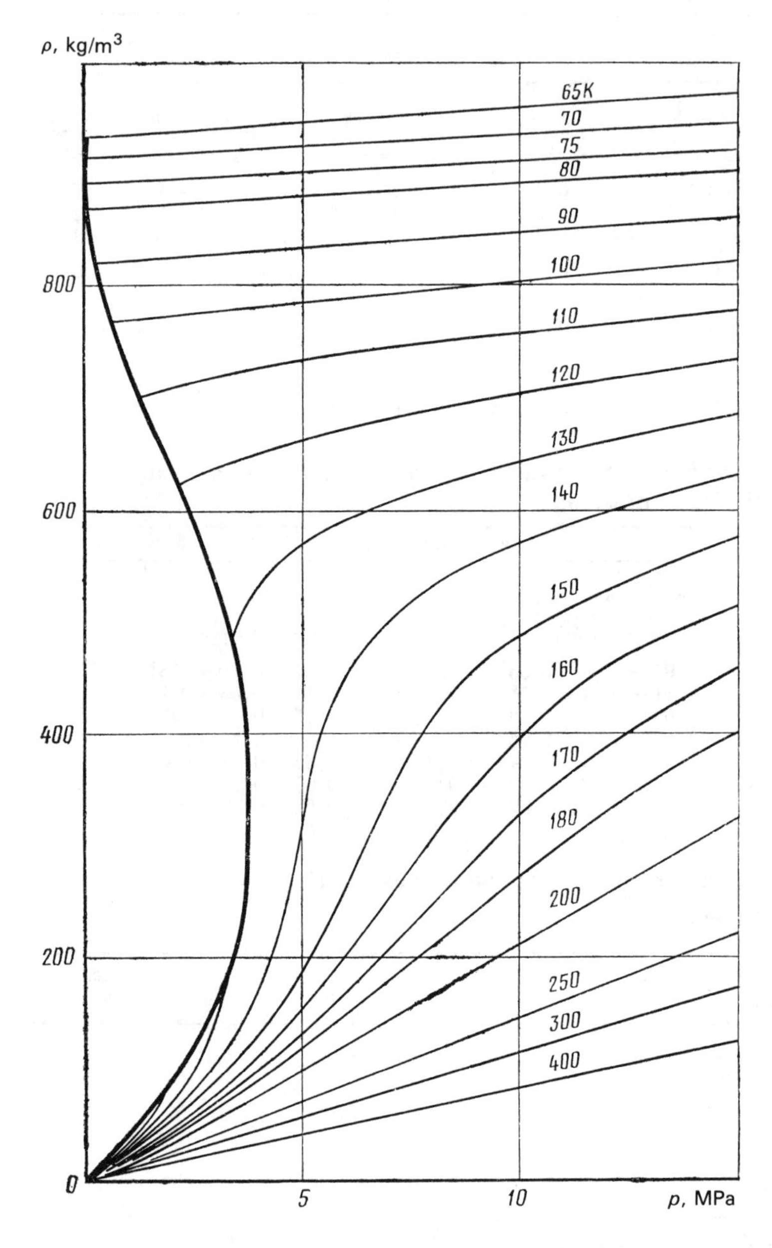

Figure 36 Pressure and temperature dependence of the density of air.

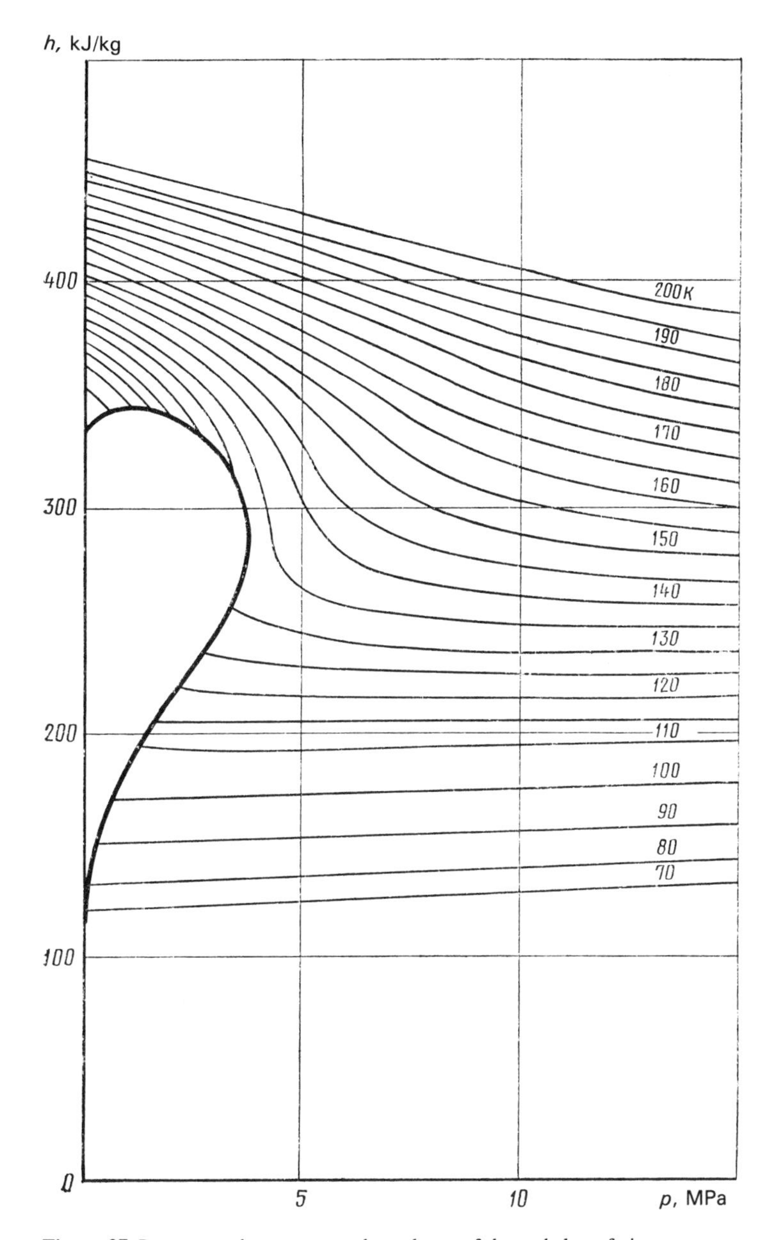

Figure 37 Pressure and temperature dependence of the enthalpy of air.

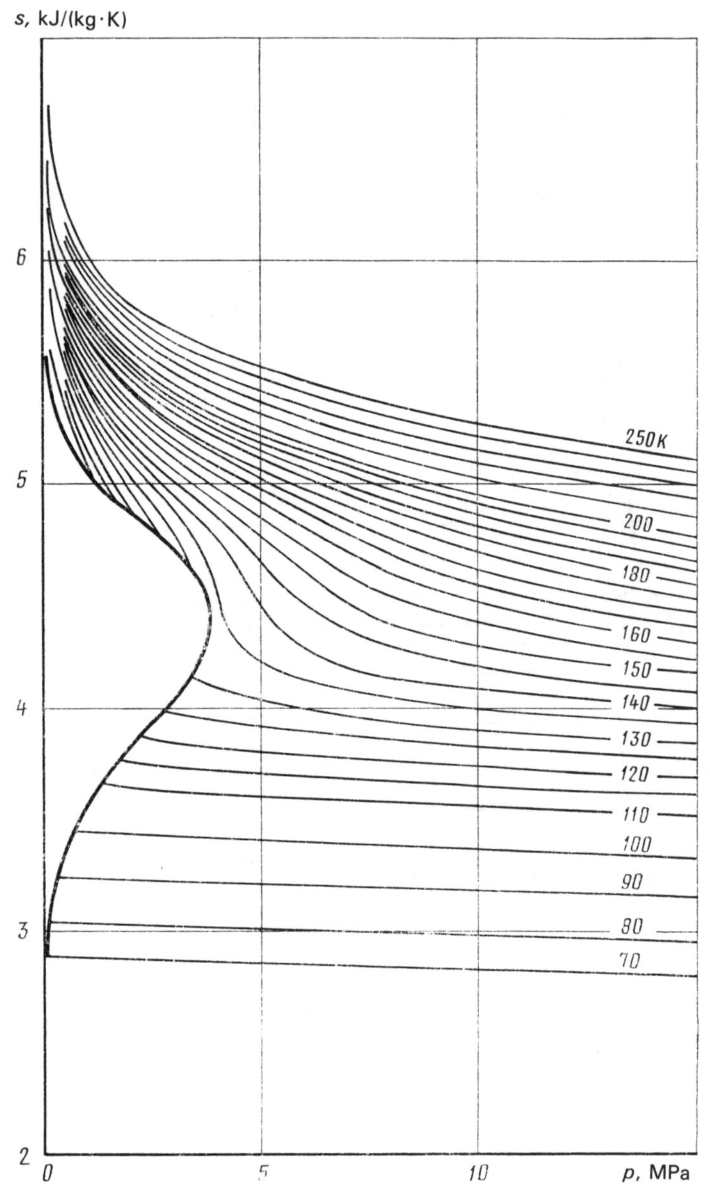

Figure 38 Pressure and temperature dependence of the entropy of air.

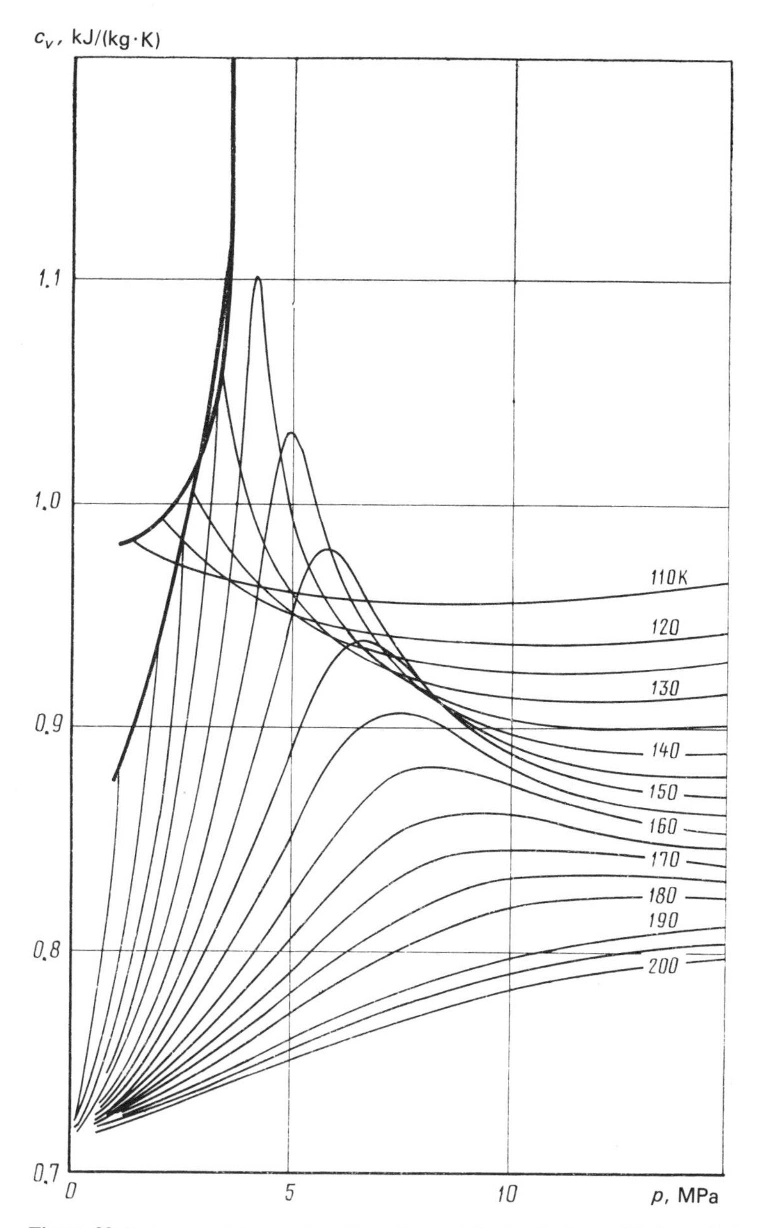

Figure 39 Pressure and temperature dependence of the isochoric specific heat of air.

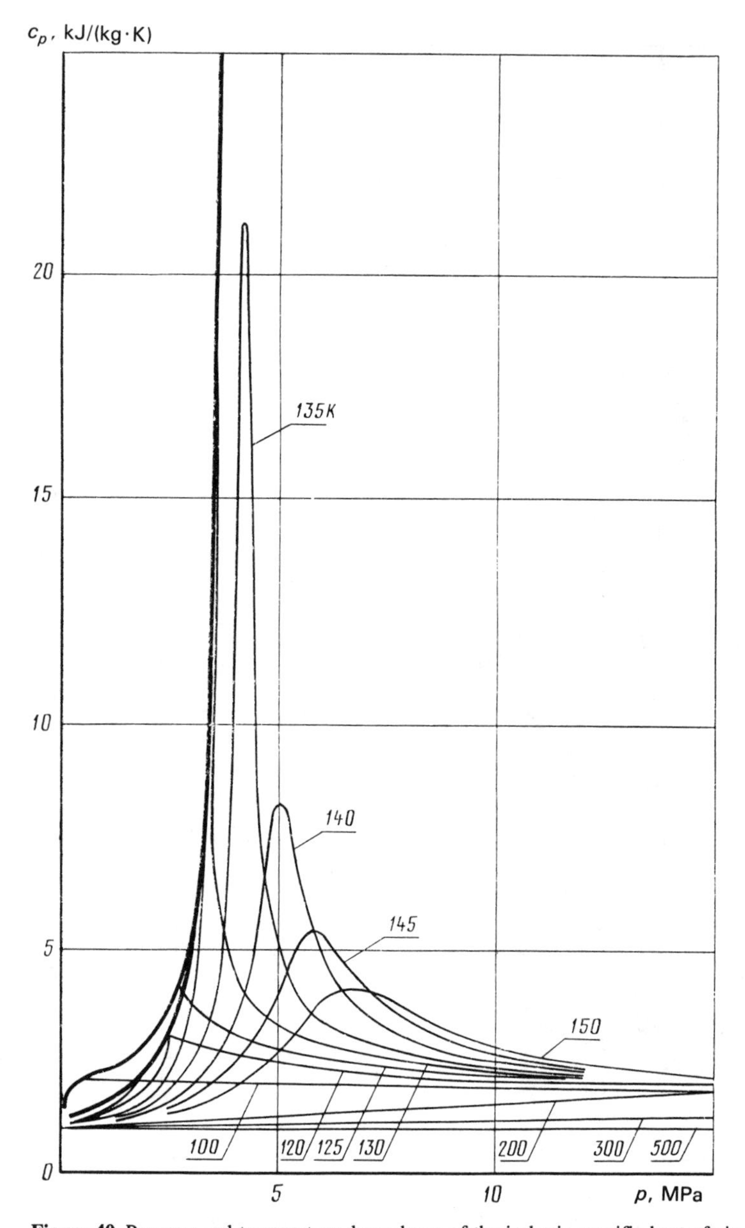

Figure 40 Pressure and temperature dependence of the isobaric specific heat of air.

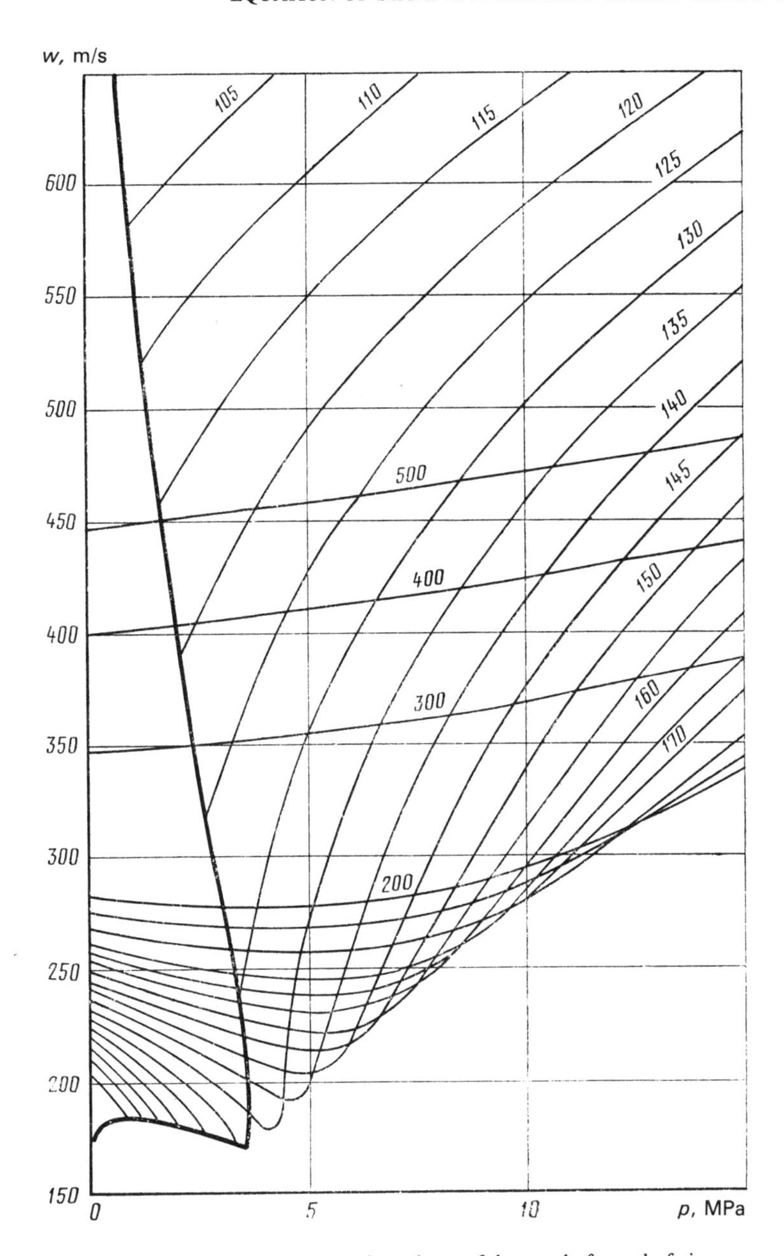

Figure 41 Pressure and temperature dependence of the speed of sound of air.

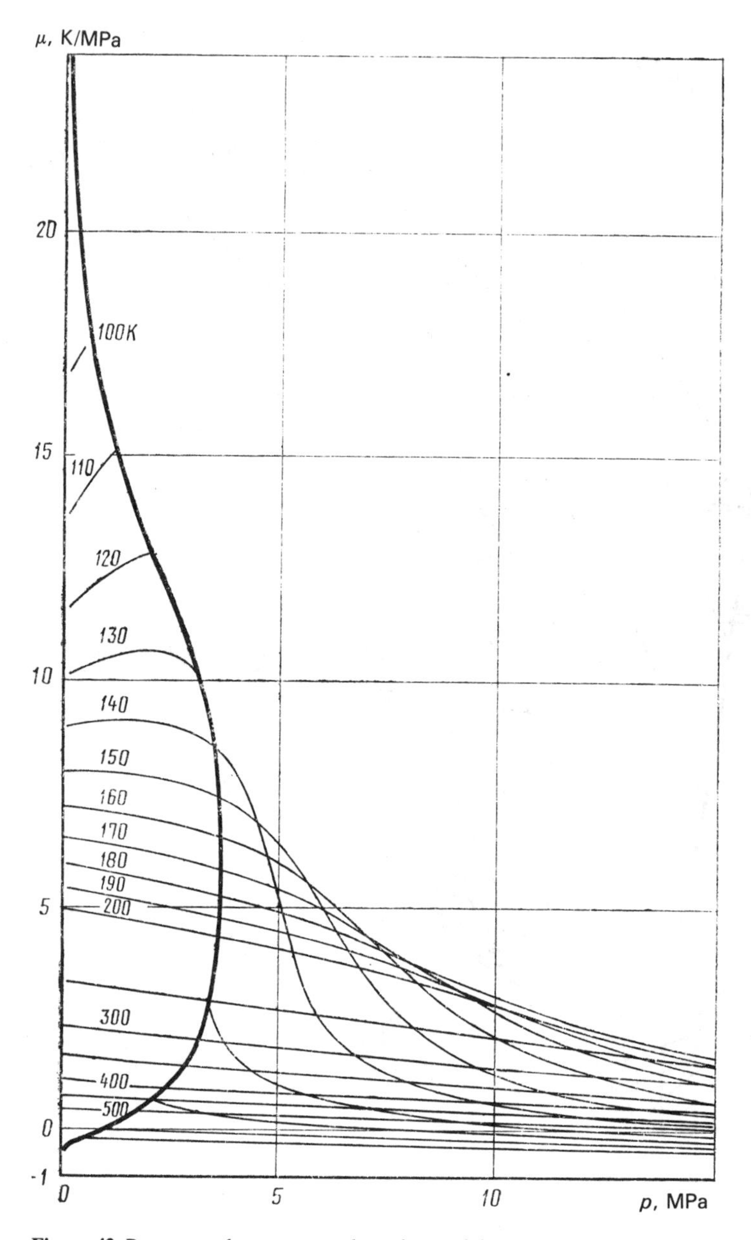

Figure 42 Pressure and temperature dependence of the adiabatic Joule–Thomson effect of air.

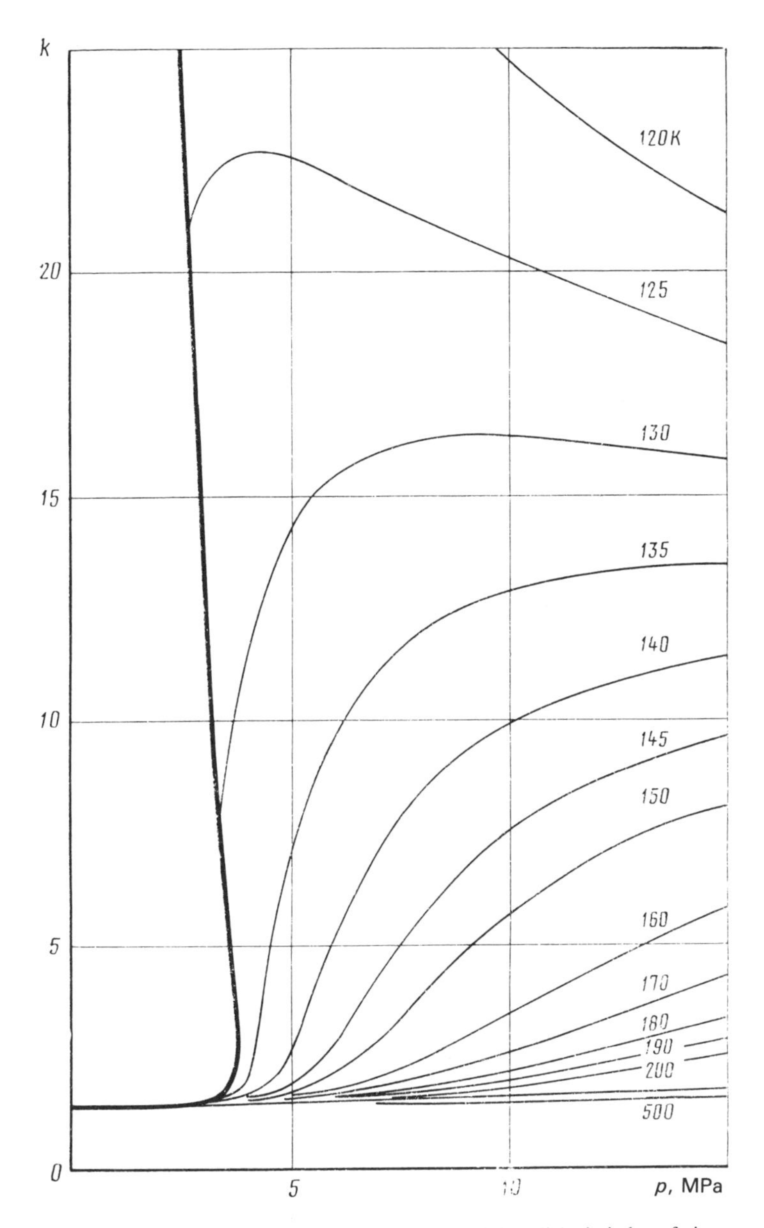

Figure 43 Pressure and temperature dependence of the adiabatic index of air.

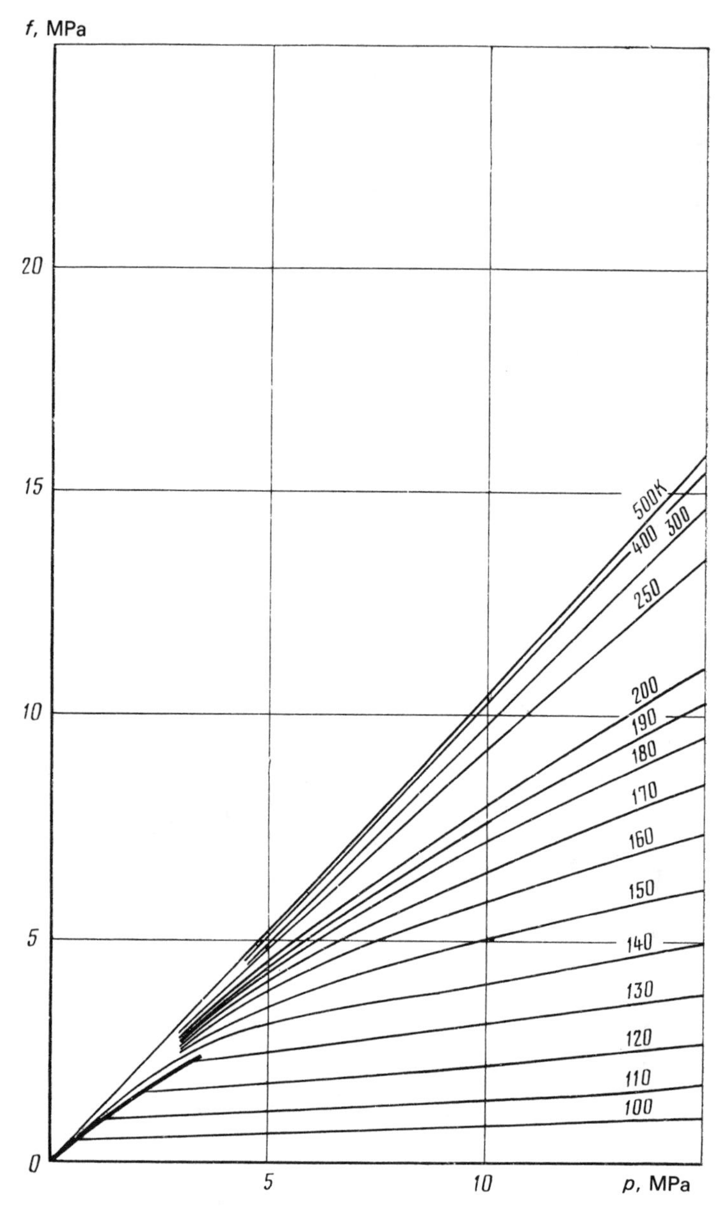

Figure 44 Pressure and temperature dependence of the fugacity of air.

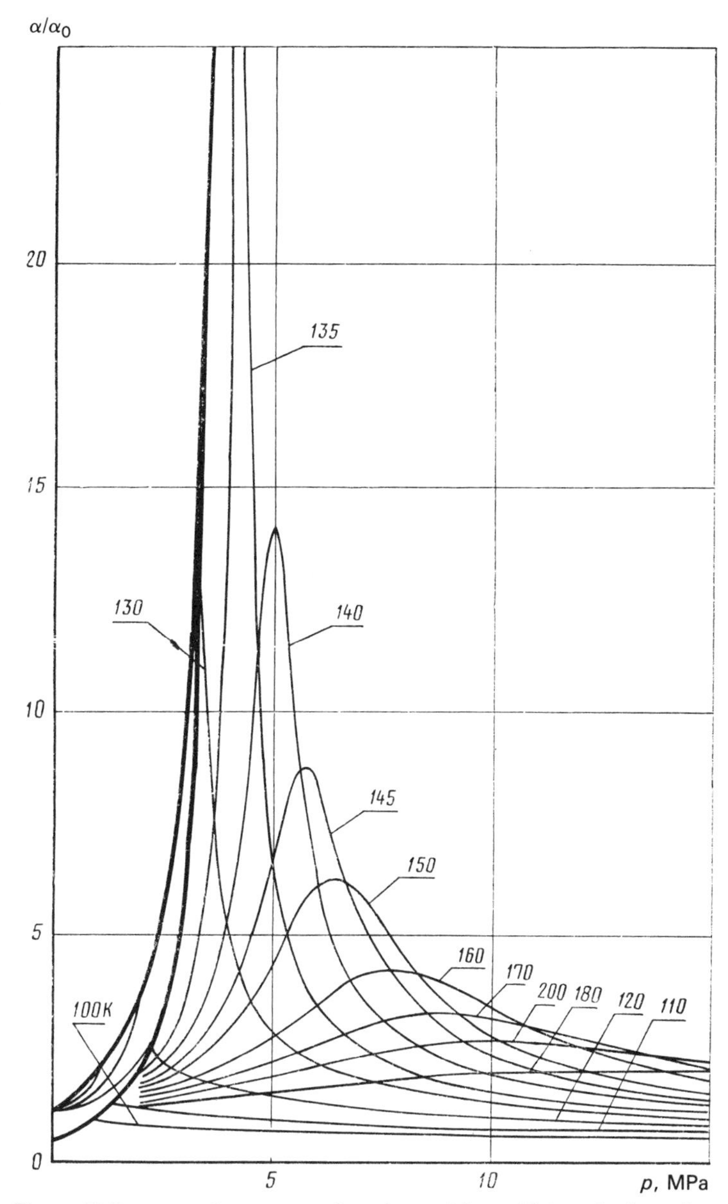

Figure 45 Pressure and temperature dependence of the coefficient of volumetric expansion of air.

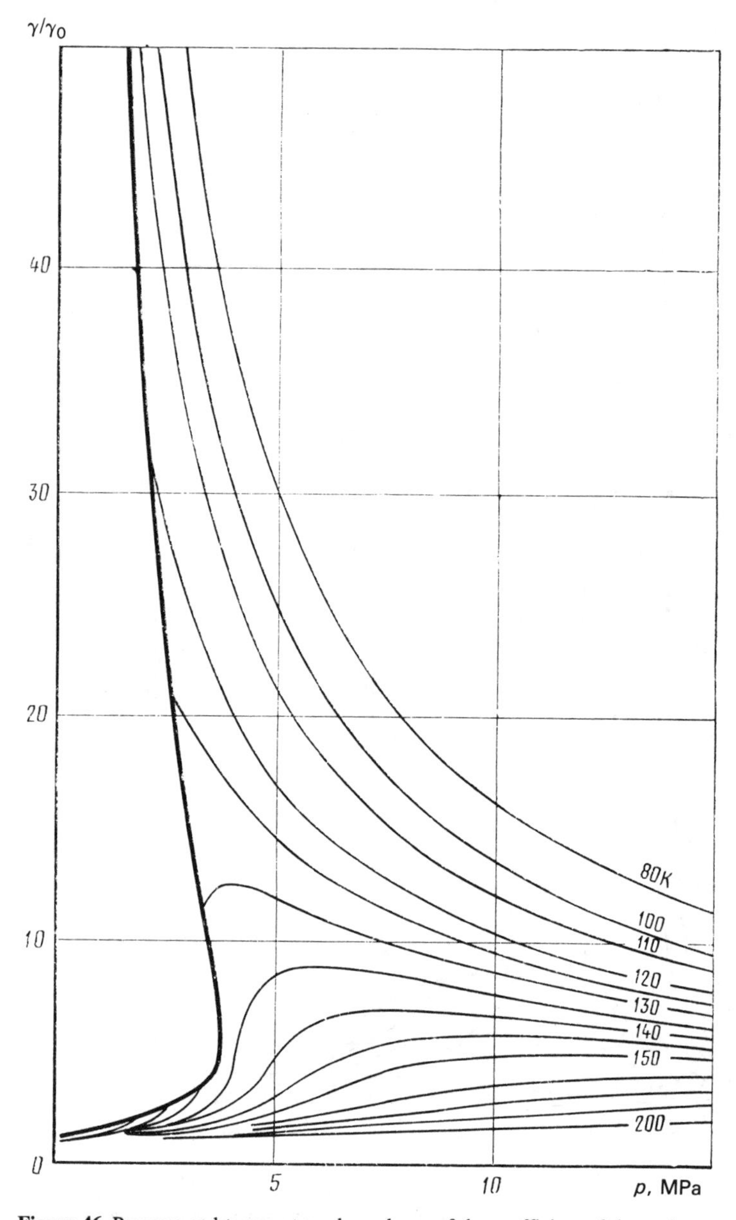

Figure 46 Pressure and temperature dependence of the coefficient of thermal pressure of air.

PART TWO

TABLES OF THERMODYNAMIC PROPERTIES OF AIR

VALUES OF FUNDAMENTAL CONSTANTS

Molecular weight	M = 28.96 kg/kmol
Gas constant	R = 287.1 J/(kg·K) = 287.1 (m^3·MPa)/(kg·K)
Temperature at the normal boiling point	$T_{n.b}$ = 78.67 ± 0.10 K; K
Temperature at the critical contact point	T_{cr} = 132.5 ± 0.1 K;
Pressure at the critical contact point	p_{cr} = 3.766 ± 0.020 MPa;
Density at the critical point contact	ρ_{cr} = 316.5 ± 6.0 kg/m^3;
Heat of sublimation at 0 K	h_0^0 = 253.4 kJ/kg.

NOMENCLATURE AND UNITS OF TABULATED QUANTITIES

T	temperature, K	τ	reduced temperature = T/T_{cr}
p	pressure, MPa	π	reduced pressure = P/P_{cr}
ρ	density, kg/m^3	ω	reduced density = ρ/ρ_{cr}
Z	compressibility		
h	enthalpy, kJ/kg		
r	heat of vaporization, kJ/kg		
s	entropy, kJ/(kg·K)		
Φ	Gibbs potential, kJ/(kg·K)		
F	Helmoltz function, kJ/(kg·K)		
μ	internal energy, kJ/kg		
c_v	specific heat at constant volume (isochoric), kJ/(kg·K)		
c_p	specific heat at constant pressure (isobaric), kJ/(kg·K)		

c_s	specific heat on the saturation curve, kJ/(kg·K)
c_λ	specific heat along the solidification curve, kJ/(kg·K)
κ	heat capacity ratio, c_p/c_v
w	speed of sound, m/s
μ	adiabatic Joule-Thomson effect K/MPa
δ	isothermal Joule–Thomson effect, m^3/kg
k	adiabatic index
f	volatility (fugacity), MPa
α/α_0	volumetric expansion coefficient
γ/γ_0	thermal pressure coefficient
$d\pi/d\tau$	first derivative of reduced pressure with respect to reduced temperature
$d^2\pi/d\tau^2$	second derivative of reduced pressure with respect to reduced temperature

Log Notation

$\lg_{10}$	$\log_{10}$
ln	$\log_e$

Subscripts

cr	critical
0	reference T
s	saturation conditions

Superscripts

0	ideal gas
(′)	at the boiling curve; liquid at saturation on the bubble point curve
(″)	at the condensation curve; gas at saturation on the dew point curve

TABLES OF DATA

Thermodynamic properties of air at boiling and condensation curves (by temperature)

Table II.1

T	p'	p''	r	$(d\pi/d\tau)'$	$(d\pi/d\tau)''$	$(d^2\pi/d\tau^2)'$	$(d^2\pi/d\tau^2)''$
70	0.03401	0.01939	200.5	0.169	0.116	2.535	2.085
71	0.03910	0.02292	200.8	0.189	0.132	2.744	2.291
72	0.04478	0.02694	201.0	0.211	0.151	2.962	2.506
73	0.05110	0.03150	200.9	0.234	0.170	3.188	2.731
74	0.05810	0.03664	200.8	0.259	0.192	3.422	2.963

Table II.1

T	p'	p''	r	$(d\pi/d\tau)'$	$(d\pi/d\tau)''$	$(d^2\pi/d\tau^2)'$	$(d^2\pi/d\tau^2)''$
75	0.06583	0.04242	200.5	0.286	0.215	3.662	3.204
76	0.07435	0.04889	200.1	0.314	0.240	3.909	3.452
77	0.08371	0.05610	199.7	0.345	0.267	4.163	3.707
78	0.09396	0.06410	199.1	0.377	0.296	4.423	3.969
79	0.10515	0.07295	198.5	0.411	0.327	4.689	4.237
80	0.11736	0.08272	197.8	0.448	0.360	4.960	4.511
81	0.13063	0.09345	197.0	0.486	0.395	5.236	4.791
82	0.14502	0.10521	196.2	0.527	0.432	5.517	5.076
83	0.16060	0.11806	195.3	0.570	0.472	5.801	5.366
84	0.17742	0.13206	194.3	0.614	0.513	6.090	5.661
85	0.19554	0.14727	193.4	0.662	0.557	6.381	5.961
86	0.21504	0.16376	192.3	0.711	0.603	6.676	6.265
87	0.23597	0.18160	191.3	0.762	0.652	6.974	6.574
88	0.25839	0.20085	190.0	0.816	0.703	7.273	6.887
89	0.28238	0.22157	188.9	0.872	0.756	7.575	7.205
90	0.30798	0.24385	187.6	0.930	0.811	7.878	7.527
91	0.33528	0.26773	186.3	0.991	0.870	8.183	7.853
92	0.36434	0.29330	185.0	1.054	0.930	8.488	8.184
93	0.39521	0.32063	183.5	1.119	0.993	8.795	8.520
94	0.42797	0.34979	182.0	1.187	1.059	9.102	8.860
95	0.46269	0.38084	180.5	1.257	1.127	9.410	9.205
96	0.49942	0.41387	179.0	1.329	1.198	9.718	9.556
97	0.53823	0.44896	177.4	1.403	1.271	10.026	9.911
98	0.57920	0.48616	175.7	1.480	1.347	10.334	10.272
99	0.62238	0.52558	173.9	1.559	1.426	10.642	10.638
100	0.66785	0.56727	172.2	1.641	1.508	10.950	11.011
101	0.71567	0.61133	170.4	1.724	1.592	11.258	11.389
102	0.76590	0.65783	168.5	1.811	1.680	11.565	11.773
103	0.81861	0.70685	166.5	1.899	1.770	11.873	12.164
104	0.87387	0.75849	164.5	1.990	1.863	12.180	12.561
105	0.93174	0.81282	162.5	2.083	1.960	12.487	12.966
106	0.99229	0.86993	160.3	2.178	2.059	12.795	13.377
107	1.0556	0.92992	158.0	2.276	2.162	13.103	13.795
108	1.1217	0.99286	155.7	2.376	2.267	13.411	14.220
109	1.1907	1.0589	153.4	2.478	2.376	13.719	14.653
110	1.2626	1.1280	150.9	2.583	2.489	14.028	15.092
111	1.3375	1.2004	148.3	2.690	2.604	14.338	15.540
112	1.4155	1.2761	145.7	2.799	2.723	14.650	15.994
113	1.4967	1.3552	143.0	2.911	2.846	14.963	16.456
114	1.5810	1.4379	140.1	3.025	2.972	15.277	16.926
115	1.6687	1.5242	137.1	3.142	3.101	15.595	17.402
116	1.7597	1.6142	134.0	3.261	3.234	15.914	17.886
117	1.8541	1.7081	130.8	3.382	3.371	16.237	18.376
118	1.9519	1.8059	127.5	3.506	3.512	16.561	18.873
119	2.0534	1.9078	123.9	3.632	3.656	16.891	19.377
120	2.1584	2.0138	120.1	3.761	3.804	17.224	19.886
121	2.2672	2.1241	116.2	3.892	3.956	17.562	20.401
122	2.3797	2.2388	112.0	4.026	4.112	17.905	20.921
123	2.4961	2.3579	107.6	4.162	4.272	18.254	21.445
124	2.6163	2.4817	103.0	4.301	4.436	18.608	21.974

Table II.1

T	p'	p''	r	$(d\pi/d\tau)'$	$(d\pi/d\tau)''$	$(d^2\pi/d\tau^2)'$	$(d^2\pi/d\tau^2)''$
125	2.7406	2.6102	97.9	4.443	4.604	18.969	22.506
126	2.8689	2.7435	92.4	4.588	4.775	19.337	23.041
127	3.0014	2.8817	86.5	4.735	4.951	19.713	23.577
128	3.1381	3.0250	79.8	4.885	5.131	20.098	24.114
129	3.2792	3.1735	72.3	5.039	5.315	20.491	24.652
130	3.4246	3.3272	63.5	5.195	5.503	20.895	25.188
131	3.5745	3.4864	52.9	5.354	5.696	21.308	25.722
132	3.7289	3.6511	38.9	5.516	5.892	21.732	26.252
132.5	3.8079	3.7356	29.8	5.599	5.991	21.949	26.516

Table II.2

T	ρ'	ρ''	h'	h''	s'	s''
70	914.44	0.9799	121.2	321.7	2.894	5.865
71	909.96	1.1422	121.8	322.6	2.902	5.830
72	905.48	1.325	122.5	323.5	2.912	5.797
73	900.99	1.533	123.4	324.3	2.924	5.765
74	896.50	1.761	124.4	325.2	2.938	5.734
75	892.01	2.017	125.5	326.0	2.953	5.704
76	887.50	2.300	126.8	326.9	2.969	5.675
77	882.97	2.611	128.0	327.7	2.985	5.648
78	878.43	2.953	129.4	328.5	3.003	5.621
79	873.86	3.325	130.8	329.3	3.021	5.595
80	869.26	3.735	132.3	330.1	3.039	5.570
81	864.64	4.179	133.9	330.9	3.058	5.546
82	859.99	4.665	135.5	331.7	3.078	5.523
83	855.30	5.191	137.1	332.4	3.097	5.501
84	850.57	5.758	138.8	333.1	3.117	5.479
85	845.80	6.370	140.5	333.9	3.137	5.458
86	840.99	7.029	142.3	334.6	3.158	5.437
87	836.14	7.736	144.0	335.3	3.178	5.417
88	831.24	8.495	145.9	335.9	3.198	5.398
89	826.29	9.313	147.7	336.6	3.219	5.378
90	821.28	10.19	149.6	337.2	3.240	5.360
91	816.22	11.12	151.5	337.8	3.260	5.342
92	811.11	12.11	153.4	338.4	3.281	5.324
93	805.93	13.17	155.4	338.9	3.302	5.307
94	800.69	14.31	157.4	339.4	3.323	5.290
95	795.38	15.51	159.4	339.9	3.344	5.273
96	790.01	16.79	161.4	340.4	3.364	5.256
97	784.56	18.15	163.5	340.9	3.385	5.240
98	779.03	19.59	165.6	341.3	3.406	5.224
99	773.43	21.12	167.7	341.6	3.427	5.208
100	767.74	22.74	169.8	342.0	3.447	5.192
101	761.96	24.46	171.9	342.3	3.468	5.177
102	756.09	26.29	174.1	342.6	3.489	5.162
103	750.12	28.22	176.3	342.8	3.509	5.146
104	744.05	30.27	178.5	343.0	3.530	5.131

Table II.2 (*Continued*)

T	ρ'	ρ''	h'	h''	s'	s''
105	737.86	32.44	180.7	343.2	3.551	5.116
106	731.56	34.74	183.0	343.3	3.571	5.111
107	725.13	37.17	185.3	343.3	3.592	5.086
108	718.56	39.75	187.6	343.3	3.613	5.071
109	711.85	42.49	189.9	343.3	3.633	5.055
110	704.99	45.39	192.3	343.2	3.654	5.040
111	697.96	48.47	194.7	343.0	3.675	5.025
112	690.75	51.74	197.1	342.8	3.696	5.010
113	683.35	55.21	199.6	342.6	3.717	4.994
114	675.74	58.91	202.1	342.2	3.738	4.978
115	667.89	62.85	204.7	341.8	3.759	4.962
116	659.80	67.04	207.3	341.3	3.780	4.946
117	651.43	71.53	209.9	340.7	3.802	4.929
118	642.75	76.33	212.6	340.1	3.824	4.912
119	633.72	81.49	215.4	339.3	3.846	4.895
120	624.31	87.03	218.3	338.4	3.868	4.877
121	614.46	93.01	221.2	337.4	3.891	4.858
122	604.11	99.49	224.2	336.2	3.914	4.839
123	593.17	106.54	227.3	334.9	3.938	4.819
124	581.54	114.26	230.5	333.5	3.962	4.758
125	569.09	122.77	233.9	331.8	3.988	4.776
126	555.61	132.21	237.5	329.9	4.014	4.752
127	540.85	142.81	241.2	327.7	4.042	4.727
128	524.39	154.89	245.3	325.1	4.072	4.699
129	505.58	168.88	249.8	322.1	4.105	4.668
130	483.24	185.46	255.0	318.5	4.143	4.634
131	454.93	205.77	261.2	314.1	4.188	4.594
132	414.34	231.32	269.7	308.6	4.250	4.547
132.5	385.55	246.43	275.7	305.5	4.294	4.520

Table II.3

T	c_v'	c_v''	c_p'	c_p''	c_s'	c_s''
90	0.906	0.856	1.881	1.261	1.862	—1.652
91	0.917	0.856	1.907	1.268	1.886	—1.634
92	0.927	0.856	1.933	1.277	1.910	—1.617
93	0.936	0.857	1.957	1.286	1.932	—1.603
94	0.943	0.857	1.981	1.295	1.953	—1.590
95	0.950	0.857	2.004	1.306	1.974	—1.580
96	0.956	0.858	2.027	1.317	1.994	—1.571
97	0.961	0.858	2.050	1.329	2.013	—1.565
98	0.965	0.859	2.073	1.342	2.033	—1.560
99	0.969	0.860	2.096	1.357	2.052	—1.558
100	0.972	0.861	2.119	1.373	2.071	—1.557
101	0.975	0.862	2.142	1.390	2.089	—1.559
102	0.977	0.864	2.166	1.409	2.109	—1.563
103	0.979	0.865	2.191	1.429	2.128	—1.569
104	0.980	0.867	2.217	1.452	2.147	—1.577

Table II.3 (*Continued*)

T	c_v'	c_v''	c_p'	c_p''	c_s'	c_s''
105	0.981	0.869	2.244	1.476	2.168	—1.587
106	0.982	0.872	2.273	1.503	2.189	—1.600
107	0.983	0.875	2.303	1.532	2.211	—1.616
108	0.983	0.878	2.336	1.564	2.234	—1.634
109	0.984	0.881	2.370	1.599	2.258	—1.655
110	0.984	0.884	2.408	1.638	2.284	—1.679
111	0.984	0.888	2.448	1.681	2.311	—1.707
112	0.985	0.893	2.492	1.729	2.340	—1.738
113	0.985	0.897	2.541	1.782	2.371	—1.773
114	0.985	0.902	2.595	1.841	2.406	—1.813
115	0.986	0.908	2.654	1.907	2.444	—1.858
116	0.986	0.914	2.721	1.981	2.485	—1.908
117	0.987	0.920	2.796	2.065	2.531	—1.965
118	0.989	0.927	2.882	2.161	2.582	—2.029
119	0.990	0.935	2.980	2.271	2.639	—2.102
120	0.992	0.943	3.094	2.398	2.705	—2.185
121	0.994	0.952	3.227	2.548	2.780	—2.280
122	0.997	0.961	3.386	2.725	2.866	—2.391
123	1.001	0.971	3.579	2.938	2.969	—2.519
124	1.005	0.982	2.817	3.199	3.091	—2.672
125	1.010	0.993	4.119	3.525	3.241	—2.854
126	1.016	1.006	4.514	3.944	3.428	—3.076
127	1.024	1.019	5.053	4.498	3.673	—3.352
128	1.033	1.034	5.831	5.264	4.006	—3.704
129	1.045	1.050	7.048	6.382	4.489	—4.166
130	1.059	1.068	9.201	8.122	5.262	—4.792
131	1.079	1.086	13.901	11.083	6.711	—5.647
132	1.106	1.104	29.272	16.445	10.181	—6.683
132.5	1.121	1.112	50.758	20.343	12.712	—7.039

Table II.4

T	w'	w''	μ'	μ''	k'	k''
90	759.7	180.1	—0.284	22.308	1539.14	1.35
91	747.7	180.6	—0.271	21.676	1360.98	1.35
92	735.8	181.0	—0.257	21.097	1205.22	1.35
93	723.9	181.5	—0.244	20.565	1068.76	1.35
94	712.2	181.9	—0.230	20.076	948.90	1.35
95	700.4	182.2	—0.216	19.624	843.38	1.35
96	688.7	182.6	—0.201	19.204	750.35	1.35
97	677.0	182.9	—0.186	18.814	668.16	1.35
98	665.3	183.2	—0.170	18.448	595.40	1.35
99	653.6	183.5	—0.153	18.106	530.92	1.35
100	641.9	183.7	—0.136	17.782	473.66	1.35
101	630.1	183.9	—0.118	17.476	422.76	1.35
102	618.3	184.0	—0.98	17.185	377.45	1.35
103	606.5	184.2	—0.78	16.906	337.07	1.35
104	594.6	184.2	—0.56	16.639	301.03	1.35

Table II.4 (*Continued*)

T	w'	w''	μ'	μ''	k'	k''
105	582.6	184.3	—0.33	16.381	268.84	1.36
106	570.6	184.3	—0.008	16.131	240.06	1.36
107	558.5	184.3	0.018	15.888	214.29	1.36
108	546.4	184.3	0.047	15.650	191.22	1.36
109	534.1	184.2	0.077	15.417	170.53	1.36
110	521.7	184.1	0.110	15.188	151.98	1.36
111	509.3	183.9	0.145	14.960	135.34	1.37
112	496.7	183.8	0.184	14.735	120.39	1.37
113	484.0	183.5	0.226	14.510	106.95	1.37
114	471.2	183.2	0.271	14.285	94.89	1.38
115	458.2	182.9	0.321	14.059	84.04	1.38
116	445.1	182.6	0.375	15.832	74.28	1.38
117	431.8	182.2	0.435	13.602	65.52	1.39
118	418.4	181.7	0.501	13.369	57.64	1.40
119	404.8	181.2	0.575	13.131	50.56	1.40
120	390.9	180.6	0.657	12.888	44.21	1.41
121	376.9	180.0	0.749	12.637	38.50	1.42
122	362.6	179.4	0.852	12.378	33.38	1.43
123	348.0	178.6	0.970	12.109	28.79	1.44
124	333.2	177.8	1.106	11.825	24.68	1.46
125	318.0	177.0	1.262	11.525	21.00	1.47
126	302.5	176.1	1.446	11.202	17.72	1.49
127	286.5	175.2	1.664	10.852	14.79	1.52
128	270.0	174.2	1.928	10.463	12.18	1.55
129	252.9	173.2	2.255	10.023	9.86	1.60
130	235.1	172.3	2,675	9.513	7.80	1.65
131	216.4	171.6	3,241	8.902	5.96	1.74
132	196.6	171.6	4,078	8.152	4.30	1.86
132,5	187.2	172.0	4,659	7.720	3.55	1.95

T **Table II.5**

T	f'	f''	α'/α_0	α''/α_0	γ'/γ_0	γ''/γ_0
90	0.258	0.227	0.562	1.286	416.49	1.183
91	0.280	0.249	0.579	1.305	378.88	1.193
92	0.304	0.271	0.597	1.326	345.06	1.204
93	0.329	0.295	0.616	1.348	314.62	1.216
94	0.355	0.320	0.635	1.372	287.16	1.228
95	0.383	0.347	0.656	1.397	262.35	1.241
96	0.412	0.375	0.678	1.424	239.92	1.255
97	0.443	0.404	0.701	1.453	219.61	1.269
98	0.475	0.436	0.726	1.485	201.19	1.285
99	0.509	0.468	0.751	1.519	184.46	1.302
100	0.544	0.502	0.779	1.555	169.25	1.319
101	0.581	0.538	0.808	1.594	155.40	1.338
102	0.619	0.575	0.839	1.636	142.78	1.358
103	0.659	0.614	0.872	1.682	131.26	1.379
104	0.700	0.655	0.907	1.731	120.74	1.401

Table II.5 (*Continued*)

T	f'	f''	α'/α_0	α''/α_0	γ'/γ_0	γ''/γ_0
105	0.743	0.697	0.945	1.784	111.11	1.425
106	0.788	0.741	0.986	1.842	102.29	1.450
107	0.834	0.787	1.030	1.904	94.20	1.477
108	0.881	0.834	1.078	1.973	86.78	1.505
109	0.931	0.883	1.130	2.047	79.97	1.535
110	0.981	0.934	1.186	2.129	73.70	1.567
111	1.034	0.986	1.248	2.219	67.92	1.602
112	1.088	1.040	1.316	2.318	62.60	1.638
113	1.143	1.096	1.392	2.427	57.69	1.677
114	1.200	1.153	1.475	2.549	53.15	1.719
115	1.259	1.212	1.569	2.684	48.96	1.764
116	1.319	1.273	1.674	2.836	45.07	1.812
117	1.381	1.336	1.793	3.009	41.48	1.864
118	1.444	1.400	1.929	3.205	38.14	1.920
119	1.509	1.466	2.085	3.430	35.04	1.980
120	1.575	1.533	2.268	3.690	32.16	2.046
121	1.642	1.602	2.484	3.994	29.47	2.117
122	1.711	1.673	2.744	4.356	26.97	2.196
123	1.782	1.745	3.060	4.790	24.63	2.282
124	1.854	1.819	3.454	5.322	22.44	2.378
125	1.927	1.894	3.958	5.987	20.38	2.486
126	2.002	1.971	4.626	6.841	18.45	2.608
127	2.078	2.049	5.547	7.971	16.62	2.747
128	2.155	2.128	6.895	9.531	14.88	2.909
129	2.234	2.209	9.036	11.803	13.21	3.101
130	2.313	2.291	12.894	15.330	11.58	3.335
131	2.394	2.374	21.497	21.300	9.94	3.629
132	2.476	2.458	50.459	32.011	8.19	4.008
132.5	2.517	2.500	92.183	39.701	7.22	4.234

Thermodynamic properties of air at boiling and condensation curves (by pressure)

Table II.6

p	T'	T''	r	$(d\pi/d\tau)'$	$(d\pi/d\tau)''$	$(d^2\pi/d\tau^2)'$	$(d^2\pi/d\tau^2)''$
0.025	67.89	71.53	202.4	0.132	0.142	2.120	2.405
0.050	72.83	76.16	203.7	0.230	0.244	3.151	3.493
0.075	76.07	79.22	202.6	0.316	0.334	3.928	4.296
0.100	78.55	81.57	201.1	0.396	0.416	4.569	4.952
0.125	80.59	83.51	199.6	0.470	0.493	5.120	5.515
0.150	82.33	85.17	198.0	0.541	0.565	5.609	6.013
0.175	83.86	86.64	196.4	0.608	0.634	6.050	6.462
0.200	85.23	87.96	195.0	0.673	0.701	6.452	6.874
0.250	87.63	90.26	192.2	0.796	0.827	7.164	7.613
0.300	89.70	92.25	189.5	0.912	0.946	7.784	8.269
0.350	91.51	94.01	187.0	1.023	1.059	8.341	8.863
0.400	93.15	95.59	184.5	1.129	1.168	8.841	9.411

Table II.6 (*Continued*)

p	T'	T''	r	$(d\pi/d\tau)'$	$(d\pi/d\tau)''$	$(d^2\pi/d\tau^2)'$	$(d^2\pi/d\tau^2)''$
0.450	94.64	97.03	182.2	1.231	1.273	9.300	9.922
0.500	96.02	98.36	179.9	1.330	1.375	9.720	10.403
0.500	97.29	99.59	177.8	1.426	1.474	10.117	10.859
0.600	98.49	100.75	175.6	1.518	1.571	10.483	11.294
0.650	99.61	101.84	173.6	1.609	1.665	10.832	11.710
0.700	100.68	102.86	171.6	1.697	1.758	11.157	12.111
0.750	101.69	103.84	169.6	1.783	1.848	11.469	12.498
0.800	102.65	104.77	167.6	1.868	1.937	11.767	12.872
0.850	103.57	105.66	165.7	1.951	2.025	12.051	13.235
0.900	104.46	106.51	163.8	2.032	2.111	12.319	13.589
0.950	105.31	107.32	161.9	2.112	2.196	12.580	13.933
1.000	106.12	108.11	160.0	2.190	2.279	12.835	14.268
1.100	107.68	109.60	156.4	2.343	2.444	13.309	14.917
1.200	109.13	110.99	152.8	2.492	2.604	13.761	15.538
1.300	110.50	112.31	149.3	2.637	2.761	14.187	16.136
1.400	111.80	113.55	145.8	2.778	2.914	14.591	16.713
1.500	113.04	114.72	142.2	2.916	3.065	14.976	17.271
1.600	114.22	115.84	138.7	3.051	3.214	15.347	17.811
1.700	115.35	116.92	135.2	3.183	3.360	15.705	18.335
1.800	116.43	117.94	131.7	3.313	3.504	16.054	18.845
1.900	117.47	118.92	128.1	3.441	3.645	16.392	19.340
2.000	118.48	119.87	124.5	3.566	3.785	16.718	19.822
2.100	119.45	120.78	120.9	3.689	3.923	17.039	20.291
2.200	120.39	121.67	117.2	3.811	4.060	17.352	20.748
2.300	121.30	122.52	113.5	3.931	4.195	17.660	21.194
2.400	122.18	123.34	109.7	4.050	4.328	17.965	21.628
2.500	123.03	124.14	105.8	4.167	4.460	18.267	22.052
2.600	123.87	124.92	101.8	4.282	4.591	18.558	22.466
2.700	124.68	125.68	97.7	4.397	4.720	18.850	22.870
2.800	125.47	126.41	96.5	4.510	4.848	19.138	23.264
2.900	126.24	127.13	89.0	4.622	4.975	19.424	23.648
3.000	126.99	127.83	84.4	4.734	5.101	19.709	24.023
3.100	127.72	128.51	79.5	4.843	5.225	19.993	24.390
3.200	128.44	129.17	74.3	4.953	5.349	20.272	24.747
3.300	129.15	129.83	68.6	5.061	5.471	20.547	25.096
3.400	129.83	130.46	62.5	5.169	5.592	20.828	25.436
3.500	130.51	131.08	55.6	5.275	5.712	21.101	25.768
3.600	131.17	131.69	48.0	5.381	5.832	21.377	26.092

Table II.7

p	ρ'	ρ''	h'	h''	s'	s''
0.025	923.91	1.239	120.7	323.1	2.886	5.812
0.050	901.75	2.348	123.3	327.0	2.922	5.671
0.075	887.18	3.413	126.9	329.5	2.970	5.590
0.100	875.92	4.451	130.2	331.3	3.013	5.533
0.125	866.54	5.471	133.2	332.8	3.051	5.490

Table II.7 (*Continued*)

p	ρ'	ρ''	h'	h''	s'	s''
0.150	858.44	6.478	136.0	334.0	3.084	5.454
0.175	851.24	7.474	138.6	335.0	3.113	5.424
0.200	844.70	8.463	140.9	335.9	3.142	5.398
0.250	833.06	10.43	145.1	337.4	3.191	5.355
0.300	822.79	12.37	149.0	338.5	3.234	5.320
0.350	813.62	14.31	152.5	339.5	3.271	5.289
0.400	805.12	16.25	155.7	340.2	3.305	5.263
0.450	797.30	18.19	158.7	340.9	3.336	5.240
0.500	789.90	20.12	161.5	341.4	3.365	5.218
0.550	782.97	22.07	164.1	341.9	3.391	5.199
0.600	776.30	24.02	166.6	342.2	3.416	5.181
0.650	769.97	25.98	168.9	342.5	3.439	5.164
0.700	763.82	27.95	171.2	342.8	3.461	5.148
0.750	757.92	29.93	173.4	343.0	3.482	5.134
0.800	752.22	31.93	175.5	343.1	3.502	5.119
0.850	746.67	33.93	177.5	343.2	3.521	5.106
0.900	741.21	35.95	179.5	343.3	3.540	5.093
0.950	735.92	38.00	181.4	343.3	3.550	5.081
1.000	730.79	40.05	183.3	343.3	3.574	5.069
1.100	720.67	44.21	186.8	343.2	3.606	5.046
1.200	710.97	48.46	190.2	343.0	3.636	5.025
1.300	701.50	52.78	193.5	342.8	3.664	5.005
1.400	692.21	57.20	196.6	342.4	3.692	4.985
1.500	683.05	61.74	199.7	341.9	3.717	4.966
1.600	674.03	66.38	202.7	341.4	3.742	4.948
1.700	665.09	71.13	205.6	340.8	3.766	4.931
1.800	656.24	76.04	208.4	340.1	3.789	4.913
1.900	647.39	81.10	211.2	339.3	3.812	4.896
2.000	638.46	86.30	214.0	338.5	3.834	4.879
2.100	629.54	91.70	216.7	337.6	3.856	4.862
2.200	624.52	97.26	219.4	336.6	3.877	4.845
2.300	611.40	103.07	222.1	335.6	3.898	4.829
2.400	602.18	109.14	224.7	334.4	3.918	4.812
2.500	592.85	115.46	227.4	333.2	3.938	4.795
2.600	583.08	122.08	230.1	331.9	3.959	4.777
2.700	573.16	129.04	232.8	330.5	3.979	4.760
2.800	562.88	136.46	232.5	329.0	4.000	4.742
2.900	552.18	144.29	238.3	327.3	4.021	4.723
3.000	541.00	152.67	241.2	325.6	4.042	4.704
3.100	529.24	161.73	244.1	323.6	4.063	4.684
3.200	516.48	171.65	247.2	321.5	4.086	4.662
3.300	502.41	182.25	250.6	319.2	4.110	4.640
3.400	487.43	194.33	254.0	316.5	4.136	4.616
3.500	469.70	207.82	258.0	313.6	4.164	4.590
3.600	449.05	222.99	262.4	310.4	4.197	4.562

Table II.8

p	c_v'	c_v''	c_p'	c_p''	c_s'	c_s''
0.600	0.967	0.862	2.084	1.385	2.043	—1.558
0.650	0.971	0.864	2.110	1.405	2.064	—1.562
0.700	0.974	0.865	2.135	1.426	2.085	—1.568
0.750	0.973	0.867	2.159	1.448	2.104	—1.575
0.800	0.978	0.869	2.182	1.470	2.121	—1.585
0.850	0.979	0.871	2.206	1.493	2.140	—1.595
0.900	0.981	0.873	2.230	1.517	2.158	—1.608
0.950	0.981	0.876	2.253	1.542	2.175	—1.621
1.000	0.982	0.878	2.276	1.568	2.192	—1.636
1.100	0.983	0.883	2.325	1.622	2.227	—1.669
1.200	0.983	0.888	2.375	1.681	2.262	—1.707
1.300	0.984	0.894	2.427	1.744	2.297	—1.749
1.400	0.984	0.900	2.483	1.813	2.334	—1.795
1.500	0.984	0.907	2.543	1.888	2.374	—1.845
1.600	0.985	0.913	2.607	1.969	2.415	—1.900
1.700	0.986	0.920	2.677	2.057	2.459	—1.960
1.800	0.987	0.927	2.752	2.155	2.505	—2.025
1.900	0.988	0.934	2.835	2.263	2.555	—2.096
2.000	0.989	0.942	2.927	2.381	2.609	—2.174
2.100	0.991	0.950	3.029	2.514	2.668	—2.259
2.200	0.993	0.958	3.143	2.661	2.733	—2.352
2.300	0.995	0.966	3.272	2.830	2.804	—2.455
2.400	0.998	0.975	3.418	3.023	2.884	—2.569
2.500	1.001	0.983	3.585	3.243	2.972	—2.696
2.600	1.004	0.992	3.783	3.498	3.074	—2.839
2.700	1.008	1.002	4.014	3.795	3.189	—3.000
2.800	1.013	1.011	4.291	4.156	3.324	—3.183
2.900	1.018	1.021	4.628	4.583	3.481	—3.394
3.000	1.024	1.032	5.047	5.110	3.670	—3.638
3.100	1.030	1.042	5.577	5.776	3.898	—3.924
3.200	1.038	1.054	6.291	6.645	4.192	—4.264
3.300	1.047	1.064	7.302	7.734	4.586	—4.670
3.400	1.057	1.076	8.716	9.293	5.094	—5.157
3.500	1.069	1.088	11.094	11.465	5.881	—5.734
3.600	1.083	1.099	15.336	14.523	7.110	—6.373

Table II.9

p	w'	w''	μ'	μ''	k'	k''
0.600	659.6	183.8	—0.162	17.54	562.91	1.35
0.650	646.5	184.0	—0.143	17.22	495.08	1.35
0.700	633.9	184.1	—0.124	16.94	438.47	1.35
0.750	631.7	184.2	—0.105	16.67	248.39	1.35
0.800	610.7	184.3	—0.085	16.43	350.63	1.36
0.850	599.7	184.3	—0.066	16.21	315.96	1.36
0.900	589.1	184.3	—0.046	16.00	285.82	1.36
0.950	578.9	184.3	—0.026	15.81	259.63	1.36
1.000	569.2	184.3	—0.005	15.62	236.76	1.36
1.100	550.3	184.1	0.037	15.27	198.37	1.36

Table II. 9 (*Continued*)

p	w'	w''	μ'	μ''	k'	k''
1.200	532.5	183.9	0.081	14.96	167.99	1.37
1.300	515.5	183.7	0.127	14.66	143.41	1.37
1.400	499.2	183.4	0.176	14.38	123.22	1.37
1.500	483.5	183.0	0.227	14.12	106.44	1.38
1.600	468.3	182.6	0.282	13.87	92.40	1.38
1.700	453.6	182.2	0.339	13.62	80.51	1.39
1.800	439.4	181.7	0.400	13.38	70.40	1.40
1.900	425.6	181.2	0.465	13.15	61.71	1.40
2.000	411.9	180.7	0.536	12.92	54.15	1.41
2.100	398.6	180.2	0.611	12.69	47.62	1.42
2.200	385.5	179.6	0.691	12.46	41.91	1.43
2.300	372.6	179.0	0.779	12.24	36.90	1.44
2.400	360.0	178.4	0.873	12.01	32.51	1.45
2.500	347.6	177.7	0.974	11.78	28.66	1.46
2.600	335.1	177.1	1.087	11.55	25.18	1.47
2.700	322.9	176.4	1.210	11.31	22.13	1.49
2.800	310.7	175.7	1.345	11.06	19.41	1.50
2.900	298.6	175.0	1.495	10.80	16.98	1.52
3.000	286.6	174.4	1.662	10.53	14.82	1.55
3.100	274.7	173.7	1.848	10.25	12.89	1.57
3.200	262.6	173.0	2.062	9.94	11.13	1.61
3.300	250.2	172.5	2.313	9.61	9.53	1.64
3.400	238.3	171.9	2.594	9.25	8.14	1.69
3.500	225.6	171.6	2.943	8.84	6.83	1.75
3.600	213.0	171.4	3.362	8.40	5.66	1.82

Table II. 10

p	f'	f''	α'/α_0	α''/α_0	γ'/γ_0	γ''/γ_0
0.600	0.492	0.529	0.738	1.584	192.82	1.333
0.650	0.530	0.569	0.768	1.629	174.97	1.354
0.700	0.569	0.609	0.798	1.675	159.72	1.376
0.750	0.607	0.648	0.829	1.722	146.32	1.397
0.800	0.645	0.687	0.860	1.771	135.15	1.419
0.850	0.682	0.726	0.892	1.821	125.12	1.441
0.900	0.720	0.764	0.925	1.873	116.22	1.463
0.950	0.757	0.802	0.958	1.926	108.32	1.486
1.000	0.793	0.839	0.991	1.981	101.26	1.508
1.100	0.866	0.913	1.062	2.096	89.11	1.554
1.200	0.937	0.986	1.137	2.219	79.11	1.602
1.300	1.007	1.057	1.217	2.350	70.73	1.650
1.400	1.077	1.127	1.302	2.491	63.62	1.700
1.500	1.145	1.196	1.395	2.646	57.50	1.751
1.600	1.213	1.264	1.495	2.813	52.20	1.805
1.700	1.280	1.330	1.604	2.992	47.57	1.859
1.800	1.345	1.396	1.723	3.192	43.49	1.916
1.900	1.410	1.461	1.854	3.413	39.87	1.976
2.000	1.475	1.524	2.001	3.655	36.62	2.037
2.100	1.538	1.587	2.164	3.926	33.72	2.102

Table II. 10 (*Continued*)

p	f'	f''	α'/α_0	α''/α_0	γ'/γ_0	γ''/γ_0
2.200	1.601	1.649	2.349	4.226	31.09	2.168
2.300	1.663	1.710	2.558	4.569	28.71	2.239
2.400	1.724	1.770	2.796	4.963	26.54	2.315
2.500	1.784	1.830	3.070	5.413	24.56	2.394
2.600	1.844	1.888	3.398	5.931	22.72	2.477
2.700	1.904	1.946	3.783	6.537	21.03	2.566
2.800	1.962	2.003	4.248	7.273	19.46	2.663
2.900	2.020	2.059	4.820	8.144	18.00	2.766
3.000	2.077	2.114	5.536	9.217	16.64	2.878
3.100	2.133	2.169	6.453	10.572	15.36	3.002
3.200	2.190	2.223	7.700	12.337	14.14	3.140
3.300	2.245	2.276	9.487	14.544	12.96	3.288
3.400	2.300	2.329	12.020	17.695	11.86	3.462
3.500	2.354	2.380	16.334	22.068	10.75	3.660
3.600	2.408	2.432	24.156	28.190	9.65	3.884

Thermodynamic properties of single-phase air

Table II.11

p	ρ	Z	h	s
		$T=70$ K		
0.1	914.62	0.0054	121.3	2.893
0.5	915.68	0.0272	121.6	2.891
1.0	916.97	0.0543	121.9	2.889
1.5	918.23	0.0813	122.3	2.886
2.0	919.46	0.1082	122.6	2.883
2.5	920.66	0.1351	123.0	2.880
3.0	921.84	0.1619	123.3	2.878
3.5	922.99	0.1887	123.7	2.875
4.0	924.12	0.2154	124.0	2.872
4.5	925.23	0.2420	124.4	2.869
5.0	926.32	0.2686	124.7	2.867
6.0	928.44	0.3216	125.4	2.861
7.0	930.49	0.3743	126.1	2.856
8.0	932.47	0.4269	126.8	2.850
9.0	934.40	0.4793	127.5	2.845
10.0	936.27	0.5315	128.2	2.839
11.0	938.09	0.5835	128.9	2.834
12.0	939.87	0.6353	129.5	2.828
13.0	941.61	0.6870	130.2	2.823
14.0	943.30	0.7385	130.9	2.818
15.0	944.97	0.7898	131.6	2.812
16.0	946.59	0.8411	132.3	2.807
17.0	948.19	0.8921	133.0	2.802
18.0	949.75	0.9430	133.6	2.796
19.0	951.29	0.9938	134.3	2.791

Table II.11 (*Continued*)

p	ρ	Z	h	s
		$T=70$ K		
20.0	952.80	1.0445	135.0	2.786
21.0	954.29	1.0950	135.7	2.781
22.0	955.75	1.1454	136.4	2.776
23.0	957.19	1.1956	137.1	2.770
24.0	958.60	1.2458	137.8	2.765
25.0	960.00	1.2958	138.5	2.760
26.0	961.38	1.3457	139.1	2.755
27.0	962.74	1.3955	139.8	2.750
28.0	964.08	1.4451	140.5	2.745
29.0	965.41	1.4947	141.2	2.740
30.0	966.72	1.5441	141.9	2.735
35.0	973.06	1.7898	145.4	2.711
40.0	979.10	2.0328	148.9	2.688
45.0	984.89	2.2735	152.4	2.665
		$T=72$ K		
0.1	905.62	0.0053	122.6	2.912
0.5	906.64	0.0267	122.9	2.910
1.0	907.89	0.0533	123.2	2.907
1.5	909.12	0.0798	123.6	2.904
2.0	910.32	0.1063	123.9	2.902
2.5	911.49	0.1327	124.3	2.899
3.0	912.65	0.1590	124.6	2.896
3.5	913.78	0.1853	125.0	2.893
4.0	914.89	0.2115	125.3	2.891
4.5	915.99	0.2377	125.7	2.888
5.0	917.06	0.2638	126.0	2.885
6.0	919.16	0.3158	126.7	2.880
7.0	921.20	0.3676	127.4	2.874
8.0	923.18	0.4192	128.1	2.869
9.0	925.11	0.4706	128.8	2.864
10.0	926.99	0.5219	129.5	2.858
11.0	928.82	0.5729	130.2	2.853
12.0	930.61	0.6238	130.9	2.848
13.0	932.36	0.6745	131.6	2.843
14.0	934.08	0.7251	132.3	2.838
15.0	935.76	0.7755	133.0	2.833
16.0	937.41	0.8257	133.7	2.827
17.0	939.03	0.8758	134.4	2.822
18.0	940.62	0.9258	135.1	2.817
19.0	942.18	0.9756	135.8	2.812
20.0	943.72	1.0252	136.5	2.807
21.0	945.23	1.0748	137.3	2.803
22.0	946.72	1.1242	138.0	2.798
23.0	948.19	1.1735	138.7	2.793
24.0	949.64	1.2226	139.4	2.788

Table II.11 (*Continued*)

p	ρ	Z	h	s
		$T=72$ K		
25.0	951.07	1.2716	140.1	2.783
26.0	952.48	1.3205	140.8	2.779
27.0	953.88	1.3693	141.5	2.774
28.0	955.25	1.4180	142.2	2.769
29.0	956.62	1.4665	142.9	2.765
30.0	957.96	1.5150	143.6	2.760
35.0	964.48	1.7555	147.2	2.737
40.0	970.71	1.9935	150.8	2.716
45.0	976.70	2.2289	154.5	2.695
50.0	982.49	2.4619	158.1	2.675
		$T=74$ K		
0.1	896.61	0.0052	124.5	2.938
0.5	897.62	0.0262	124.7	2.935
1.0	898.87	0.0524	125.1	2.933
1.5	900.09	0.0784	125.4	2.930
2.0	901.29	0.1044	125.8	2.927
2.5	902.46	0.1304	126.1	2.924
3.0	903.62	0.1563	126.5	2.922
3.5	904.76	0.1821	126.8	2.919
4.0	905.87	0.2078	127.2	2.916
4.5	906.97	0.2335	127.5	2.913
5.0	908.06	0.2592	127.9	2.911
6.0	910.17	0.3103	128.6	2.905
7.0	912.23	0.3612	129.3	2.900
8.0	914.24	0.4119	130.0	2.895
9.0	916.19	0.4624	130.7	2.889
10.0	918.10	0.5127	131.4	2.884
11.0	919.96	0.5628	132.1	2.879
12.0	921.78	0.6128	132.8	2.874
13.0	923.57	0.6625	133.5	2.869
14.0	925.32	0.7122	134.2	2.864
15.0	927.03	0.7616	134.9	2.859
16.0	928.72	0.8109	135.7	2.854
17.0	930.38	0.8601	136.4	2.849
18.0	932.00	0.9091	137.1	2.844
19.0	933.61	0.9579	137.8	2.839
20.0	935.18	1.0066	138.5	2.834
21.0	936.74	1.0552	139.2	2.830
22.0	938.27	1.1036	140.0	2.825
23.0	939.78	1.1520	140.7	2.820
24.0	941.27	1.2001	141.4	2.816
25.0	942.74	1.2482	142.1	2.811
26.0	944.19	1.2961	142.8	2.807
27.0	945.62	1.3439	143.6	2.802
28.0	947.04	1.3916	144.3	2.798
29.0	948.44	1.4392	145.0	2.793

Table II.11 (*Continued*)

p	ρ	Z	h	s
		$T=74$ K		
30.0	949.83	1.4867	145.8	2.789
35.0	956.56	1.7222	149.4	2.767
40.0	963.00	1.9551	153.1	2.747
45.0	969.21	2.1854	156.8	2.727
50.0	975.23	2.4132	160.6	2.709
		$T=75$ K		
0.1	892.09	0.0052	125.6	2.953
0.5	893.11	0.0260	125.8	2.950
1.0	894.36	0.0519	126.2	2.947
1.5	895.59	0.0778	126.5	2.945
2.0	896.79	0.1036	126.9	2.942
2.5	897.98	0.1293	127.2	2.939
3.0	899.14	0.1550	127.6	2.936
3.5	900.28	0.1805	127.9	2.933
4.0	901.41	0.2061	128.3	2.931
4.5	902.52	0.2316	128.6	2.928
5.0	903.61	0.2570	129.0	2.925
6.0	905.75	0.3076	129.7	2.920
7.0	907.83	0.3581	130.4	2.915
8.0	909.85	0.4083	131.1	2.909
9.0	911.83	0.4584	131.8	2.904
10.0	913.75	0.5082	132.5	2.899
11.0	915.64	0.5579	133.2	2.894
12.0	917.48	0.6074	133.9	2.889
13.0	919.29	0.6567	134.6	2.884
14.0	921.06	0.7059	135.3	2.879
15.0	922.80	0.7549	136.1	2.874
16.0	924.51	0.8037	136.8	2.869
17.0	926.19	0.8524	137.5	2.864
18.0	927.84	0.9010	138.2	2.859
19.0	929.46	0.9494	138.9	2.854
20.0	931.06	0.9976	139.6	2.849
21.0	932.64	1.0457	140.4	2.845
22.0	934.19	1.0937	141.1	2.840
23.0	935.73	1.1415	141.8	2.836
24.0	937.24	1.1892	142.5	2.831
25.0	938.73	1.2368	143.3	2.826
26.0	940.21	1.2843	144.0	2.822
27.0	941.66	1.3316	144.7	2.818
28.0	943.11	1.3788	145.5	2.813
29.0	944.53	1.4259	146.2	2.809
30.0	945.94	1.4729	146.9	2.805
35.0	952.79	1.7060	150.6	2.784
40.0	959.34	1.9364	154.3	2.763
45.0	965.66	2.1642	158.1	2.744
50.0	971.80	2.3895	161.9	2.726

Table II.11 (*Continued*)

p	ρ	Z	h	s
		$T=75$ K		
55.0	977.80	2.6123	165.7	2.709
60.0	983.69	2.8327	169.6	2.692
65.0	989.51	3.0507	173.5	2.676
70.0	995.29	3.2663	177.4	2.661
75.0	1001.08	3.4794	181.3	2.647
		$T=76$ K		
0.1	887.56	0.0052	126.8	2.968
0.5	888.59	0.0258	127.0	2.966
1.0	889.85	0.0515	127.4	2.963
1.5	891.09	0.0771	127.7	2.960
2.0	892.30	0.1027	128.1	2.958
2.5	893.50	0.1282	128.4	2.955
3.0	894.67	0.1537	128.8	2.952
3.5	895.83	0.1791	129.1	2.949
4.0	896.97	0.2044	129.5	2.946
4.5	898.09	0.2296	129.8	2.944
5.0	899.19	0.2548	130.2	2.941
6.0	901.35	0.3051	130.9	2.936
7.0	903.46	0.3551	131.6	2.930
8.0	905.51	0.4049	132.3	2.925
9.0	907.51	0.4545	133.0	2.920
10.0	909.46	0.5039	133.7	2.914
11.0	911.37	0.5532	134.4	2.909
12.0	913.24	0.6022	135.1	2.904
13.0	915.07	0.6511	135.8	2.899
14.0	916.87	0.6998	136.5	2.894
15.0	918.63	0.7483	137.2	2.889
16.0	920.37	0.7967	138.7	2.884
17.0	922.07	0.8450	138.7	2.880
18.0	923.75	0.8930	139.4	2.875
19.0	925.40	0.9410	140.1	2.870
20.0	927.02	0.9888	140.8	2.865
21.0	928.62	1.0364	141.6	2.861
22.0	930.20	1.0839	142.3	2.856
23.0	931.76	1.1313	143.0	2.852
24.0	933.29	1.1785	143.7	2.847
25.0	934.81	1.2257	144.5	2.843
26.0	936.31	1.2726	145.2	2.838
27.0	937.79	1.3195	145.9	2.834
28.0	939.26	1.3662	146.7	2.829
29.0	940.71	1.4129	147.4	2.825
30.0	942.14	1.4593	148.2	2.821
35.0	949.11	1.6901	151.9	2.800
40.0	955.78	1.9180	155.6	2.780
45.0	962.22	2.1433	159.4	2.762
50.0	968.48	2.3661	163.2	2.744

p	ρ	Z	h	s
		$T=76$ K		
55.0	974.60	2.5864	167.1	2.727
60.0	980.61	2.8042	171.0	2.711
65.0	986.56	3.0195	174.9	2.695
70.0	992.49	3.2324	178.8	2.681
75.0	998.42	3.4427	182.8	2.667
		$T=78$ K		
0.1	878.44	0.0051	129.4	3.003
0.5	879.49	0.0254	129.7	3.000
1.0	880.79	0.0507	130.0	2.997
1.5	882.06	0.0759	130.4	2.994
2.0	883.31	0.1011	130.7	2.992
2.5	884.54	0.1262	131.0	2.989
3.0	885.74	0.1512	131.4	2.986
3.5	886.93	0.1762	131.7	2.983
4.0	888.10	0.2011	132.1	2.980
4.5	889.26	0.2260	132.4	2.977
5.0	890.39	0.2508	132.8	2.975
6.0	892.62	0.3002	133.4	2.969
7.0	894.78	0.3493	134.1	2.964
8.0	896.90	0.3983	134.8	2.958
9.0	898.96	0.4471	135.5	2.953
10.0	900.97	0.4956	136.2	2.948
11.0	902.94	0.5440	137.0	2.943
12.0	904.87	0.5922	137.7	2.937
13.0	906.77	0.6402	138.4	2.932
14.0	908.62	0.6880	139.1	2.927
15.0	910.45	0.7357	139.8	2.923
16.0	912.24	0.7832	140.5	2.918
17.0	914.00	0.8306	141.2	2.913
18.0	915.73	0.8778	142.0	2.908
19.0	917.43	0.9248	142.7	2.903
20.0	919.11	0.9717	143.4	2.899
21.0	920.77	1.0185	144.1	2.894
22.0	922.40	1.0651	144.9	2.890
23.0	924.01	1.1115	145.6	2.885
24.0	925.60	1.1579	146.3	2.881
25.0	927.17	1.2041	147.1	2.876
26.0	928.72	1.2501	147.8	2.872
27.0	930.26	1.2961	148.5	2.868
28.0	931.78	1.3419	149.3	2.863
29.0	933.28	1.3876	150.0	2.859
30.0	934.76	1.4332	150.8	2.855
35.0	941.98	1.6592	154.5	2.835
40.0	948.90	1.8824	158.3	2.815
45.0	955.59	2.1029	162.1	2.797
50.0	962.09	2.3207	166.0	2.780

p	ρ	Z	h	s
		$T=78$ K		
55.0	968.46	2.5360	169.9	2.763
60.0	974.73	2.7488	173.8	2.748
65.0	980.95	2.9590	177.8	2.733
70.0	987.15	3.1666	181.8	2.719
75.0	993.38	3.3715	185.9	2.707
		$T=80$ K		
0.1 *	869.22	0.0050	132.3	3.039
0.5	870.31	0.0250	132.6	3.037
1.0	871.65	0.0499	132.9	3.034
1.5	872.97	0.0748	133.2	3.031
2.0	874.26	0.0996	133.6	3.028
2.5	875.53	0.1243	133.9	3.025
3.0	876.78	0.1490	134.2	3.022
3.5	878.02	0.1736	134.6	3.019
4.0	879.23	0.1981	134.9	3.016
4.5	880.43	0.2225	135.3	3.013
5.0	881.60	0.2469	135.6	3.011
6.0	883.91	0.2955	136.3	3.005
7.0	886.16	0.3439	137.0	2.999
8.0	888.35	0.3921	137.7	2.994
9.0	890.48	0.4400	138.4	2.989
10.0	892.57	0.4878	139.1	2.983
11.0	894.61	0.5353	139.8	2.978
12.0	896.62	0.5827	140.5	2.973
13.0	898.58	0.6299	141.2	2.968
14.0	900.50	0.6769	141.9	2.963
15.0	902.39	0.7237	142.6	2.958
16.0	904.24	0.7704	143.3	2.953
17.0	906.07	0.8169	144.0	2.948
18.0	907.86	0.8632	144.7	2.943
19.0	909.63	0.9094	145.5	2.939
20.0	911.37	0.9555	146.2	2.934
21.0	913.09	1.0013	146.9	2.929
22.0	914.78	1.0471	147.7	2.925
23.0	916.45	1.0927	148.4	2.920
24.0	918.10	1.1381	149.1	2.916
25.0	919.73	1.1835	149.9	2.912
26.0	921.34	1.2287	150.6	2.907
27.0	922.93	1.2737	151.3	2.903
28.0	924.50	1.3186	152.1	2.899
29.0	926.05	1.3634	152.8	2.895

*This point lies in the two-phase region.

Table II.11 (*Continued*)

p	ρ	Z	h	s
		$T=80$ K		
30,0	927,59	1.4081	153,6	2.890
35,0	935,08	1.6297	157.3	2.870
40,0	942,26	1.8483	161.2	2.851
45,0	949,20	2,0641	165,0	2.833
50,0	955,95	2.2772	168.9	2.816
55,0	962.57	2,4877	172,8	2.800
60,0	969,10	2.6956	176,8	2.785
65,0	975,58	2.9009	180,8	2.771
70,0	982,06	3,1034	184.8	2.757
75,0	988.58	3.3031	188.9	2.745
80,0	995.22	3.4998	193,0	2.733
85,0	1002,04	3,6933	197,2	2.723
90,0	1009,17	3,8829	201,4	2.713
95.0	1016.77	4,0680	205,6	2.705
100,0	1025.18	4.2469	210,0	2,698
		$T=85$ K		
0,1	4,25	0.9652	335,4	5.582
0.5	846.74	0,0242	140.7	3.135
1,0	848,25	0.0483	141.0	3.132
1.5	849,74	0,0723	141.3	3.129
2,0	851,19	0,0963	141.6	3.125
2,5	852,62	0,1202	141.9	3.122
3,0	854,03	0,1439	142.2	3.119
3,5	855,41	0.1677	142.6	3,116
4,0	856,77	0,1913	142,9	3,113
4,5	858,10	0,2149	143,2	3,110
5,0	859.42	0,2384	143.5	3,106
6,0	862,00	0,2852	144,2	3,100
7,0	864.50	0,3318	144.8	3.095
8,0	866,94	0,3781	145.5	3.089
9.0	869.31	0.4242	146.2	3.083
10,0	871,63	0.4701	146.8	3.077
11.0	873.90	0,5158	147.5	3.072
12,0	876,11	0,5613	148.2	3,067
13.0	878,28	0,6065	148.9	3,061
14.0	880.41	0.6516	149,6	3,056
15,0	882,49	0.6965	150.3	3,051
16,0	884,54	0.7412	151,0	3,046
17,0	886,55	0,7858	151.7	3,041
18,0	888,53	0,8301	152,4	3,036
19,0	890,48	0,8743	153,1	3,031
20,0	892,39	0,9184	153,8	3,026
21.0	894.28	0.9623	154.5	3,022
22,0	896,14	1,0660	155,3	3.017
23.0	897.97	1,0496	156.0	3.012
24,0	899,78	1,0930	156,7	3,008

Table II.11 (*Continued*)

p	ρ	Z	h	s
		$T=85$ K		
25.0	901.57	1.1363	157.4	3.003
26.0	903.33	1.1794	158.2	2.999
27.0	905.08	1.2224	158.9	2.995
28.0	906.80	1.2653	159.7	2.990
29.0	908.50	1.3080	160.4	2.986
30.0	910.19	1.3506	161.1	2.982
35.0	918.37	1.5617	164.9	2.962
40.0	926.21	1.7697	168.7	2.943
45.0	933.79	1.9747	172.6	2.925
50.0	941.16	2.1770	176.5	2.908
55.0	948.38	2.3764	180.4	2.892
60.0	955.51	2.5731	184.3	2.877
65.0	962.59	2.7671	188.3	2.862
70.0	969.69	2.9581	192.4	2.849
75.0	976.86	3.1461	196.4	2.836
80.0	984.19	3.3309	200.5	2.824
85.0	991.78	3.5120	204.6	2.813
90.0	999.80	3.6887	208.7	2.802
95.0	1008.52	3.8600	212.9	2.793
100.0	1018.54	4.0232	217.1	2.784
		$T=90$ K		
0.1	3.98	0.9715	341.0	5.646
0.5	821.97	0.0235	149.7	3.238
1.0	823.73	0.0470	150.0	3.234
1.5	825.46	0.0703	150.2	3.231
2.0	827.14	0.0936	150.5	3.227
2.5	828.79	0.1167	150.8	3.224
3.0	830.41	0.1398	151.1	3.220
3.5	832.00	0.1628	151.4	3.217
4.0	833.56	0.1857	151.7	3.213
4.5	835.09	0.2085	152.0	3.210
5.0	836.60	0.2313	152.3	3.206
6.0	839.53	0.2766	152.9	3.200
7.0	842.37	0.3216	153.5	3.193
8.0	845.13	0.3663	154.1	3.187
9.0	847.81	0.4108	154.7	3.181
10.0	850.42	0.4551	155.4	3.175
11.0	852.96	0.4991	156.0	3.169
12.0	855.45	0.5429	156.7	3.163
13.0	857.87	0.5865	157.3	3.158
14.0	860.24	0.6298	158.0	3.152
15.0	862.56	0.6730	158.7	3.147
16.0	864.84	0.7160	159.3	3.141
17.0	867.07	0.7588	160.0	3.136
18.0	869.26	0.8014	160.7	3.131
19.0	871.41	0.8438	161.4	3.126

p	ρ	Z	h	s
		$T=90$ K		
20.0	873.52	0.8861	162.1	3.121
21.0	875.60	0.9282	162.8	3.116
22.0	877.65	0.9701	163.5	3.111
23.0	879.67	1.0119	164.2	3.106
24.0	881.65	1.0535	164.9	3.102
25.0	883.61	1.0950	165.6	3.097
26.0	885.55	1.1363	166.4	3.093
27.0	887.45	1.1774	167.1	3.088
28.0	889.34	1.2185	167.8	3.084
29.0	891.20	1.2594	168.5	3.079
30.0	893.04	1.3001	169.3	3.075
35.0	901.94	1.5018	173.0	3.054
40.0	910.44	1.7003	176.7	3.035
45.0	918.62	1.8958	180.5	3.016
50.0	926.56	2.0884	184.4	2.998
55.0	934.31	2.2782	188.2	2.982
60.0	941.94	2.4652	192.1	2.966
65.0	949.50	2.6494	196.1	2.951
70.0	957.05	2.8307	200.0	2.936
75.0	964.66	3.0089	204.0	2.923
80.0	972.40	3.1840	207.9	2.909
85.0	980.39	3.3554	211.9	2.896
90.0	988.78	3.5226	215.9	2.884
95.0	997.83	3.6846	219.8	2.872
100.0	1008.07	3.8391	223.7	2.859
		$T=95$ K		
0.1	3.76	0.9761	346.4	5.704
0.5	795.54	0.0230	159.4	3.343
1.0	797.66	0.0460	159.6	3.339
1.5	799.72	0.0688	159.9	3.335
2.0	801.72	0.0915	160.1	3.331
2.5	803.68	0.1141	160.3	3.327
3.0	805.59	0.1365	160.6	3.323
3.5	807.45	0.1589	160.8	3.319
4.0	809.28	0.1812	161.1	3.315
4.5	811.06	0.2034	161.3	3.311
5.0	812.81	0.2255	161.6	3.307
6.0	816.21	0.2695	162.1	3.300
7.0	819.49	0.3132	162.7	3.293
8.0	822.65	0.3565	163.3	3.286
9.0	825.71	0.3996	163.8	3.279
10.0	828.67	0.4424	164.4	3.273
11.0	831.55	0.4850	165.0	3.267
12.0	834.35	0.5273	165.6	3.260
13.0	837.08	0.5694	166.3	3.254
14.0	839.74	0.6113	166.9	3.248

p	ρ	Z	h	s
		$T=95$ K		
15.0	842.34	0.6529	167.5	3.243
16.0	844.87	0.6943	168.2	3.237
17.0	847.35	0.7356	168.8	3.231
18.0	849.79	0.7766	169.5	3.226
19.0	852.17	0.8175	170.1	3.220
20.0	854.50	0.8581	170.8	3.215
21.0	856.80	0.8986	171.5	3.210
22.0	859.05	0.9390	172.2	3.205
23.0	861.27	0.9791	172.8	3.200
24.0	863.45	1.0191	173.5	3.195
25.0	865.59	1.0589	174.2	3.190
26.0	867.70	1.0986	174.9	3.185
27.0	869.79	1.1381	175.6	3.180
28.0	871.84	1.1775	176.3	3.176
29.0	873.86	1.2167	177.0	3.171
30.0	875.86	1.2558	177.8	3.167
35.0	885.50	1.4492	181.4	3.145
40.0	894.64	1.6393	185.1	3.125
45.0	903.39	1.8263	188.8	3.105
50.0	911.82	2.0105	192.6	3.087
55.0	920.02	2.1918	196.4	3.070
60.0	928.04	2.3704	200.2	3.053
65.0	935.94	2.5463	204.0	3.037
70.0	943.78	2.7194	207.9	3.021
75.0	951.61	2.8897	211.7	3.006
80.0	959.50	3.0570	215.6	2.992
85.0	967.54	3.2210	219.4	2.977
90.0	975.83	3.3815	223.2	2.963
95.0	984.53	3.5378	226.9	2.948
100.0	993.90	3.6889	230.5	2.934
		$T=100$ K		
0.1	3.56	0.9796	351.6	5.758
0.5	19.64	0.8867	343.6	5.240
1.0	769.48	0.0453	169.9	3.444
1.5	772.02	0.0677	170.0	3.439
2.0	774.47	0.0899	170.2	3.434
2.5	776.85	0.1121	170.4	3.430
3.0	779.16	0.1341	170.6	3.425
3.5	781.40	0.1560	170.8	3.421
4.0	783.59	0.1778	171.0	3.416
4.5	785.71	0.1995	171.2	3.412
5.0	787.79	0.2211	171.4	3.408
6.0	791.79	0.2639	171.8	3.400
7.0	795.61	0.3065	172.3	3.392
8.0	799.28	0.3486	172.8	3.384
9.0	802.81	0.3905	173.3	3.377

Table II.11 (*Continued*)

p	ρ	Z	h	s
		$T = 100$ K		
10.0	806.21	0.4320	173.8	3.370
11.0	809.50	0.4733	174.4	3.363
12.0	812.68	0.5143	174.9	3.356
13.0	815.77	0.5551	175.5	3.349
14.0	818.77	0.5956	176.1	3.343
15.0	821.68	0.6358	176.7	3.337
16.0	824.52	0.6759	177.3	3.331
17.0	827.29	0.7157	177.9	3.325
18.0	830.00	0.7554	178.5	3.319
19.0	832.64	0.7948	179.2	3.313
20.0	835.23	0.8341	179.8	3.307
21.0	837.76	0.8731	180.4	3.302
22.0	840.24	0.9120	181.1	3.296
23.0	842.67	0.9507	181.7	3.291
24.0	845.06	0.9892	182.4	3.286
25.0	847.41	1.0276	183.1	3.281
26.0	849.71	1.0658	183.8	3.276
27.0	851.98	1.1038	184.4	3.271
28.0	854.21	1.1417	185.1	3.266
29.0	856.41	1.1795	185.8	3.261
30.0	858.57	1.2171	186.5	3.256
35.0	868.96	1.4029	190.0	3.234
40.0	878.74	1.5855	193.6	3.212
45.0	888.03	1.7650	197.3	3.192
50.0	896.92	1.9417	201.0	3.173
55.0	905.51	2.1156	204.7	3.155
60.0	913.84	2.2869	208.4	3.138
65.0	921.99	2.4556	212.2	3.121
70.0	930.00	2.6217	216.0	3.104
75.0	937.92	2.7852	219.7	3.088
80.0	945.81	2.9461	223.5	3.073
85.0	953.73	3.1043	227.2	3.057
90.0	961.74	3.2595	230.9	3.042
95.0	969.92	3.4116	234.5	3.026
100.0	978.38	3.5601	238.0	3.010

p	ρ	Z	h	s	c_v	c_p
			$T = 105$ K			
0.1	3.38	0.9823	356.8	5.809	0.728	1.035
0.5	18.34	0.9042	349.9	5.302	0.794	1.220
1.0	738.32	0.0449	180.7	3.550	0.981	2.240
1.5	741.59	0.0671	180.8	3.544	0.977	2.213
2.0	744.71	0.0891	180.8	3.538	0.974	2.189

p	ρ	Z	h	s	c_v	c_p
			$T=105$ K			
2,5	747,71	0.1109	180.9	3,533	0,971	2,166
3,0	750,59	0,1326	181,0	3.527	0.968	2,146
3,5	753,37	0,1541	181.2	3,522	0,966	2,126
4,0	756,05	0.1755	181.3	3,517	0,965	2.109
4,5	758,64	0,1968	181.4	3,512	0.963	2,092
5,0	761,15	0,2179	181,6	3,507	0.962	2.077
6,0	765,96	0,2599	181,9	3,498	0.961	2,049
7,0	770,50	0,3014	182,3	3,489	0,960	2,025
8,0	774,83	0,3425	182,7	3.480	0,959	2.003
9,0	778,95	0,3833	183,1	3,472	0,960	1,983
10,0	782,89	0,4237	183,6	3,464	0,960	1,965
11,0	786,68	0,4638	184,0	3,457	0,961	1,949
12,0	790,33	0.5037	184,5	3,449	0,963	1.934
13,0	793,84	0,5432	185,0	3,442	0,965	1,921
14,0	797,24	0,5825	185,5	3.435	0,967	1,908
15,0	800,54	0,6216	186,1	3,428	0,969	1,896
16,0	803,73	0,6604	186,6	3,422	0,971	1,886
17,0	806,83	0,6989	187,2	3,415	0,973	1,876
18,0	809,85	0.7373	187,8	3,409	0.976	1.866
19,0	812,79	0,7755	188,4	3,403	0.978	1.857
20,0	815,65	0,8134	189,0	3,397	0.981	1,849
21,0	818,45	0,8511	189,6	3,391	0.984	1.841
22,0	821,19	0,8887	190,2	3,385	0,987	1,834
23,0	823,86	0,9261	190,8	3,380	0,989	1,827
24,0	826,48	0,9633	191,5	3,374	0,992	1.820
25,0	829,04	1,0003	192.1	3,369	0.995	1,814
26,0	831.56	1.0372	192,7	3,363	0,998	1.808
27,0	834,03	1,0739	193,4	3,358	1,001	1.802
28,0	836,45	1,1104	194,1	3,353	1.004	1.797
29,0	838,84	1,1468	194,7	3,348	1,007	1,792
30,0	841,18	1,1831	195,4	3,343	1,009	1,787
35,0	852,35	1,3622	198,8	3,319	1,023	1,765
40,0	862,78	1,5379	202,3	3,297	1,035	1,747
45,0	872,60	1,7107	205,9	3,276	1,045	1,732
50,0	881,94	1,8807	209,5	3,257	1,054	1,718
55,0	890,88	2,0480	213,2	3,238	1,060	1,706
60,0	899,50	2,2127	216,8	3,220	1,063	1,694
65,0	907,85	2,3751	220,5	3,202	1,064	1,683
70,0	915,99	2,5350	224,3	3,185	1,061	1,673
75,0	923,97	2,6927	228,0	3,169	1,056	1,663
80,0	931,82	2,8480	231,6	3,152	1,046	1,654
85,0	939,59	3,0009	235,3	3,136	1,033	1,646
90,0	947,33	3,1515	238,9	3,120	1,015	1,640
95,0	955,08	3,2996	242,5	3,104	0,992	1,636
100,0	962,88	3,4451	246,0	3,088	0,963	1,635

p	ρ	Z	h	s	c_v	c_p
			$T=110$ K			
0.1	3.22	0.9846	362.0	5.857	0.724	1.028
0.5	17.25	0.9178	355.8	5.357	0.768	1.167
1.0	38.74	0.8175	346.2	5.096	0.854	1.495
1.5	707.09	0.0672	192.2	3.651	0.981	2.385
2.0	711.31	0.0890	192.1	3.643	0.977	2.342
2.5	715.28	0.1107	192.1	3.636	0.973	2.304
3.0	719.04	0.1321	192.1	3.630	0.970	2.270
3.5	722.62	0.1534	192.1	3.623	0.967	2.240
4.0	726.03	0.1745	192.1	3.617	0.965	2.213
4.5	729.29	0.1954	192.1	3.611	0.963	2.188
5.0	732.42	0.2162	192.2	3.606	0.961	2.165
6.0	738.33	0.2573	192.3	3.595	0.959	2.125
7.0	743.84	0.2980	192.6	3.585	0.958	2.091
8.0	749.02	0.3382	192.8	3.575	0.957	2.061
9.0	753.90	0.3780	193.1	3.566	0.957	2.034
10.0	758.54	0.4174	193.5	3.557	0.958	2.011
11.0	762.95	0.4565	193.9	3.548	0.959	1.990
12.0	767.16	0.4953	194.3	3.540	0.960	1.971
13.0	771.20	0.5338	194.7	3.532	0.962	1.954
14.0	775.08	0.5719	195.2	3.525	0.963	1.938
15.0	778.82	0.6099	195.6	3.517	0.966	1.924
16.0	782.43	0.6475	196.1	3.510	0.968	1.911
17.0	785.92	0.6849	196.6	3.503	0.970	1.899
18.0	789.30	0.7221	197.2	3.496	0.972	1.888
19.0	792.58	0.7591	197.7	3.490	0.975	1.877
20.0	795.76	0.7958	198.3	3.483	0.977	1.867
21.0	798.86	0.8324	198.8	3.477	0.980	1.858
22.0	801.88	0.8687	199.4	3.471	0.983	1.850
23.0	804.83	0.9049	200.0	3.465	0.985	1.842
24.0	807.70	0.9409	200.6	3.459	0.988	1.834
25.0	810.51	0.9767	201.2	3.453	0.991	1.827
26.0	813.26	1.0123	201.8	3.448	0.994	1.820
27.0	815.94	1.0478	202.4	3.442	0.996	1.814
28.0	818.58	1.0831	203.1	3.437	0.999	1.808
29.0	821.16	1.1183	203.7	3.432	1.001	1.802
30.0	823.69	1.1533	204.3	3.426	1.004	1.797
35.0	835.71	1.3261	207.6	3.402	1.016	1.774
40.0	846.81	1.4957	211.1	3.379	1.027	1.755
45.0	857.19	1.6623	214.6	3.357	1.036	1.740
50.0	866.98	1.8261	218.1	3.337	1.044	1.727
55.0	876.28	1.9874	221.7	3.317	1.049	1.716
60.0	885.18	2.1463	225.4	3.299	1.052	1.707
65.0	893.74	2.3029	229.0	3.281	1.052	1.698
70.0	902.01	2.4573	232.7	3.264	1.050	1.692
75.0	910.04	2.6096	236.3	3.247	1.045	1.686

p	ρ	Z	h	s	c_v	c_p
			$T=110$ K			
80.0	917.88	2.7598	240.0	3.230	1.037	1.681
85.0	925.55	2.9080	243.6	3.214	1.026	1.678
90.0	933.10	3.0541	247.2	3.198	1.012	1.677
95.0	940.55	3.1983	250.8	3.182	0.994	1.679
100.0	947.93	3.3404	254.3	3.165	0.973	1.683
			$T=115$ K			
0.1	3.07	0.9864	367.1	5.902	0.721	1.023
0.5	16.31	0.9286	361.6	5.408	0.752	1.132
1.0	35.82	0.8455	353.3	5.159	0.809	1.362
1.5	61.35	0.7406	342.4	4.971	0.902	1.866
2.0	672.00	0.0901	204.4	3.752	0.981	2.595
2.5	677.74	0.1117	204.1	3.743	0.975	2.519
3.0	683.01	0.1330	203.8	3.734	0.969	2.456
3.5	687.89	0.1541	203.6	3.726	0.965	2.402
4.0	692.46	0.1750	203.5	3.719	0.962	2.356
4.5	696.76	0.1956	203.4	3.711	0.959	2.315
5.0	700.81	0.2161	203.3	3.704	0.956	2.279
6.0	708.33	0.2566	203.2	3.691	0.953	2.218
7.0	715.19	0.2964	203.2	3.679	0.951	2.168
8.0	721.52	0.3358	203.3	3.668	0.949	2.126
9.0	727.41	0.3747	203.5	3.657	0.949	2.090
10.0	732.93	0.4132	203.7	3.647	0.949	2.059
11.0	738.12	0.4514	203.9	3.638	0.950	2.031
12.0	743.05	0.4891	204.2	3.629	0.951	2.007
13.0	747.72	0.5266	204.6	3.620	0.953	1.985
14.0	752.19	0.5637	204.9	3.611	0.954	1.965
15.0	756.46	0.6006	205.3	3.603	0.956	1.948
16.0	760.56	0.6372	205.7	3.596	0.958	1.932
17.0	764.50	0.6735	206.2	3.588	0.960	1.917
18.0	768.30	0.7096	206.7	3.581	0.963	1.903
19.0	771.97	0.7455	207.1	3.574	0.965	1.891
20.0	775.53	0.7811	207.6	3.567	0.968	1.879
21.0	778.97	0.8165	208.2	3.560	0.970	1.869
22.0	782.31	0.8517	208.7	3.553	0.973	1.859
23.0	785.56	0.8868	209.2	3.547	0.975	1.849
24.0	788.72	0.9216	209.8	3.541	0.978	1.841
25.0	791.80	0.9563	210.4	3.535	0.980	1.833
26.0	794.80	0.9908	210.9	3.529	0.983	1.825
27.0	797.74	1.0251	211.5	3.523	0.985	1.818
28.0	800.60	1.0593	212.1	3.517	0.988	1.811
29.0	803.40	1.0933	212.7	3.512	0.991	1.805
30.0	806.15	1.1271	213.3	3.506	0.993	1.799
35.0	819.07	1.2942	216.5	3.481	1.005	1.773
40.0	830.90	1.4581	219.8	3.457	1.015	1.754
45.0	841.87	1.6190	223.3	3.434	1.024	1.738
50.0	852.13	1.7772	226.7	3.413	1.031	1.725

Table II.11 (*Continued*)

p	ρ	Z	h	s	c_v	c_p
			$T=115$ K			
55.0	861.81	1.9329	230.3	3.394	1.035	1.715
60.0	871.01	2.0864	233.9	3.375	1.038	1.707
65.0	879.79	2.2377	237.5	3.356	1.039	1.701
70.0	888.22	2.3870	241.1	3.339	1.038	1.696
75.0	896.34	2.5343	244.8	3.322	1.034	1.693
80.0	904.20	2.6797	248.4	3.305	1.029	1.691
85.0	911.84	2.8234	252.0	3.289	1.020	1.691
90.0	919.28	2.9653	255.6	3.273	1.010	1.693
95.0	926.55	3.1054	259.2	3.257	0.997	1.697
100.0	933.68	3.2439	262.8	3.241	0.982	1.703
			$T=120$ K			
0.1	2.94	0.9880	372.2	5.946	0.720	1.020
0.5	15.48	0.9375	367.2	5.456	0.742	1.108
1.0	33.47	0.8671	359.9	5.215	0.782	1.280
1.5	55.53	0.7840	351.0	5.044	0.843	1.592
2.0	85.96	0.6753	338.8	4.882	0.939	2.360
2.5	630.88	0.1150	217.6	3.858	0.984	2.953
3.0	639.34	0.1362	216.9	3.845	0.974	2.798
3.5	646.77	0.1571	216.3	3.834	0.966	2.680
4.0	653.43	0.1777	215.8	3.823	0.960	2.587
4.5	659.49	0.1981	215.4	3.814	0.955	2.512
5.0	665.07	0.2182	215.1	3.805	0.951	2.448
6.0	675.07	0.2580	214.6	3.788	0.946	2.347
7.0	683.91	0.2971	214.3	3.774	0.942	2.270
8.0	691.85	0.3356	214.1	3.760	0.940	2.208
9.0	699.10	0.3737	214.1	3.748	0.938	2.157
10.0	705.78	0.4113	214.1	3.736	0.938	2.114
11.0	711.99	0.4484	214.2	3.725	0.938	2.078
12.0	717.80	0.4852	214.4	3.715	0.939	2.046
13.0	723.27	0.5217	214.6	3.705	0.940	2.018
14.0	728.45	0.5578	214.8	3.696	0.942	1.993
15.0	733.36	0.5937	215.1	3.687	0.943	1.971
16.0	738.05	0.6292	215.5	3.678	0.945	1.951
17.0	742.52	0.6645	215.8	3.670	0.947	1.933
18.0	746.82	0.6996	216.2	3.662	0.949	1.917
19.0	750.94	0.7344	216.6	3.654	0.952	1.902
20.0	754.92	0.7690	217.1	3.647	0.954	1.889
21.0	758.75	0.8034	217.5	3.640	0.957	1.876
22.0	762.46	0.8375	218.0	3.633	0.959	1.864
23.0	766.05	0.8715	218.5	3.626	0.961	1.854
24.0	769.53	0.9053	219.0	3.619	0.964	1.844
25.0	772.92	0.9388	219.5	3.613	0.966	1.834
26.0	776.20	0.9723	220.1	3.607	0.969	1.826
27.0	779.41	1.0055	220.6	3.600	0.971	1.817
28.0	782.53	1.0386	221.2	3.594	0.974	1.810
29.0	785.57	1.0715	221.7	3.589	0.976	1.803

p	ρ	Z	h	s	c_v	c_p
			$T=120$ K			
30.0	788.54	1.1043	222.3	3.583	0.979	1.796
35.0	802.45	1.2660	225.4	3.556	0.990	1.768
40.0	815.07	1.4245	228.6	3.531	1.000	1.746
45.0	826.66	1.5800	231.9	3.508	1.009	1.730
50.0	837.44	1.7330	235.4	3.487	1.015	1.717
55.0	847.53	1.8836	238.9	3.466	1.021	1.707
60.0	857.05	2.0320	242.4	3.447	1.024	1.699
65.0	866.09	2.1784	246.0	3.429	1.026	1.694
70.0	874.70	2.3229	249.6	3.411	1.025	1.690
75.0	882.96	2.4655	253.2	3.394	1.024	1.688
80.0	890.89	2.6065	256.9	3.377	1.020	1.687
85.0	898.55	2.7458	260.5	3.361	1.014	1.689
90.0	905.95	2.8835	264.1	3.345	1.007	1.692
95.0	913.14	3.0198	267.7	3.329	0.998	1.697
100.0	920.13	3.1545	271.3	3.313	0.987	1.704
			$T=125$ K			
0.1	2.82	0.9893	377.3	5.987	0.719	1.017
0.5	14.75	0.9449	372.7	5.501	0.735	1.091
1.0	31.51	0.8844	366.2	5.266	0.765	1.225
1.5	51.22	0.8160	358.6	5.106	0.807	1.441
2.0	75.89	0.7343	349.1	4.966	0.867	1.845
2.5	111.48	0.6249	335.9	4.816	0.962	2.932
3.0	578.58	0.1445	232.7	3.974	0.997	3.745
3.5	593.10	0.1644	231.0	3.954	0.980	3.311
4.0	604.65	0.1843	229.7	3.937	0.967	3.048
4.5	614.37	0.2041	228.7	3.923	0.958	2.867
5.0	622.82	0.2237	227.9	3.910	0.950	2.733
6.0	637.11	0.2624	226.8	3.888	0.940	2.543
7.0	649.04	0.3005	226.0	3.869	0.934	2.414
8.0	659.36	0.3381	225.4	3.852	0.930	2.318
9.0	668.51	0.3751	225.1	3.837	0.927	2.244
10.0	676.76	0.4117	224.8	3.824	0.926	2.184
11.0	684.29	0.4479	224.7	3.811	0.926	2.134
12.0	691.24	0.4837	224.7	3.799	0.926	2.092
13.0	697.70	0.5192	224.7	3.788	0.926	2.056
14.0	703.74	0.5543	224.9	3.778	0.928	2.025
15.0	709.43	0.5892	225.0	3.768	0.929	1.997
16.0	714.81	0.6237	225.3	3.758	0.931	1.973
17.0	719.92	0.6580	225.5	3.749	0.933	1.951
18.0	724.79	0.6920	225.8	3.741	0.935	1.931
19.0	729.44	0.7258	226.2	3.732	0.937	1.913
20.0	733.90	0.7594	226.5	3.724	0.939	1.897
21.0	738.18	0.7927	226.9	3.716	0.941	1.882
22.0	742.30	0.8258	227.3	3.709	0.944	1.869
23.0	746.28	0.8588	227.8	3.702	0.946	1.856
24.0	750.12	0.8915	228.2	3.695	0.948	1.845

p	ρ	Z	h	s	c_v	c_p
			$T = 125$ K			
25.0	753.84	0.9241	228.7	3.688	0.951	1.834
26.0	757.45	0.9565	229.2	3.681	0.953	1.824
27.0	760.95	0.9887	229.7	3.675	0.956	1.815
28.0	764.35	1.0208	230.2	3.668	0.958	1.806
29.0	767.66	1.0527	230.7	3.662	0.960	1.798
30.0	770.88	1.0844	231.3	3.656	0.963	1.791
35.0	785.87	1.2410	234.2	3.628	0.974	1.759
40.0	799.33	1.3944	237.3	3.602	0.984	1.735
45.0	811.60	1.5450	240.5	3.579	0.992	1.717
50.0	822.92	1.6930	243.9	3.557	0.999	1.704
55.0	833.46	1.8388	247.4	3.536	1.005	1.693
60.0	843.34	1.9825	250.9	3.516	1.009	1.686
65.0	852.66	2.1242	254.4	3.498	1.011	1.680
70.0	861.50	2.2641	258.0	3.480	1.012	1.677
75.0	869.92	2.4024	261.6	3.462	1.012	1.675
80.0	877.98	2.5390	265.3	3.446	1.010	1.675
85.0	885.71	2.6742	268.9	3.429	1.007	1.677
90.0	893.14	2.8079	272.5	3.414	1.002	1.680
95.0	900.31	2.9403	276.2	3.398	0.996	1.685
100.0	907.25	3.0714	279.8	3.383	0.989	1.692
			$T = 130$ K			
0.1	2.71	0.9905	382.4	6.027	0.718	1.016
0.5	14.08	0.9511	378.1	5.543	0.731	1.078
1.0	29.82	0.8986	372.2	5.313	0.753	1.187
1.5	47.80	0.8409	365.5	5.160	0.784	1.347
2.0	69.10	0.7754	357.7	5.033	0.825	1.605
2.5	96.05	0.6974	348.0	4.911	0.882	2.085
3.0	135.46	0.5934	334.4	4.773	0.968	3.378
3.5	493.54	0.1900	253.4	4.129	1.046	7.602
4.0	531.53	0.2016	247.8	4.078	0.999	4.566
4.5	553.03	0.2180	244.8	4.049	0.975	3.738
5.0	568.69	0.2356	242.9	4.027	0.960	3.320
6.0	591.78	0.2717	240.2	3.993	0.940	2.875
7.0	609.07	0.3079	238.5	3.968	0.929	2.631
8.0	623.11	0.3440	237.4	3.946	0.922	2.472
9.0	635.02	0.3797	236.5	3.927	0.917	2.358
10.0	645.43	0.4151	236.0	3.911	0.915	2.272
11.0	654.72	0.4502	235.6	3.896	0.913	2.204
12.0	663.13	0.4849	235.3	3.882	0.913	2.148
13.0	670.83	0.5192	235.1	3.870	0.913	2.101
14.0	677.95	0.5533	235.1	3.858	0.914	2.061
15.0	684.57	0.5871	235.1	3.847	0.915	2.027
16.0	690.79	0.6206	235.2	3.836	0.916	1.997
17.0	696.64	0.6538	235.3	3.826	0.918	1.971
18.0	702.17	0.6868	235.5	3.817	0.920	1.947
19.0	707.43	0.7196	235.8	3.807	0.922	1.926

p	ρ	Z	h	s	c_v	c_p
			$T=130$ K			
20.0	712.44	0.7521	236.0	3.799	0.924	1.907
21.0	717.23	0.7845	236.3	3.790	0.926	1.890
22.0	721.82	0.8166	236.7	3.782	0.928	1.874
23.0	726.23	0.8485	237.1	3.775	0.930	1.859
24.0	730.48	0.8803	237.4	3.767	0.933	1.846
25.0	734.57	0.9119	237.9	3.760	0.935	1.834
26.0	738.53	0.9433	238.3	3.753	0.937	1.822
27.0	742.36	0.9745	238.8	3.746	0.939	1.812
28.0	746.06	1.0056	239.2	3.739	0.942	1.802
29.0	749.66	1.0365	239.7	3.733	0.944	1.793
30.0	753.16	1.0672	240.2	3.726	0.946	1.784
35.0	769.31	1.2190	243.0	3.697	0.957	1.749
40.0	783.68	1.3676	245.9	3.670	0.967	1.723
45.0	796.68	1.5134	249.1	3.646	0.975	1.703
50.0	808.58	1.6568	252.4	3.623	0.983	1.689
55.0	819.60	1.7980	255.8	3.602	0.989	1.677
60.0	829.87	1.9371	259.3	3.582	0.993	1.669
65.0	839.52	2.0745	262.8	3.563	0.996	1.663
70.0	848.62	2.2101	266.4	3.545	0.999	1.659
75.0	857.25	2.3441	270.0	3.528	0.999	1.657
80.0	865.46	2.4767	273.6	3.511	0.999	1.657
85.0	873.30	2.6078	277.3	3.495	0.998	1.659
90.0	880.81	2.7377	280.9	3.479	0.995	1.662
95.0	888.03	2.8663	284.5	3.464	0.992	1.666
100.0	894.97	2.9937	288.2	3.449	0.987	1.673
			$T=131$ K			
0.1	2.68	0.9907	383.4	6.035	0.718	1.015
0.5	13.96	0.9523	379.2	5.552	0.730	1.076
1.0	29.51	0.9011	373.4	5.322	0.752	1.180
1.5	47.19	0.8452	366.9	5.171	0.780	1.333
2.0	67.97	0.7823	359.3	5.045	0.818	1.571
2.5	93.83	0.7084	350.0	4.927	0.871	1.998
3.0	130.05	0.6133	337.6	4.797	0.946	3.014
3.5*	417.15	0.2231	267.6	4.238	1.120	59.609
4.0	508.93	0.2090	252.7	4.117	1.014	5.480
4.5	536.81	0.2229	248.7	4.079	0.983	4.097
5.0	555.40	0.2394	246.3	4.053	0.964	3.521
6.0	581.42	0.2744	243.2	4.016	0.941	2.970
7.0	600.27	0.3101	241.2	3.988	0.928	2.688
8.0	615.29	0.3457	239.9	3.965	0.921	2.510
9.0	627.91	0.3811	238.9	3.946	0.916	2.386
10.0	638.85	0.4162	238.2	3.928	0.913	2.293
11.0	648.55	0.4510	237.8	3.913	0.911	2.220
12.0	657.30	0.4854	237.4	3.899	0.910	2.160
13.0	665.28	0.5196	237.2	3.886	0.910	2.111
14.0	672.65	0.5534	237.1	3.874	0.911	2.069

*This point lies in the two-phase region.

Table II.11 (*Continued*)

p	ρ	Z	h	s	c_v	c_p
			$T=131$ K			
15.0	679.48	0.5870	237.1	3.862	0.912	2.034
16.0	685.88	0.6203	237.2	3.851	0.913	2.002
17.0	691.90	0.6533	237.3	3.841	0.915	1.975
18.0	697.58	0.6861	237.5	3.831	0.917	1.950
19.0	702.97	0.7186	237.7	3.822	0.919	1.929
20.0	708.10	0.7510	237.9	3.813	0.921	1.909
21.0	713.00	0.7831	238.2	3.805	0.923	1.891
22.0	717.69	0.8150	238.6	3.797	0.925	1.875
23.0	722.19	0.8468	238.9	3.789	0.927	1.860
24.0	726.52	0.8783	239.3	3.781	0.929	1.846
25.0	730.70	0.9097	239.7	3.774	0.932	1.834
26.0	734.72	0.9409	240.1	3.767	0.934	1.822
27.0	738.62	0.9719	240.6	3.760	0.936	1.811
28.0	742.39	1.0028	241.0	3.753	0.939	1.801
29.0	746.05	1.0335	241.5	3.746	0.941	1.792
30.0	749.61	1.0641	242.0	3.740	0.943	1.783
35.0	766.00	1.2149	244.7	3.710	0.954	1.747
40.0	780.56	1.3625	247.7	3.683	0.964	1.721
45.0	793.71	1.5075	250.8	3.659	0.972	1.701
50.0	805.74	1.6500	254.1	3.636	0.979	1.685
55.0	816.85	1.7902	257.5	3.615	0.985	1.674
60.0	827.21	1.9285	260.9	3.595	0.990	1.665
65.0	836.92	2.0650	264.5	3.576	0.993	1.659
70.0	846.08	2.1998	268.0	3.558	0.996	1.655
75.0	854.75	2.3330	271.6	3.541	0.997	1.653
80.0	863.00	2.4648	275.3	3.524	0.997	1.653
85.0	870.87	2.5951	278.9	3.508	0.996	1.655
90.0	878.40	2.7242	282.6	3.492	0.994	1.658
95.0	885.63	2.8521	286.2	3.476	0.990	1.662
100.0	892.58	2.9788	289.9	3.461	0.986	1.668
			$T=132$ K			
0.1	2.66	0.9909	384.4	6.043	0.718	1.015
0.5	13.84	0.9534	380.3	5.560	0.730	1.074
1.0	29.20	0.9036	374.6	5.331	0.750	1.174
1.5	46.60	0.8494	368.2	5.181	0.777	1.319
2.0	66.90	0.7889	360.8	5.057	0.813	1.541
2.5	91.80	0.7186	352.0	4.942	0.861	1.923
3.0	125.49	0.6308	340.5	4.819	0.928	2.752
3.5	186.07	0.4964	321.3	4.649	1.041	6.526
4.0	478.74	0.2205	259.0	4.164	1.036	7.404
4.5	518.13	0.2292	253.1	4.112	0.992	4.612
5.0	540.82	0.2440	249.9	4.081	0.969	3.777
6.0	570.50	0.2775	246.2	4.039	0.943	3.079
7.0	591.13	0.3125	243.9	4.009	0.928	2.750
8.0	607.26	0.3476	242.4	3.984	0.920	2.551
9.0	620.64	0.3826	241.3	3.964	0.914	2.415

p	ρ	Z	h	s	c_v	c_p
			$T=132$ K			
10.0	632.14	0.4174	240.5	3.946	0.911	2.314
11.0	642.29	0.4519	240.0	3.930	0.909	2.236
12.0	651.40	0.4861	239.6	3.915	0.908	2.173
13.0	659.68	0.5200	239.4	3.902	0.908	2.121
14.0	667.30	0.5536	239.2	3.889	0.908	2.078
15.0	674.35	0.5869	239.2	3.878	0.909	2.040
16.0	680.94	0.6200	239.2	3.867	0.910	2.008
17.0	687.12	0.6528	239.3	3.856	0.912	1.979
18.0	692.96	0.6854	239.4	3.846	0.914	1.954
19.0	698.48	0.7178	239.6	3.837	0.916	1.931
20.0	703.74	0.7499	239.8	3.828	0.918	1.911
21.0	708.75	0.7818	240.1	3.819	0.920	1.893
22.0	713.54	0.8136	240.4	3.811	0.922	1.876
23.0	718.14	0.8451	240.8	3.803	0.924	1.861
24.0	722.55	0.8765	241.1	3.795	0.926	1.847
25.0	726.81	0.9076	241.5	3.788	0.929	1.834
26.0	730.91	0.9386	241.9	3.780	0.931	1.822
27.0	734.88	0.9695	242.4	3.773	0.933	1.811
28.0	738.72	1.0002	242.8	3.767	0.935	1.801
29.0	742.44	1.0307	243.3	3.760	0.938	1.791
30.0	746.05	1.0611	243.8	3.753	0.940	1.782
35.0	762.70	1.2109	246.5	3.723	0.951	1.745
40.0	777.45	1.3576	249.4	3.696	0.960	1.718
45.0	790.75	1.5016	252.5	3.672	0.969	1.698
50.0	802.90	1.6432	255.8	3.649	0.976	1.682
55.0	814.12	1.7827	259.1	3.628	0.982	1.670
60.0	824.56	1.9201	262.6	3.608	0.987	1.662
65.0	834.34	2.0557	266.1	3.589	0.990	1.656
70.0	843.55	2.1897	269.7	3.570	0.993	1.652
75.0	852.27	2.3221	273.3	3.553	0.994	1.650
80.0	860.55	2.4530	276.9	3.536	0.994	1.649
85.0	868.45	2.5826	280.6	3.520	0.994	1.650
90.0	876.01	2.7110	284.2	3.504	0.992	1.653
95.0	883.25	2.8381	287.9	3.489	0.989	1.658
100.0	890.21	2.9641	291.5	3.474	0.986	1.664
			$T=133$ K			
0.1	2.64	0.9911	385.4	6.050	0.718	1.015
0.5	13.72	0.9544	381.3	5.568	0.729	1.072
1.0	28.91	0.9059	375.7	5.340	0.748	1.169
1.5	46.03	0.8534	369.5	5.190	0.774	1.306
2.0	65.88	0.7951	362.4	5.069	0.807	1.514
2.5	89.91	0.7282	353.9	4.956	0.852	1.858
3.0	121.53	0.6465	343.1	4.839	0.913	2.552
3.5	172.40	0.5317	326.9	4.691	1.006	4.846
4.0	428.27	0.2446	269.0	4.239	1.076	14.342
4.5	495.85	0.2377	258.1	4.149	1.005	5.411

p	ρ	Z	h	s	c_v	c_p
			$T=133$ K			
5.0	524.61	0.2496	253.9	4.110	0.975	4.111
6.0	558.93	0.2811	249.3	4.062	0.945	3.205
7.0	581.64	0.3152	246.7	4.030	0.928	2.819
8.0	598.99	0.3498	245.0	4.004	0.919	2.595
9.0	613.20	0.3844	243.7	3.982	0.913	2.446
10.0	625.32	0.4188	242.9	3.963	0.909	2.337
11.0	635.94	0.4530	242.2	3.947	0.907	2.253
12.0	645.43	0.4869	241.8	3.932	0.906	2.187
13.0	654.02	0.5206	241.5	3.918	0.905	2.132
14.0	661.90	0.5539	241.3	3.905	0.906	2.086
15.0	669.18	0.5870	241.2	3.893	0.907	2.047
16.0	675.97	0.6199	241.2	3.882	0.908	2.013
17.0	682.32	0.6525	241.3	3.871	0.909	1.984
18.0	688.31	0.6849	241.4	3.861	0.911	1.958
19.0	693.98	0.7170	241.5	3.851	0.913	1.934
20.0	699.36	0.7489	241.8	3.842	0.915	1.913
21.0	704.48	0.7807	242.0	3.834	0.917	1.894
22.0	709.38	0.8122	242.3	3.825	0.919	1.877
23.0	714.07	0.8435	242.6	3.817	0.921	1.861
24.0	718.58	0.8747	243.0	3.809	0.923	1.847
25.0	722.92	0.9057	243.4	3.802	0.925	1.834
26.0	727.10	0.9365	243.8	3.794	0.928	1.822
27.0	731.13	0.9671	244.2	3.787	0.930	1.810
28.0	735.04	0.9976	244.6	3.780	0.932	1.800
29.0	738.83	1.0279	245.1	3.773	0.934	1.790
30.0	742.50	1.0581	245.6	3.767	0.937	1.781
35.0	759.39	1.2070	248.2	3.737	0.947	1.743
40.0	774.33	1.3528	251.1	3.709	0.957	1.716
45.0	787.79	1.4960	254.2	3.684	0.965	1.695
50.0	800.07	1.6367	257.4	3.662	0.973	1.679
55.0	811.39	1.7752	260.8	3.640	0.979	1.667
60.0	821.91	1.9118	264.3	3.620	0.984	1.658
65.0	831.76	2.0466	267.8	3.601	0.988	1.652
70.0	841.03	2.1797	271.3	3.583	0.990	1.648
75.0	849.80	2.3113	274.9	3.565	0.992	1.646
80.0	858.12	2.4415	278.6	3.549	0.992	1.645
85.0	866.05	2.5703	282.2	3.533	0.992	1.646
90.0	873.63	2.6979	285.9	3.517	0.990	1.649
95.0	880.89	2.8243	289.5	3.502	0.988	1.653
100.0	887.87	2.9496	293.2	3.487	0.985	1.659
			$T=134$ K			
0.1	2.62	0.9913	386.4	6.058	0.718	1.014
0.5	13.60	0.9555	382.4	5.576	0.729	1.070
1.0	28.62	0.9082	376.9	5.349	0.747	1.164
1.5	45.48	0.8572	370.8	5.200	0.771	1.295
2.0	64.90	0.8010	363.9	5.080	0.802	1.489

Table II.11 (*Continued*)

p	ρ	Z	h	s	c_v	c_p
			$T=134$ K			
2.5	88.16	0.7371	355.7	4.970	0.844	1.802
3.0	118.04	0.6607	345.6	4.857	0.899	2.394
3.5	162.73	0.5591	331.3	4.723	0.978	4.001
4.0	293.81	0.3539	296.6	4.446	1.115	26.302
4.5	467.75	0.2501	264.1	4.195	1.022	6.790
5.0	506.25	0.2567	258.2	4.143	0.984	4.562
6.0	546.61	0.2853	252.6	4.087	0.947	3.353
7.0	571.75	0.3182	249.6	4.051	0.929	2.896
8.0	590.48	0.3522	247.6	4.023	0.918	2.643
9.0	605.60	0.3863	246.2	4.001	0.911	2.478
10.0	618.37	0.4204	245.2	3.981	0.907	2.361
11.0	629.49	0.4542	244.5	3.964	0.905	2.271
12.0	639.38	0.4879	244.0	3.948	0.903	2.201
13.0	648.30	0.5212	243.6	3.934	0.903	2.143
14.0	656.45	0.5544	243.4	3.921	0.903	2.095
15.0	663.97	0.5872	243.3	3.908	0.904	2.054
16.0	670.96	0.6199	243.2	3.897	0.905	2.019
17.0	677.49	0.6522	243.2	3.886	0.906	1.988
18.0	683.64	0.6844	243.3	3.876	0.908	1.961
19.0	689.45	0.7163	243.5	3.866	0.910	1.937
20.0	694.96	0.7481	243.7	3.857	0.912	1.915
21.0	700.20	0.7796	243.9	3.848	0.914	1.896
22.0	705.20	0.8109	244.2	3.839	0.916	1.878
23.0	710.00	0.8420	244.5	3.831	0.918	1.862
24.0	714.59	0.8730	244.8	3.823	0.920	1.847
25.0	719.01	0.9038	245.2	3.815	0.922	1.834
26.0	723.27	0.9344	245.6	3.808	0.925	1.821
27.0	727.38	0.9649	246.0	3.801	0.927	1.810
28.0	731.36	0.9952	246.4	3.794	0.929	1.799
29.0	735.21	1.0253	246.9	3.787	0.931	1.789
30.0	738.94	1.0553	247.4	3.780	0.933	1.780
35.0	756.09	1.2033	249.9	3.750	0.944	1.741
40.0	771.22	1.3482	252.8	3.722	0.954	1.713
45.0	784.84	1.4904	255.9	3.697	0.962	1.692
50.0	797.24	1.6302	259.1	3.674	0.970	1.676
55.0	808.67	1.7679	262.5	3.653	0.976	1.664
60.0	819.28	1.9036	265.9	3.632	0.981	1.655
65.0	829.20	2.0376	269.4	3.613	0.985	1.648
70.0	838.53	2.1699	273.0	3.595	0.987	1.644
75.0	847.35	2.3007	276.6	3.578	0.989	1.641
80.0	855.71	2.4301	280.2	3.561	0.990	1.641
85.0	863.67	2.5582	283.9	3.545	0.990	1.642
90.0	871.27	2.6850	287.5	3.529	0.989	1.645
95.0	878.55	2.8107	291.2	3.514	0.987	1.649
100.0	885.54	2.9353	294.8	3.499	0.984	1.654

p	ρ	Z	h	s	c_v	c_p
			$T=135$ K			
0.1	2.60	0.9915	387.5	6.066	0.717	1.014
0.5	13.49	0.9565	383.5	5.584	0.728	1.069
1.0	28.34	0.9104	378.1	5.357	0.746	1.159
1.5	44.95	0.8609	372.1	5.210	0.768	1.284
2.0	63.97	0.8067	365.3	5.091	0.798	1.466
2.5	86.52	0.7455	357.5	4.983	0.836	1.752
3.0	114.91	0.6736	347.9	4.875	0.886	2.265
3.5	155.19	0.5819	335.0	4.751	0.956	3.478
4.0	236.41	0.4365	312.0	4.561	1.064	9.853
4.5	429.34	0.2704	272.1	4.254	1.046	9.447
5.0	484.99	0.2660	263.0	4.179	0.994	5.192
6.0	533.43	0.2902	256.0	4.113	0.950	3.527
7.0	561.43	0.3217	252.5	4.073	0.929	2.981
8.0	581.71	0.3548	250.3	4.043	0.917	2.694
9.0	597.81	0.3884	248.7	4.019	0.910	2.513
10.0	611.29	0.4221	247.6	3.999	0.905	2.385
11.0	622.94	0.4556	246.8	3.981	0.903	2.290
12.0	633.25	0.4889	246.2	3.965	0.901	2.215
13.0	642.51	0.5220	245.8	3.950	0.901	2.154
14.0	650.95	0.5549	245.5	3.936	0.901	2.104
15.0	658.71	0.5875	245.3	3.924	0.901	2.061
16.0	665.91	0.6199	245.2	3.912	0.902	2.025
17.0	672.63	0.6521	245.2	3.901	0.904	1.993
18.0	678.95	0.6840	245.3	3.890	0.905	1.965
19.0	684.90	0.7157	245.4	3.880	0.907	1.940
20.0	690.54	0.7473	245.6	3.871	0.909	1.918
21.0	695.90	0.7786	245.8	3.862	0.911	1.898
22.0	701.02	0.8097	246.1	3.853	0.913	1.880
23.0	705.91	0.8406	246.4	3.845	0.915	1.863
24.0	710.60	0.8714	246.7	3.837	0.917	1.848
25.0	715.10	0.9020	247.0	3.829	0.919	1.834
26.0	719.44	0.9324	247.4	3.821	0.921	1.821
27.0	723.63	0.9627	247.8	3.814	0.924	1.809
28.0	727.67	0.9928	248.2	3.807	0.926	1.798
29.0	731.58	1.0227	248.7	3.800	0.928	1.788
30.0	735.38	1.0526	249.1	3.793	0.930	1.779
35.0	752.78	1.1996	251.7	3.763	0.941	1.740
40.0	768.12	1.3436	254.5	3.735	0.950	1.711
45.0	781.89	1.4849	257.6	3.710	0.959	1.689
50.0	794.42	1.6239	260.8	3.687	0.966	1.673
55.0	805.95	1.7607	264.1	3.665	0.973	1.660
60.0	816.65	1.8956	267.6	3.645	0.978	1.651
65.0	826.65	2.0287	271.1	3.626	0.982	1.644
70.0	836.04	2.1603	274.6	3.607	0.985	1.640
75.0	844.90	2.2903	278.2	3.590	0.986	1.637
80.0	853.31	2.4189	281.8	3.573	0.987	1.637
85.0	861.30	2.5462	285.5	3.557	0.988	1.638
90.0	868.93	2.6723	289.2	3.541	0.987	1.640
95.0	876.23	2.7973	292.8	3.526	0.985	1.644
100.0	883.23	2.9212	296.5	3.511	0.983	1.649

Table II.11 (*Continued*)

p	ρ	Z	h	s	c_v	c_p
			$T = 136$ K			
0,1	2.58	0.9916	388.5	6.073	0.717	1.014
0,5	13.37	0.9574	384.5	5.592	0.728	1.067
1.0	28.07	0.9125	379.2	5.366	0.744	1.154
1.5	44.44	0.8645	373.4	5.219	0.766	1.273
2.0	63.08	0.8121	366.8	5.102	0.794	1.445
2.5	84.97	0.7535	359.2	4.996	0.829	1.707
3.0	112.07	0.6856	350.1	4.891	0.876	2.158
3.5	149.00	0.6016	338.3	4.775	0.937	3.119
4.0	212.32	0.4825	319.9	4.619	1.026	6.478
4.5	373.93	0.3082	283.6	4.339	1.071	13.461
5.0	459.75	0.2785	268.7	4.220	1.006	6.087
6.0	519.24	0.2959	259.7	4.139	0.954	3.732
7.0	550.64	0.3256	255.5	4.095	0.930	3.075
8.0	572.65	0.3578	253.0	4.063	0.917	2.750
9.0	589.84	0.3908	251.2	4.038	0.909	2.550
10.0	604.08	0.4240	250.0	4.016	0.904	2.411
11.0	616.29	0.4571	249.1	3.998	0.901	2.309
12.0	627.04	0.4901	248.4	3.981	0.899	2.230
13.0	636.67	0.5229	247.9	3.966	0.898	2.166
14.0	645.40	0.5556	247.6	3.952	0.898	2.113
15.0	653.42	0.5879	247.4	3.939	0.899	2.069
16.0	660.83	0.6201	247.3	3.927	0.900	2.031
17.0	667.75	0.6520	247.2	3.916	0.901	1.998
18.0	674.23	0.6837	247.3	3.905	0.902	1.969
19.0	680.33	0.7153	247.4	3.895	0.904	1.943
20.0	686.10	0.7466	247.5	3.885	0.906	1.920
21.0	691.59	0.7777	247.7	3.876	0.908	1.900
22.0	696.82	0.8086	247.9	3.867	0.910	1.881
23.0	701.81	0.8393	248.2	3.858	0.912	1.864
24.0	706.59	0.8699	248.5	3.850	0.914	1.848
25.0	711.19	0.9003	248.9	3.842	0.916	1.834
26.0	715.61	0.9305	249.2	3.835	0.918	1.821
27.0	719.87	0.9606	249.6	3.827	0.921	1.809
28.0	723.98	0.9905	250.0	3.820	0.923	1.798
29.0	727.96	1.0203	250.5	3.813	0.925	1.787
30.0	731.81	1.0499	250.9	3.807	0.927	1.778
35.0	749.48	1.1960	253.4	3.775	0.938	1.738
40.0	765.02	1.3391	256.2	3.747	0.947	1.708
45.0	778.95	1.4796	259.3	3.722	0.956	1.686
50.0	791.61	1.6177	262.5	3.699	0.963	1.670
55.0	803.25	1.7536	265.8	3.677	0.969	1.657
60.0	814.04	1.8877	269.2	3.657	0.975	1.647
65.0	824.10	2.0200	272.7	3.638	0.979	1.641
70.0	833.56	2.1508	276.3	3.620	0.982	1.636
75.0	842.47	2.2800	279.9	3.602	0.984	1.633

p	ρ	Z	h	s	c_v	c_p
			$T=136$ K			
80.0	850.92	2.4079	283.5	3.585	0.985	1.632
85.0	858.95	2.5344	287.1	3.569	0.985	1.633
90.0	866.60	2.6598	290.8	3.553	0.985	1.635
95.0	873.92	2.7841	294.5	3.538	0.984	1.639
100.0	880.94	2.9072	298.1	3.523	0.982	1.644
			$T=137$ K			
0.1	2.56	0.9918	389.5	6.080	0.717	1.014
0.5	13.26	0.9584	385.6	5.599	0.727	1.065
1.0	27.80	0.9146	380.4	5.374	0.743	1.149
1.5	43.94	0.8679	374.6	5.229	0.764	1.264
2.0	62.22	0.8172	368.2	5.112	0.790	1.425
2.5	83.52	0.7610	360.9	5.008	0.823	1.667
3.0	109.49	0.6966	352.2	4.906	0.866	2.068
3.5	143.75	0.6190	341.3	4.797	0.921	2.853
4.0	197.04	0.5161	325.5	4.660	0.997	5.025
4.5	313.42	0.3650	297.2	4.438	1.072	12.581
5.0	429.18	0.2962	275.3	4.269	1.021	7.304
6.0	503.89	0.3027	263.5	4.167	0.958	3.974
7.0	539.33	0.3300	258.7	4.118	0.931	3.180
8.0	563.29	0.3611	255.8	4.084	0.917	2.809
9.0	581.68	0.3934	253.8	4.057	0.908	2.588
10.0	596.72	0.4261	252.4	4.034	0.902	2.439
11.0	609.54	0.4588	251.4	4.015	0.899	2.329
12.0	620.76	0.4915	250.7	3.997	0.897	2.245
13.0	630.75	0.5240	250.1	3.982	0.896	2.178
14.0	639.80	0.5563	249.7	3.967	0.896	2.123
15.0	648.08	0.5885	249.5	3.954	0.896	2.076
16.0	655.72	0.6204	249.3	3.942	0.897	2.037
17.0	662.83	0.6521	249.2	3.930	0.898	2.003
18.0	669.48	0.6836	249.2	3.919	0.900	1.973
19.0	675.74	0.7149	249.3	3.909	0.901	1.946
20.0	681.65	0.7460	249.4	3.899	0.903	1.923
21.0	687.26	0.7769	249.6	3.890	0.905	1.901
22.0	692.60	0.8076	249.8	3.881	0.907	1.882
23.0	697.70	0.8381	250.1	3.872	0.909	1.865
24.0	702.58	0.8685	250.4	3.864	0.911	1.849
25.0	707.26	0.8987	250.7	3.856	0.913	1.834
26.0	711.76	0.9287	251.1	3.848	0.915	1.821
27.0	716.10	0.9586	251.4	3.841	0.918	1.809
28.0	720.29	0.9883	251.8	3.833	0.920	1.797
29.0	724.33	1.0179	252.3	3.826	0.922	1.786
30.0	728.25	1.0473	252.7	3.820	0.924	1.777
35.0	746.18	1.1925	255.2	3.788	0.934	1.736
40.0	761.92	1.3347	257.9	3.760	0.944	1.706
45.0	776.01	1.4743	260.9	3.735	0.953	1.684
50.0	788.81	1.6116	264.1	3.711	0.960	1.667

Table II.11 (*Continued*)

p	ρ	Z	h	s	c_v	c_p
			$T=137$ K			
55.0	800.55	1.7467	267.4	3.689	0.966	1.654
60.0	811.43	1.8800	270.9	3.669	0.972	1.644
65.0	821.57	2.0115	274.3	3.650	0.976	1.637
70.0	831.08	2.1414	277.9	3.631	0.979	1.632
75.0	840.05	2.2699	281.5	3.614	0.981	1.629
80.0	848.54	2.3970	285.1	3.597	0.983	1.628
85.0	856.61	2.5228	288.8	3.581	0.983	1.629
90.0	864.29	2.6474	292.4	3.565	0.983	1.631
95.0	871.64	2.7710	296.1	3.550	0.982	1.634
100.0	878.87	2.8935	299.8	3.535	0.981	1.639
			$T=138$ K			
0.1	2.54	0.9920	390.5	6.088	0.717	1.013
0.5	13.16	0.9593	386.7	5.607	0.727	1.064
1.0	27.54	0.9166	381.5	5.383	0.742	1.145
1.5	43.46	0.8712	375.9	5.238	0.762	1.255
2.0	61.40	0.8222	369.6	5.122	0.786	1.407
2.5	82.14	0.7682	362.6	5.020	0.818	1.631
3.0	107.10	0.7070	354.3	4.921	0.857	1.990
3.5	139.19	0.6347	344.0	4.817	0.907	2.649
4.0	185.86	0.5432	330.1	4.694	0.973	4.203
4.5	272.03	0.4175	307.9	4.516	1.048	9.032
5.0	392.80	0.3213	283.3	4.327	1.034	8.561
6.0	487.19	0.3108	267.6	4.197	0.963	4.257
7.0	527.45	0.3350	261.9	4.142	0.933	3.296
8.0	553.62	0.3647	258.6	4.104	0.917	2.873
9.0	573.30	0.3962	256.4	4.076	0.907	2.629
10.0	589.23	0.4284	254.9	4.052	0.901	2.467
11.0	602.68	0.4607	253.7	4.032	0.897	2.350
12.0	614.39	0.4930	252.9	4.014	0.895	2.261
13.0	624.78	0.5252	252.3	3.998	0.894	2.190
14.0	634.15	0.5572	251.8	3.983	0.893	2.132
15.0	642.70	0.5891	251.5	3.969	0.894	2.084
16.0	650.57	0.6207	251.3	3.957	0.894	2.043
17.0	657.88	0.6522	251.2	3.945	0.896	2.008
18.0	664.71	0.6835	251.2	3.934	0.897	1.977
19.0	671.13	0.7146	251.3	3.923	0.899	1.949
20.0	677.18	0.7454	251.4	3.913	0.900	1.925
21.0	682.92	0.7761	251.5	3.904	0.902	1.903
22.0	688.38	0.8066	251.7	3.894	0.904	1.884
23.0	693.58	0.8370	251.9	3.886	0.906	1.866
24.0	698.56	0.8672	252.2	3.877	0.908	1.849
25.0	703.33	0.8972	252.5	3.869	0.910	1.835
26.0	707.91	0.9270	252.9	3.861	0.912	1.821
27.0	712.33	0.9567	253.2	3.854	0.915	1.808
28.0	716.59	0.9862	253.6	3.847	0.917	1.796
29.0	720.70	1.0156	254.0	3.839	0.919	1.786

p	ρ	Z	h	s	c_v	c_p
			T=138 K			
30.0	724.68	1.0449	254.5	3.833	0.921	1.776
35.0	742.88	1.1892	256.9	3.801	0.931	1.734
40.0	758.83	1.3305	259.6	3.772	0.941	1.704
45.0	773.08	1.4692	262.6	3.747	0.949	1.681
50.0	786.01	1.6056	265.8	3.723	0.957	1.663
55.0	797.86	1.7399	269.1	3.701	0.963	1.650
60.0	808.83	1.8723	272.5	3.681	0.969	1.640
65.0	819.05	2 0031	276.0	3.662	0.973	1.633
70.0	828.63	2.1322	279.5	3.643	0.976	1.628
75.0	837.65	2.2599	283.1	3.626	0.979	1.625
80.0	846.18	2.3862	286.7	3.609	0.980	1.624
85.0	854.28	2.5113	290.4	3.593	0.981	1.624
90.0	862.00	2.6353	294.1	3.577	0.981	1.626
95.0	869.37	2.7581	297.7	3.562	0.981	1.630
100.0	876.42	2.8799	301.4	3.547	0.979	1.634
			T=139 K			
0.1	2.53	0.9922	391.5	6.095	0.717	1.013
0.5	13.05	0.9602	387.7	5.615	0.726	1.062
1.0	27.28	0.9185	382.7	5.391	0.741	1.141
1.5	42.99	0.8744	377.1	5.247	0.760	1.246
2.0	60.61	0.8269	371.0	5.133	0.783	1.391
2.5	80.84	0.7749	364.2	5.032	0.812	1.599
3.0	104.90	0.7166	356.2	4.935	0.849	1.922
3.5	135.15	0.6489	346.6	4.836	0.894	2.486
4.0	177.06	0.5661	334.0	4.722	0.952	3.670
4.5	245.78	0.4588	315.7	4.573	1.021	6.766
5.0	353.73	0.3542	292.2	4.391	1.039	8.972
6.0	469.00	0.3206	272.0	4.229	0.968	4.579
7.0	514.96	0.3406	265.2	4.166	0.934	3.423
8.0	543.61	0.3688	261.5	4.125	0.917	2.942
9.0	564.72	0.3994	259.1	4.095	0.906	2.672
10.0	581.59	0.4309	257.3	4.070	0.899	2.497
11.0	595.72	0.4627	256.1	4.049	0.895	2.372
12.0	607.94	0.4946	255.2	4.030	0.893	2.277
13.0	618.74	0.5265	254.5	4.013	0.892	2.203
14.0	628.44	0.5582	254.0	3.998	0.891	2.142
15.0	637.27	0.5898	253.6	3.984	0.891	2.092
16.0	645.39	0.6212	253.4	3.971	0.892	2.049
17.0	652.91	0.6525	253.2	3.959	0.893	2.013
18.0	659.92	0.6835	253.2	3.948	0.894	1.981
19.0	666.50	0.7143	253.2	3.937	0.896	1.952
20.0	672.70	0.7450	253.3	3.927	0.898	1.927
21.0	678.56	0.7755	253.4	3.917	0.899	1.905
22.0	684.14	0.8058	253.6	3.908	0.901	1.885
23.0	689.45	0.8359	253.8	3.899	0.903	1.867
24.0	694.53	0.8659	254.1	3.891	0.905	1.850

p	ρ	Z	h	s	c_v	c_p
			$T=139$ K			
25.0	699.39	0.8957	254.4	3.882	0.907	1.835
26.0	704.06	0.9254	254.7	3.875	0.910	1.821
27.0	708.55	0.9549	255.0	3.867	0.912	1.808
28.0	712.88	0.9842	255.4	3.859	0.914	1.796
29.0	717.07	1.0134	255.8	3.852	0.916	1.785
30.0	721.11	1.0425	256.2	3.845	0.918	1.775
35.0	739.58	1.1859	258.6	3.813	0.928	1.732
40.0	755.73	1.3263	261.3	3.785	0.938	1.701
45.0	770.15	1.4642	264.3	3.759	0.946	1.678
50.0	783.21	1.5997	267.5	3.735	0.954	1.660
55.0	795.18	1.7332	270.7	3.713	0.960	1.647
60.0	806.24	1.8648	274.1	3.693	0.966	1.637
65.0	816.53	1.9948	277.6	3.673	0.970	1.629
70.0	826.18	2.1231	281.1	3.655	0.973	1.624
75.0	835.25	2.2501	284.7	3.638	0.976	1.621
80.0	843.83	2.3757	288.4	3.621	0.978	1.620
85.0	851.97	2.5000	292.0	3.605	0.979	1.620
90.0	859.78	2.6232	295.7	3.589	0.979	1.622
95.0	867.11	2.7454	299.4	3.574	0.979	1.625
100.0	874.18	2.8665	303.1	3.559	0.978	1.629
			$T=140$ K			
0.1	2.51	0.9923	392.5	6.102	0.717	1.013
0.5	12.94	0.9611	388.8	5.622	0.726	1.061
1.0	27.03	0.9204	383.8	5.399	0.740	1.137
1.5	42.53	0.8774	378.4	5.256	0.758	1.238
2.0	59.84	0.8315	372.4	5.142	0.780	1.375
2.5	79.60	0.7814	365.8	5.043	0.807	1.569
3.0	102.84	0.7257	358.1	4.949	0.841	1.863
3.5	131.54	0.6620	349.0	4.853	0.883	2.352
4.0	169.80	0.5861	337.5	4.747	0.935	3.293
4.5	227.59	0.4919	321.7	4.616	0.996	5.431
5.0	318.26	0.3909	300.8	4.453	1.033	8.217
6.0	449.23	0.3323	276.8	4.263	0.974	4.926
7.0	501.81	0.3471	268.7	4.191	0.936	3.560
8.0	533.25	0.3732	264.5	4.147	0.917	3.014
9.0	555.92	0.4028	261.8	4.114	0.905	2.717
10.0	573.79	0.4336	259.9	4.088	0.898	2.527
11.0	588.64	0.4649	258.5	4.066	0.894	2.394
12.0	601.41	0.4964	257.5	4.046	0.891	2.294
13.0	612.64	0.5279	256.7	4.029	0.889	2.215
14.0	622.69	0.5594	256.1	4.014	0.889	2.152
15.0	631.81	0.5907	255.7	3.999	0.889	2.100
16.0	640.17	0.6218	255.4	3.986	0.890	2.055
17.0	647.90	0.6528	255.3	3.974	0.891	2.018
18.0	655.10	0.6836	255.2	3.962	0.892	1.985
19.0	661.84	0.7142	255.2	3.951	0.893	1.956

p	ρ	Z	h	s	c_v	c_p
			$T = 140$ K			
20.0	668.19	0.7447	255.2	3.941	0.895	1.930
21.0	674.19	0.7750	255.3	3.931	0.897	1.907
22.0	679.89	0.8051	255.5	3.922	0.899	1.886
23.0	685.31	0.8350	255.7	3.913	0.901	1.868
24.0	690.49	0.8648	255.9	3.904	0.903	1.851
25.0	695.44	0.8944	256.2	3.896	0.905	1.835
26.0	700.20	0.9238	256.5	3.888	0.907	1.821
27.0	704.77	0.9531	256.9	3.880	0.909	1.808
28.0	709.18	0.9823	257.2	3.872	0.911	1.795
29.0	713.43	1.0113	257.6	3.865	0.913	1.784
30.0	717.54	1.0402	258.0	3.858	0.915	1.774
35.0	736.28	1.1827	260.4	3.826	0.925	1.731
40.0	752.65	1.3222	263.0	3.797	0.935	1.699
45.0	767.23	1.4592	266.0	3.771	0.943	1.675
50.0	780.43	1.5940	269.1	3.747	0.951	1.657
55.0	792.50	1.7266	272.4	3.725	0.957	1.644
60.0	803.66	1.8575	275.8	3.704	0.963	1.633
65.0	814.03	1.9866	279.2	3.685	0.967	1.626
70.0	823.74	2.1142	282.8	3.667	0.971	1.620
75.0	832.87	2.2404	286.4	3.649	0.974	1.617
80.0	841.50	2.3653	290.0	3.632	0.976	1.615
85.0	849.67	2.4889	293.6	3.616	0.977	1.615
90.0	857.45	2.6114	297.3	3.601	0.977	1.617
95.0	864.87	2.7328	301.0	3.585	0.977	1.620
100.0	871.96	2.8532	304.7	3.571	0.977	1.624
			$T = 145$ K			
0.1	2.42	0.9931	397.6	6.138	0.717	1.012
0.5	12.45	0.9650	394.1	5.660	0.724	1.055
1.0	25.86	0.9289	389.4	5.439	0.736	1.121
1.5	40.43	0.8913	384.5	5.298	0.750	1.204
2.0	56.41	0.8517	379.1	5.190	0.768	1.312
2.5	74.17	0.8097	373.3	5.096	0.788	1.455
3.0	94.28	0.7644	366.8	5.010	0.813	1.650
3.5	117.59	0.7150	359.6	4.927	0.841	1.930
4.0	145.50	0.6604	351.2	4.843	0.874	2.350
4.5	180.25	0.5997	341.3	4.754	0.911	3.013
5.0	225.05	0.5337	329.5	4.655	0.949	4.024
6.0	340.22	0.4236	304.0	4.454	0.978	5.384
7.0	426.57	0.3942	288.4	4.328	0.946	4.247
8.0	476.03	0.4037	280.5	4.259	0.919	3.412
9.0	508.51	0.4251	275.9	4.214	0.903	2.963
10.0	532.53	0.4511	272.9	4.179	0.893	2.692
11.0	551.62	0.4790	270.7	4.152	0.886	2.5.0
12.0	567.51	0.5079	269.1	4.128	0.882	2.380
13.0	581.16	0.5373	267.9	4.108	0.880	2.281
14.0	593.16	0.5670	267.0	4.090	0.879	2.203

p	ρ	Z	h	s	c_v	c_p
			$T=145$ K			
15.0	603.88	0.5967	266.3	4.074	0.878	2.140
16.0	613.59	0.6264	265.8	4.059	0.878	2.087
17.0	622.47	0.6560	265.4	4.045	0.879	2.043
18.0	630.68	0.6856	265.1	4.032	0.880	2.005
19.0	638.30	0.7150	265.0	4.020	0.881	1.972
20.0	645.44	0.7443	264.9	4.009	0.882	1.943
21.0	652.14	0.7735	264.9	3.998	0.884	1.917
22.0	658.47	0.8026	264.9	3.988	0.886	1.893
23.0	664.47	0.8315	265.0	3.978	0.887	1.873
24.0	670.18	0.8602	265.2	3.969	0.889	1.854
25.0	675.62	0.8889	265.4	3.960	0.891	1.836
26.0	680.82	0.9174	265.6	3.951	0.893	1.821
27.0	685.81	0.9457	265.9	3.943	0.895	1.806
28.0	690.60	0.9739	266.2	3.935	0.897	1.793
29.0	695.22	1.0020	266.5	3.928	0.899	1.781
30.0	699.66	1.0300	266.9	3.920	0.901	1.769
35.0	719.81	1.1680	269.0	3.886	0.911	1.723
40.0	737.26	1.3033	271.5	3.856	0.920	1.688
45.0	752.70	1.4361	274.3	3.829	0.928	1.663
50.0	766.59	1.5668	277.4	3.805	0.936	1.643
55.0	779.24	1.6955	280.6	3.782	0.943	1.628
60.0	790.87	1.8224	283.9	3.761	0.948	1.616
65.0	801.64	1.9477	287.3	3.742	0.953	1.608
70.0	811.69	2.0716	290.8	3.723	0.957	1.601
75.0	821.12	2.1941	294.4	3.706	0.961	1.597
80.0	829.99	2.3154	298.0	3.689	0.964	1.594
85.0	838.38	2.4354	301.6	3.673	0.966	1.594
90.0	846.33	2.5545	305.3	3.657	0.967	1.594
95.0	853.90	2.6725	309.0	3.642	0.969	1.596
100.0	861.12	2.7896	312.7	3.627	0.969	1.599
			$T=150$ K			
0.1	2.34	0.9938	402.7	6.172	0.717	1.011
0.5	11.99	0.9685	399.3	5.695	0.723	1.050
1.0	24.80	0.9363	395.0	5.477	0.733	1.107
1.5	38.57	0.9031	390.4	5.339	0.745	1.178
2.0	53.46	0.8687	385.6	5.233	0.759	1.266
2.5	69.78	0.8327	380.4	5.144	0.775	1.376
3.0	87.65	0.7947	374.7	5.064	0.793	1.518
3.5	107.72	0.7545	368.6	4.989	0.814	1.704
4.0	130.52	0.7117	361.8	4.915	0.837	1.953
4.5	156.88	0.6661	354.3	4.842	0.863	2.291
5.0	187.83	0.6181	345.9	4.766	0.889	2.745
6.0	265.34	0.5251	327.0	4.610	0.932	3.840
7.0	348.67	0.4662	309.8	4.474	0.936	4.132
8.0	412.44	0.4504	298.3	4.380	0.918	3.647
9.0	456.23	0.4581	291.3	4.318	0.901	3.174

p	ρ	Z	h	s	c_v	c_p
			$T = 150$ K			
10.0	487.78	0.4761	286.8	4.273	0.889	2.847
11.0	512.05	0.4988	283.6	4.239	0.881	2.624
12.0	531.71	0.5241	281.3	4.211	0.875	2.465
13.0	548.22	0.5506	279.5	4.187	0.872	2.346
14.0	562.47	0.5780	278.2	4.166	0.870	2.253
15.0	575.03	0.6057	277.1	4.147	0.869	2.179
16.0	586.26	0.6337	276.3	4.130	0.868	2.118
17.0	596.44	0.6618	275.7	4.115	0.868	2.067
18.0	605.76	0.6900	275.2	4.100	0.869	2.024
19.0	614.36	0.7181	274.9	4.087	0.870	1.987
20.0	622.35	0.7462	274.6	4.075	0.871	1.954
21.0	629.81	0.7743	274.5	4.063	0.872	1.926
22.0	636.83	0.8022	274.4	4.052	0.874	1.900
23.0	643.45	0.8300	274.4	4.142	0.875	1.877
24.0	649.72	0.8578	274.5	4.032	0.877	1.857
25.0	655.68	0.8854	274.6	4.022	0.879	1.838
26.0	661.35	0.9129	274.7	4.013	0.881	1.821
27.0	666.78	0.9403	274.9	4.004	0.883	1.805
28.0	671.98	0.9676	275.1	3.996	0.884	1.791
29.0	676.97	0.9947	275.4	3.988	0.886	1.778
30.0	681.77	1.0218	275.7	3.980	0.888	1.765
35.0	703.39	1.1554	277.6	3.944	0.898	1.715
40.0	721.96	1.2865	279.9	3.913	0.907	1.679
45.0	738.28	1.4154	282.6	3.886	0.915	1.651
50.0	752.89	1.5421	285.6	3.860	0.922	1.630
55.0	766.14	1.6670	288.7	3.837	0.929	1.614
60.0	778.27	1.7902	291.9	3.816	0.935	1.601
65.0	789.47	1.9119	295.3	3.796	0.940	1.591
70.0	799.88	2.0321	298.8	3.777	0.945	1.583
75.0	809.61	2.1511	302.3	3.759	0.948	1.578
80.0	818.75	2.2689	305.9	3.742	0.952	1.575
85.0	827.37	2.3856	309.6	3.726	0.955	1.573
90.0	835.53	2.5013	313.2	3.711	0.957	1.572
95.0	843.27	2.6160	316.9	3.696	0.959	1.573
100.0	850.63	2.7298	320.7	3.681	0.961	1.575
			$T = 155$ K			
0.1	2.26	0.9943	407.7	6.205	0.717	1.010
0.5	11.56	0.9716	404.6	5.730	0.722	1.045
1.0	23.84	0.9428	400.5	5.513	0.731	1.096
1.5	36.91	0.9133	396.3	5.377	0.741	1.157
2.0	50.90	0.8831	391.8	5.274	0.752	1.230
2.5	65.95	0.8518	387.1	5.188	0.765	1.319
3.0	82.27	0.8195	382.1	5.112	0.779	1.428
3.5	100.10	0.7858	376.7	5.042	0.795	1.562
4.0	119.75	0.7506	371.0	4.975	0.812	1.731
4.5	141.60	0.7141	364.8	4.911	0.831	1.942

Table II.11 (*Continued*)

p	ρ	Z	h	s	c_v	c_p
			$T=155$ K			
5.0	166.07	0.6766	358.1	4.847	0.850	2.205
6.0	223.91	0.6022	343.5	4.719	0.886	2.859
7.0	290.24	0.5420	328.6	4.597	0.906	3.387
8.0	352.90	0.5094	316.2	4.497	0.905	3.439
9.0	403.12	0.5017	307.4	4.423	0.894	3.198
10.0	441.32	0.5092	301.2	4.368	0.883	2.925
11.0	470.88	0.5249	296.9	4.326	0.875	2.701
12.0	494.57	0.5452	293.8	4.293	0.869	2.529
13.0	514.20	0.5681	291.4	4.264	0.864	2.398
14.0	530.93	0.5926	289.5	4.240	0.862	2.294
15.0	545.49	0.6179	288.1	4.219	0.860	2.212
16.0	558.38	0.6439	287.0	4.200	0.859	2.144
17.0	569.96	0.6703	286.1	4.183	0.859	2.088
18.0	580.48	0.6968	285.4	4.167	0.859	2.041
19.0	590.12	0.7235	284.8	4.153	0.860	2.000
20.0	599.02	0.7503	284.4	4.139	0.861	1.964
21.0	607.30	0.7771	284.1	4.126	0.862	1.933
22.0	615.04	0.8038	283.9	4.115	0.863	1.905
23.0	622.31	0.8305	283.8	4.103	0.865	1.881
24.0	629.17	0.8572	283.8	4.093	0.866	1.859
25.0	635.67	0.8838	283.8	4.083	0.868	1.838
26.0	641.85	0.9103	283.8	4.073	0.870	1.820
27.0	647.73	0.9367	283.9	4.064	0.871	1.804
28.0	653.35	0.9630	284.1	4.055	0.873	1.788
29.0	658.73	0.9893	284.3	4.046	0.875	1.774
30.0	663.90	1.0154	284.5	4.038	0.877	1.761
35.0	687.04	1.1448	286.1	4.001	0.886	1.708
40.0	706.76	1.2718	288.3	3.968	0.894	1.670
45.0	723.99	1.3967	290.8	3.940	0.902	1.641
50.0	739.34	1.5197	293.7	3.914	0.910	1.618
55.0	753.19	1.6409	296.7	3.890	0.916	1.601
60.0	765.84	1.7605	299.9	3.868	0.922	1.587
65.0	777.48	1.8787	303.2	3.848	0.928	1.576
70.0	788.27	1.9955	306.7	3.829	0.932	1.567
75.0	798.33	2.1111	310.2	3.811	0.937	1.561
80.0	807.76	2.2256	313.7	3.794	0.941	1.556
85.0	816.63	2.3390	317.4	3.777	0.944	1.553
90.0	825.00	2.4515	321.1	3.762	0.947	1.552
95.0	832.93	2.5630	324.8	3.747	0.950	1.552
100.0	840.46	2.6737	328.5	3.732	0.953	1.553
			$T=160$ K			
0.1	2.19	0.9949	412.8	6.237	0.716	1.010
0.5	11.17	0.9743	409.8	5.763	0.722	1.042
1.0	22.95	0.9484	406.0	5.547	0.729	1.087
1.5	35.41	0.9221	402.0	5.414	0.737	1.140
2.0	48.63	0.8954	397.9	5.313	0.747	1.203

p	ρ	Z	h	s	c_v	c_p
			$T=160$ K			
2.5	62.69	0.8681	393.6	5.229	0.757	1.276
3.0	77.74	0.8401	389.1	5.156	0.769	1.362
3.5	93.91	0.8113	384.3	5.090	0.781	1.466
4.0	111.39	0.7818	379.3	5.028	0.795	1.589
4.5	130.35	0.7515	374.0	4.969	0.809	1.736
5.0	151.00	0.7208	368.4	4.912	0.824	1.911
6.0	197.88	0.6601	356.3	4.800	0.852	2.333
7.0	251.17	0.6067	343.9	4.694	0.874	2.758
8.0	305.84	0.5694	332.4	4.600	0.882	2.998
9.0	355.35	0.5514	322.9	4.522	0.881	2.999
10.0	396.61	0.5489	315.8	4.461	0.874	2.866
11.0	430.03	0.5569	310.5	4.412	0.867	2.701
12.0	457.25	0.5713	306.5	4.373	0.862	2.550
13.0	479.86	0.5898	303.4	4.341	0.857	2.422
14.0	499.05	0.6107	301.1	4.313	0.855	2.318
15.0	515.65	0.6333	299.2	4.290	0.853	2.232
16.0	530.25	0.6569	297.7	4.268	0.852	2.161
17.0	543.28	0.6812	296.6	4.249	0.851	2.102
18.0	555.04	0.7060	295.6	4.232	0.851	2.051
19.0	565.75	0.7311	294.9	4.216	0.851	2.008
20.0	575.61	0.7564	294.3	4.202	0.852	1.970
21.0	584.72	0.7818	293.8	4.188	0.853	1.937
22.0	593.21	0.8073	293.5	4.175	0.854	1.908
23.0	601.16	0.8329	293.2	4.163	0.856	1.882
24.0	608.63	0.8584	293.0	4.152	0.857	1.859
25.0	615.69	0.8840	293.0	4.141	0.858	1.838
26.0	622.37	0.9094	292.9	4.131	0.860	1.818
27.0	628.72	0.9349	293.0	4.121	0.862	1.811
28.0	634.78	0.9603	293.0	4.111	0.863	1.785
29.0	640.56	0.9856	293.2	4.102	0.865	1.770
30.0	646.10	1.0108	293.3	4.094	0.867	1.757
35.0	670.79	1.1359	294.7	4.055	0.875	1.702
40.0	691.67	1.2589	296.6	4.021	0.883	1.662
45.0	709.83	1.3801	299.0	3.992	0.891	1.632
50.0	725.92	1.4994	301.7	3.965	0.898	1.608
55.0	740.40	1.6171	304.7	3.941	0.905	1.589
60.0	753.58	1.7333	307.8	3.918	0.911	1.574
65.0	765.67	1.8481	311.1	3.898	0.916	1.562
70.0	776.86	1.9616	314.5	3.878	0.921	1.553
75.0	787.26	2.0739	317.9	3.860	0.926	1.546
80.0	796.98	2.1852	321.5	3.843	0.930	1.540
85.0	806.11	2.2955	325.1	3.826	0.934	1.536
90.0	814.72	2.4048	328.8	3.811	0.937	1.534
95.0	822.86	2.5133	332.5	3.796	0.941	1.532
100.0	830.57	2.6210	336.2	3.781	0.944	1.532

p	ρ	Z	h	s	c_v	c_p
			$T = 165$ K			
0.1	2.12	0.9953	417.8	6.269	0.716	1.009
0.5	10.81	0.9767	415.0	5.795	0.721	1.039
1.0	22.14	0.9534	411.4	5.581	0.727	1.080
1.5	34.05	0.9299	407.7	5.448	0.735	1.127
2.0	46.59	0.9061	403.8	5.349	0.743	1.180
2.5	59.83	0.8820	399.9	5.268	0.751	1.242
3.0	73.85	0.8576	395.7	5.197	0.761	1.313
3.5	88.73	0.8327	391.4	5.134	0.771	1.396
4.0	104.58	0.8074	387.0	5.075	0.782	1.491
4.5	121.50	0.7818	382.3	5.020	0.793	1.601
5.0	139.60	0.7561	377.4	4.967	0.805	1.727
6.0	179.56	0.7054	367.2	4.867	0.828	2.022
7.0	224.11	0.6594	356.5	4.772	0.847	2.338
8.0	270.92	0.6233	346.3	4.686	0.859	2.588
9.0	316.22	0.6008	337.2	4.610	0.863	2.703
10.0	356.90	0.5915	329.7	4.546	0.862	2.694
11.0	391.83	0.5926	323.8	4.494	0.858	2.615
12.0	421.32	0.6013	319.2	4.451	0.854	2.511
13.0	446.25	0.6150	315.5	4.415	0.850	2.408
14.0	467.57	0.6321	312.7	4.385	0.847	2.315
15.0	486.04	0.6515	310.4	4.358	0.845	2.234
16.0	502.26	0.6725	308.6	4.335	0.844	2.165
17.0	516.70	0.6945	307.1	4.314	0.844	2.105
18.0	529.68	0.7174	305.9	4.295	0.843	2.054
19.0	541.47	0.7407	304.9	4.278	0.844	2.010
20.0	552.27	0.7645	304.1	4.262	0.844	1.971
21.0	562.23	0.7885	303.5	4.248	0.845	1.937
22.0	571.48	0.8127	303.0	4.234	0.846	1.907
23.0	580.10	0.8370	302.6	4.221	0.847	1.881
24.0	588.19	0.8613	302.3	4.209	0.848	1.856
25.0	595.80	0.8858	302.1	4.197	0.850	1.835
26.0	603.00	0.9102	302.0	4.187	0.851	1.815
27.0	609.83	0.9346	301.9	4.176	0.853	1.797
28.0	616.32	0.9590	301.9	4.166	0.854	1.781
29.0	622.51	0.9834	302.0	4.157	0.856	1.765
30.0	628.42	1.0077	302.1	4.148	0.857	1.751
35.0	654.67	1.1286	303.2	4.107	0.865	1.695
40.0	676.72	1.2478	304.9	4.072	0.873	1.654
45.0	695.81	1.3652	307.2	4.042	0.881	1.623
50.0	712.66	1.4811	309.8	4.014	0.888	1.598
55.0	727.77	1.5953	312.6	3.989	0.894	1.579
60.0	741.48	1.7082	315.7	3.967	0.900	1.563
65.0	754.03	1.8197	318.9	3.946	0.906	1.550
70.0	765.61	1.9301	322.2	3.926	0.911	1.540
75.0	776.36	2.0393	325.6	3.907	0.916	1.532

Table II.11 (*Continued*)

p	ρ	Z	h	s	c_v	c_p
			$T=165$ K			
80.0	786.40	2.1475	329.2	3.890	0.920	1.525
85.0	795.80	2.2547	332.7	3.874	0.924	1.521
90.0	804.65	2.3611	336.4	3.858	0.928	1.517
95.0	813.01	2.4667	340.1	3.843	0.932	1.515
100.0	820.92	2.5715	343.8	3.828	0.935	1.514
			$T=170$ K			
0.1	2.06	0.9958	422.8	6.299	0.716	1.009
0.5	10.47	0.9789	420.2	5.826	0.721	1.036
1.0	21.39	0.9578	416.8	5.613	0.726	1.073
1.5	32.81	0.9367	413.3	5.482	0.733	1.115
2.0	44.76	0.9155	409.7	5.384	0.740	1.162
2.5	57.29	0.8941	406.0	5.305	0.747	1.215
3.0	70.44	0.8726	402.2	5.236	0.755	1.275
3.5	84.28	0.8509	398.3	5.175	0.763	1.343
4.0	98.86	0.8290	394.2	5.119	0.772	1.419
4.5	114.25	0.8070	390.0	5.066	0.781	1.505
5.0	130.50	0.7850	385.7	5.017	0.791	1.601
6.0	165.68	0.7420	376.7	4.924	0.809	1.820
7.0	204.19	0.7024	367.5	4.838	0.826	2.059
8.0	244.79	0.6696	358.4	4.758	0.838	2.274
9.0	285.32	0.6463	350.0	4.686	0.845	2.418
10.0	323.53	0.6333	342.6	4.624	0.847	2.478
11.0	357.96	0.6296	336.5	4.570	0.847	2.470
12.0	388.20	0.6333	331.5	4.525	0.844	2.420
13.0	414.47	0.6426	327.4	4.487	0.842	2.353
14.0	437.30	0.6559	324.2	4.454	0.840	2.282
15.0	457.27	0.6721	321.5	4.425	0.838	2.214
16.0	474.89	0.6903	319.4	4.399	0.837	2.152
17.0	490.59	0.7100	317.6	4.377	0.837	2.096
18.0	504.71	0.7307	316.1	4.357	0.837	2.047
19.0	517.52	0.7522	315.0	4.338	0.837	2.004
20.0	529.22	0.7743	314.0	4.321	0.837	1.966
21.0	540.00	0.7968	313.2	4.305	0.838	1.933
22.0	549.98	0.8196	312.5	4.291	0.839	1.962
23.0	559.27	0.8426	312.0	4.277	0.840	1.875
24.0	567.96	0.8658	311.6	4.264	0.841	1.851
25.0	576.13	0.8891	311.3	4.252	0.842	1.829
26.0	583.83	0.9124	311.1	4.241	0.844	1.809
27.0	591.12	0.9358	310.9	4.230	0.845	1.791
28.0	598.05	0.9593	310.8	4.219	0.846	1.774
29.0	604.64	0.9827	310.8	4.209	0.848	1.759
30.0	610.93	1.0061	310.8	4.200	0.849	1.745
35.0	638.72	1.1227	311.6	4.157	0.857	1.688
40.0	661.94	1.2381	313.2	4.121	0.865	1.646
45.0	681.94	1.3520	315.3	4.090	0.872	1.614
50.0	699.54	1.4644	317.7	4.062	0.879	1.589

p	ρ	Z	h	s	c_v	c_p
			$T=170$ K			
55.0	715.28	1.5754	320.5	4.036	0.885	1.569
60.0	729.53	1.6851	323.4	4.013	0.891	1.553
65.0	742.54	1.7935	326.6	3.992	0.896	1.540
70.0	754.53	1.9008	329.9	3.972	0.901	1.529
75.0	765.64	2.0070	333.3	3.953	0.906	1.520
80.0	775.99	2.1123	336.7	3.935	0.911	1.512
85.0	785.68	2.2166	340.3	3.919	0.915	1.507
90.0	794.78	2.3201	343.9	3.903	0.919	1.502
95.0	803.36	2.4229	347.6	3.888	0.923	1.499
100.0	811.48	2.5249	351.4	3.873	0.927	1.497
			$T=175$ K			
0.1	2.00	0.9961	427.9	6.328	0.716	1.008
0.5	10.15	0.9808	425.4	5.856	0.720	1.033
1.0	20.70	0.9617	422.1	5.644	0.725	1.067
1.5	31.67	0.9427	418.8	5.514	0.731	1.105
2.0	43.10	0.9237	415.5	5.418	0.737	1.147
2.5	55.00	0.9047	412.0	5.339	0.743	1.193
3.0	67.42	0.8856	408.5	5.272	0.750	1.244
3.5	80.39	0.8665	404.9	5.213	0.757	1.301
4.0	93.95	0.8474	401.2	5.159	0.765	1.364
4.5	108.13	0.8283	397.4	5.109	0.772	1.434
5.0	122.96	0.8093	393.5	5.062	0.780	1.510
6.0	154.62	0.7723	385.5	4.975	0.795	1.681
7.0	188.75	0.7381	377.3	4.894	0.809	1.866
8.0	224.62	0.7089	369.2	4.820	0.821	2.043
9.0	260.91	0.6866	361.5	4.753	0.829	2.184
10.0	296.09	0.6722	354.5	4.692	0.833	2.271
11.0	328.92	0.6656	348.4	4.640	0.835	2.305
12.0	358.73	0.6658	343.3	4.594	0.835	2.298
13.0	385.36	0.6714	339.0	4.554	0.834	2.266
14.0	408.99	0.6813	335.4	4.519	0.832	2.221
15.0	429.95	0.6944	332.5	4.488	0.831	2.171
16.0	448.62	0.7099	330.0	4.461	0.831	2.121
17.0	465.35	0.7271	328.0	4.437	0.830	2.073
18.0	480.43	0.7457	326.3	4.416	0.830	2.030
19.0	494.14	0.7653	324.9	4.396	0.830	1.990
20.0	506.67	0.7857	323.8	4.378	0.831	1.954
21.0	518.20	0.8066	322.8	4.361	0.831	1.922
22.0	528.86	0.8280	322.0	4.346	0.832	1.893
23.0	538.78	0.8497	321.4	4.331	0.833	1.866
24.0	548.05	0.8716	320.8	4.318	0.834	1.842
25.0	556.75	0.8937	320.4	4.305	0.835	1.821
26.0	564.95	0.9160	320.1	4.293	0.837	1.801
27.0	572.69	0.9384	319.9	4.282	0.838	1.783
28.0	580.04	0.9608	319.7	4.271	0.839	1.766
29.0	587.02	0.9833	319.6	4.260	0.841	1.751

p	ρ	Z	h	s	c_v	c_p
			$T=175$ K			
30.0	593.67	1.0058	319.5	4.250	0.842	1.737
35.0	622.97	1.1182	320.0	4.206	0.849	1.680
40.0	647.34	1.2299	321.4	4.169	0.857	1.638
45.0	668.25	1.3403	323.3	4.137	0.864	1.606
50.0	686.60	1.4494	325.6	4.108	0.870	1.581
55.0	702.96	1.5573	328.3	4.082	0.876	1.561
60.0	717.73	1.6639	331.2	4.058	0.882	1.544
65.0	731.21	1.7693	334.3	4.036	0.888	1.530
70.0	743.60	1.8736	337.5	4.016	0.893	1.519
75.0	755.07	1.9770	340.8	3.997	0.897	1.509
80.0	765.74	2.0794	344.3	3.979	0.902	1.501
85.0	775.72	2.1809	347.8	3.962	0.906	1.495
90.0	785.08	2.2817	351.4	3.946	0.911	1.490
95.0	793.90	2.3817	355.1	3.931	0.915	1.486
100.0	802.23	2.4810	358.8	3.916	0.919	1.483
			$T=180$ K			
0.1	1.94	0.9965	432.9	6.356	0.716	1.008
0.5	9.85	0.9825	430.5	5.885	0.720	1.031
1.0	20.05	0.9652	427.4	5.674	0.724	1.062
1.5	30.62	0.9480	424.3	5.545	0.730	1.096
2.0	41.57	0.9310	421.2	5.450	0.735	1.134
2.5	52.93	0.9140	417.9	5.373	0.740	1.174
3.0	64.72	0.8970	414.7	5.307	0.746	1.219
3.5	76.95	0.8801	411.3	5.249	0.753	1.268
4.0	89.66	0.8633	407.9	5.197	0.759	1.321
4.5	102.86	0.8466	404.4	5.149	0.765	1.379
5.0	116.56	0.8301	400.9	5.103	0.772	1.441
6.0	145.49	0.7980	393.6	5.021	0.785	1.578
7.0	176.31	0.7683	386.2	4.945	0.797	1.727
8.0	208.52	0.7424	378.9	4.875	0.807	1.872
9.0	241.28	0.7218	371.9	4.812	0.815	1.998
10.0	273.55	0.7074	365.4	4.754	0.820	2.091
11.0	304.36	0.6994	359.6	4.702	0.823	2.146
12.0	333.05	0.6972	354.5	4.656	0.825	2.167
13.0	359.31	0.7001	350.1	4.616	0.825	2.163
14.0	383.09	0.7072	346.3	4.580	0.825	2.141
15.0	404.53	0.7175	343.2	4.549	0.824	2.110
16.0	423.85	0.7305	340.5	4.521	0.824	2.074
17.0	441.31	0.7454	338.3	4.495	0.824	2.037
18.0	457.14	0.7619	336.4	4.472	0.824	2.001
19.0	471.58	0.7796	334.8	4.452	0.824	1.966
20.0	484.82	0.7983	333.5	4.433	0.825	1.934
21.0	497.01	0.8176	332.4	4.415	0.825	1.905
22.0	508.29	0.8375	331.4	4.399	0.826	1.878
23.0	518.79	0.8579	330.7	4.384	0.827	1.853
24.0	528.59	0.8786	330.0	4.370	0.828	1.830

Table II.11 (*Continued*)

p	ρ	Z	h	s	c_v	c_p
			$T = 180$ K			
25.0	537.79	0.8995	329.5	4.356	0.829	1.809
26.0	546.44	0.9207	329.1	4.344	0.830	1.790
27.0	554.62	0.9420	328.7	4.332	0.832	1.772
28.0	562.36	0.9635	328.5	4.320	0.833	1.756
29.0	569.71	0.9850	328.3	4.310	0.834	1.741
30.0	576.72	1.0066	328.2	4.299	0.836	1.727
35.0	607.48	1.1149	328.4	4.253	0.843	1.671
40.0	632.96	1.2229	329.6	4.215	0.850	1.629
45.0	654.76	1.3299	331.3	4.182	0.856	1.598
50.0	673.83	1.4359	333.5	4.152	0.863	1.573
55.0	690.80	1.5407	336.1	4.126	0.869	1.552
60.0	706.10	1.6443	338.9	4.102	0.874	1.536
65.0	720.03	1.7468	341.9	4.079	0.880	1.521
70.0	732.83	1.8484	345.1	4.059	0.885	1.510
75.0	744.65	1.9490	348.4	4.039	0.889	1.500
80.0	755.64	2.0486	351.8	4.021	0.894	1.491
85.0	765.91	2.1475	355.3	4.004	0.899	1.484
90.0	775.54	2.2456	358.8	3.988	0.903	1.478
95.0	784.59	2.3430	362.5	3.973	0.907	1.474
100.0	793.14	2.4397	366.2	3.958	0.911	1.470
			$T = 185$ K			
0.1	1.89	0.9968	438.0	6.384	0.716	1.008
0.5	9.57	0.9841	435.7	5.913	0.720	1.029
1.0	19.44	0.9684	432.7	5.703	0.724	1.058
1.5	29.64	0.9528	429.8	5.575	0.728	1.089
2.0	40.17	0.9375	426.8	5.481	0.733	1.123
2.5	51.04	0.9222	423.8	5.405	0.738	1.159
3.0	62.27	0.9071	420.7	5.340	0.743	1.198
3.5	73.87	0.8921	417.6	5.284	0.749	1.240
4.0	85.85	0.8772	414.4	5.233	0.754	1.286
4.5	98.23	0.8625	411.2	5.186	0.760	1.335
5.0	111.01	0.8480	407.9	5.142	0.765	1.387
6.0	137.77	0.8200	401.3	5.063	0.776	1.501
7.0	165.98	0.7940	394.6	4.991	0.786	1.622
8.0	195.31	0.7712	387.9	4.925	0.796	1.744
9.0	225.18	0.7525	381.5	4.864	0.803	1.853
10.0	254.85	0.7388	375.5	4.809	0.809	1.942
11.0	283.61	0.7302	369.9	4.759	0.813	2.005
12.0	310.89	0.7267	365.0	4.714	0.815	2.042
13.0	336.32	0.7277	360.6	4.674	0.816	2.056
14.0	359.77	0.7326	356.8	4.638	0.817	2.053
15.0	381.25	0.7408	353.6	4.606	0.817	2.038
16.0	400.84	0.7515	350.8	4.577	0.818	2.016
17.0	418.73	0.7644	348.4	4.550	0.818	1.990
18.0	435.07	0.7789	346.3	4.527	0.818	1.963
19.0	450.06	0.7948	344.6	4.505	0.819	1.935

p	ρ	Z	h	s	c_v	c_p
			$T=185$ K			
20.0	463.85	0.8118	343.1	4.485	0.819	1.908
21.0	476.59	0.8296	341.9	4.467	0.820	1.882
22.0	488.40	0.8481	340.8	4.450	0.821	1.858
23.0	499.40	0.8671	339.9	4.434	0.822	1.835
24.0	509.68	0.8866	339.1	4.419	0.823	1.814
25.0	519.32	0.9064	338.5	4.406	0.824	1.794
26.0	528.40	0.9264	338.0	4.392	0.825	1.776
27.0	536.97	0.9467	337.6	4.380	0.826	1.759
28.0	545.08	0.9671	337.2	4.368	0.827	1.744
29.0	552.79	0.9877	337.0	4.357	0.829	1.729
30.0	560.12	1.0084	336.8	4.346	0.830	1.716
35.0	592.28	1.1126	336.7	4.299	0.837	1.661
40.0	618.83	1.2170	337.7	4.260	0.843	1.620
45.0	641.48	1.3208	339.3	4.225	0.850	1.589
50.0	661.25	1.4236	341.4	4.195	0.856	1.565
55.0	678.82	1.5255	343.8	4.168	0.862	1.544
60.0	694.63	1.6263	346.5	4.143	0.867	1.528
65.0	709.01	1.7261	349.5	4.121	0.873	1.513
70.0	722.20	1.8249	352.6	4.100	0.877	1.501
75.0	734.38	1.9228	355.8	4.080	0.882	1.491
80.0	745.69	2.0199	359.2	4.062	0.887	1.482
85.0	756.25	2.1162	362.7	4.045	0.891	1.475
90.0	766.14	2.2117	366.2	4.028	0.896	1.468
95.0	775.44	2.3066	369.8	4.013	0.900	1.463
100.0	784.21	2.4008	373.5	3.998	0.904	1.459
			$T=190$ K			
0.1	1.84	0.9971	443.0	6.411	0.716	1.007
0.5	9.30	0.9855	440.8	5.940	0.719	1.028
1.0	18.88	0.9712	438.0	5.731	0.723	1.054
1.5	28.73	0.9571	435.2	5.604	0.727	1.083
2.0	38.87	0.9433	432.4	5.511	0.732	1.113
2.5	49.30	0.9296	429.5	5.436	0.736	1.146
3.0	60.04	0.9160	426.6	5.372	0.741	1.181
3.5	71.08	0.9026	423.7	5.316	0.745	1.218
4.0	82.45	0.8894	420.8	5.266	0.750	1.258
4.5	94.13	0.8764	417.8	5.221	0.755	1.300
5.0	106.14	0.8636	414.8	5.179	0.760	1.344
6.0	131.10	0.8390	408.6	5.102	0.769	1.440
7.0	157.21	0.8163	402.5	5.033	0.778	1.542
8.0	184.22	0.7961	396.4	4.970	0.787	1.644
9.0	211.70	0.7794	390.5	4.912	0.794	1.739
10.0	239.13	0.7666	384.9	4.859	0.799	1.821
11.0	265.97	0.7582	379.7	4.811	0.803	1.884
12.0	291.76	0.7540	374.9	4.767	0.806	1.928
13.0	316.16	0.7538	370.6	4.727	0.808	1.954
14.0	338.98	0.7571	366.9	4.691	0.810	1.964

Table II.11 (*Continued*)

p	ρ	Z	h	s	c_v	c_p
			$T=190$ K			
15.0	360.16	0.7635	363.6	4.659	0.811	1.962
16.0	379.73	0.7724	360.7	4.630	0.811	1.952
17.0	397.76	0.7835	358.2	4.603	0.812	1.937
18.0	414.39	0.7963	356.0	4.579	0.812	1.918
19.0	429.73	0.8105	354.2	4.556	0.813	1.897
20.0	443.91	0.8259	352.6	4.536	0.814	1.875
21.0	457.07	0.8423	351.2	4.517	0.815	1.854
22.0	469.31	0.8594	350.0	4.499	0.816	1.833
23.0	480.72	0.8771	349.0	4.483	0.817	1.813
24.0	491.41	0.8953	348.2	4.468	0.818	1.794
25.0	501.44	0.9140	347.4	4.453	0.819	1.776
26.0	510.89	0.9329	346.8	4.440	0.820	1.760
27.0	519.82	0.9522	346.3	4.427	0.821	1.744
28.0	528.27	0.9717	345.9	4.415	0.822	1.729
29.0	536.30	0.9913	345.6	4.403	0.824	1.716
30.0	543.94	1.0111	345.4	4.392	0.825	1.703
35.0	577.39	1.1112	345.0	4.343	0.831	1.650
40.0	604.96	1.2121	345.8	4.303	0.838	1.611
45.0	628.43	1.3127	347.2	4.268	0.844	1.580
50.0	648.89	1.4126	349.2	4.237	0.850	1.556
55.0	667.03	1.5116	351.5	4.209	0.856	1.536
60.0	683.34	1.6096	354.2	4.184	0.861	1.520
65.0	698.15	1.7068	357.0	4.161	0.866	1.506
70.0	711.73	1.8030	360.1	4.140	0.871	1.494
75.0	724.25	1.8984	363.3	4.120	0.876	1.483
80.0	735.88	1.9930	366.6	4.101	0.880	1.474
85.0	746.72	2.0868	370.0	4.084	0.884	1.466
90.0	756.87	2.1799	373.5	4.067	0.889	1.460
95.0	766.41	2.2723	377.1	4.052	0.893	1.454
100.0	775.41	2.3642	380.8	4.037	0.897	1.449
			$T=195$ K			
0.1	1.79	0.9973	448.0	6.437	0.716	1.007
0.5	9.05	0.9867	445.9	5.967	0.719	1.026
1.0	18.34	0.9737	443.3	5.758	0.723	1.051
1.5	27.88	0.9610	440.6	5.632	0.726	1.077
2.0	37.67	0.9485	437.9	5.539	0.730	1.105
2.5	47.70	0.9361	435.2	5.465	0.734	1.134
3.0	58.00	0.9240	432.5	5.402	0.738	1.166
3.5	68.55	0.9120	429.8	5.348	0.743	1.199
4.0	79.37	0.9002	427.0	5.299	0.747	1.234
4.5	90.45	0.8887	424.2	5.254	0.751	1.271
5.0	101.79	0.8774	421.4	5.213	0.755	1.309
6.0	125.25	0.8557	415.7	5.139	0.764	1.391
7.0	149.62	0.8357	410.0	5.072	0.772	1.478
8.0	174.71	0.8179	404.4	5.012	0.779	1.565
9.0	200.21	0.8029	399.0	4.956	0.785	1.648

p	ρ	Z	h	s	c_v	c_p
			$T = 195$ K			
10.0	225.72	0.7913	393.7	4.905	0.791	1.722
11.0	250.84	0.7833	388.8	4.858	0.795	1.783
12.0	275.19	0.7789	384.3	4.816	0.798	1.829
13.0	298.47	0.7780	380.2	4.777	0.801	1.861
14.0	320.50	0.7802	376.5	4.741	0.803	1.879
15.0	341.18	0.7853	373.2	4.709	0.804	1.887
16.0	360.49	0.7928	370.3	4.679	0.805	1.886
17.0	378.46	0.8023	367.7	4.652	0.806	1.880
18.0	395.16	0.8136	365.5	4.628	0.807	1.868
19.0	410.68	0.8264	363.6	4.605	0.808	1.854
20.0	425.11	0.8404	361.9	4.584	0.809	1.838
21.0	438.55	0.8553	360.4	4.565	0.810	1.821
22.0	451.10	0.8711	359.1	4.547	0.811	1.805
23.0	462.85	0.8876	358.0	4.530	0.812	1.788
24.0	473.86	0.9047	357.1	4.514	0.813	1.771
25.0	484.22	0.9222	356.3	4.499	0.814	1.756
26.0	493.99	0.9401	355.6	4.485	0.815	1.741
27.0	503.23	0.9584	355.0	4.472	0.816	1.727
28.0	511.98	0.9769	354.5	4.459	0.818	1.713
29.0	520.29	0.9956	354.1	4.447	0.819	1.700
30.0	528.21	1.0145	353.8	4.436	0.820	1.688
35.0	562.86	1.1107	353.2	4.386	0.827	1.638
40.0	591.39	1.2081	353.8	4.344	0.833	1.601
45.0	615.64	1.3056	355.1	4.309	0.839	1.571
50.0	636.74	1.4026	356.9	4.277	0.845	1.548
55.0	655.43	1.4989	359.2	4.249	0.850	1.529
60.0	672.22	1.5943	361.7	4.223	0.855	1.512
65.0	687.46	1.6889	364.5	4.200	0.860	1.499
70.0	701.41	1.7826	367.5	4.178	0.865	1.487
75.0	714.27	1.8756	370.7	4.158	0.870	1.476
80.0	726.20	1.9677	373.9	4.140	0.874	1.467
85.0	737.32	2.0592	377.3	4.122	0.878	1.459
90.0	747.73	2.1499	380.8	4.105	0.883	1.452
95.0	757.52	2.2401	384.4	4.089	0.887	1.446
100.0	766.74	2.3296	388.0	4.074	0.891	1.440
			$T = 200$ K			
0.1	1.75	0.9976	453.1	6.462	0.716	1.007
0.5	8.81	0.9879	451.1	5.993	0.719	1.025
1.0	17.84	0.9761	448.5	5.785	0.722	1.048
1.5	27.09	0.9645	446.0	5.659	0.726	1.072
2.0	36.54	0.9531	443.4	5.567	0.729	1.097
2.5	46.22	0.9420	440.9	5.494	0.733	1.124
3.0	56.11	0.9311	438.3	5.432	0.737	1.153
3.5	66.23	0.9204	435.7	5.378	0.740	1.182
4.0	76.56	0.9099	433.1	5.330	0.744	1.213
4.5	87.12	0.8996	430.5	5.286	0.748	1.246

Table II.11 (*Continued*)

p	ρ	Z	h	s	c_v	c_p
			$T=200$ K			
5.0	97.89	0.8895	427.9	5.246	0.752	1.280
6.0	120.06	0.8704	422.6	5.173	0.759	1.352
7.0	142.96	0.8527	417.3	5.109	0.766	1.426
8.0	166.45	0.8370	412.1	5.050	0.773	1.502
9.0	190.28	0.8238	407.0	4.997	0.779	1.575
10.0	214.14	0.8133	402.1	4.948	0.784	1.641
11.0	237.72	0.8059	397.5	4.902	0.788	1.697
12.0	260.73	0.8015	393.2	4.861	0.792	1.743
13.0	282.91	0.8003	389.3	4.823	0.794	1.778
14.0	304.08	0.8018	385.7	4.788	0.797	1.801
15.0	324.15	0.8059	382.4	4.756	0.798	1.815
16.0	343.05	0.8123	379.6	4.726	0.800	1.822
17.0	360.79	0.8206	377.0	4.699	0.801	1.822
18.0	377.40	0.8306	374.7	4.674	0.802	1.817
19.0	392.94	0.8421	372.7	4.651	0.803	1.809
20.0	407.48	0.8548	370.9	4.630	0.804	1.798
21.0	421.09	0.8685	369.4	4.610	0.805	1.786
22.0	433.85	0.8831	368.1	4.592	0.806	1.773
23.0	445.83	0.8985	366.9	4.575	0.807	1.760
24.0	457.10	0.9144	365.9	4.558	0.809	1.746
25.0	467.72	0.9309	365.0	4.543	0.810	1.733
26.0	477.75	0.9478	364.2	4.529	0.811	1.720
27.0	487.24	0.9651	363.6	4.515	0.812	1.707
28.0	496.25	0.9826	363.1	4.502	0.813	1.695
29.0	504.81	1.0005	362.6	4.490	0.815	1.684
30.0	512.97	1.0185	362.2	4.479	0.816	1.673
35.0	548.71	1.1109	361.4	4.427	0.822	1.626
40.0	578.13	1.2049	361.8	4.385	0.828	1.590
45.0	603.12	1.2994	362.9	4.348	0.834	1.562
50.0	624.83	1.3936	364.7	4.316	0.840	1.539
55.0	644.05	1.4872	366.8	4.288	0.845	1.520
60.0	661.29	1.5801	369.3	4.262	0.850	1.505
65.0	676.93	1.6723	372.0	4.238	0.855	1.491
70.0	691.24	1.7636	374.9	4.216	0.860	1.480
75.0	704.43	1.8542	378.0	4.196	0.864	1.469
80.0	716.66	1.9441	381.3	4.177	0.868	1.460
85.0	728.06	2.0332	384.6	4.159	0.873	1.452
90.0	738.72	2.1218	388.0	4.142	0.877	1.445
95.0	748.74	2.2097	391.6	4.126	0.881	1.439
100.0	758.18	2.2970	395.2	4.111	0.885	1.433
			$T=210$ K			
0.1	1.66	0.9980	463.2	6.512	0.716	1.007
0.5	8.38	0.9899	461.3	6.043	0.719	1.022
1.0	16.92	0.9801	459.0	5.836	0.721	1.042
1.5	25.63	0.9705	456.7	5.711	0.724	1.063
2.0	34.51	0.9612	454.4	5.620	0.728	1.085

Table II.11 (*Continued*)

p	ρ	Z	h	s	c_v	c_p
			$T=210$ K			
2.5	43.55	0.9522	452.0	5.548	0.731	1.108
3.0	52.75	0.9433	449.7	5.487	0.734	1.131
3.5	62.11	0.9347	447.4	5.435	0.737	1.156
4.0	71.63	0.9262	445.1	5.388	0.740	1.181
4.5	81.30	0.9180	442.7	5.346	0.743	1.207
5.0	91.12	0.9101	440.4	5.307	0.746	1.234
6.0	111.19	0.8950	435.8	5.238	0.752	1.290
7.0	131.75	0.8812	431.1	5.176	0.758	1.348
8.0	152.69	0.8690	426.6	5.121	0.763	1.407
9.0	173.85	0.8586	422.2	5.071	0.768	1.464
10.0	195.05	0.8504	417.9	5.025	0.773	1.517
11.0	216.09	0.8443	413.8	4.982	0.777	1.565
12.0	236.76	0.8406	409.9	4.943	0.780	1.607
13.0	256.91	0.8393	406.3	4.906	0.783	1.641
14.0	276.38	0.8402	403.0	4.872	0.786	1.668
15.0	295.08	0.8431	400.0	4.841	0.788	1.689
16.0	312.94	0.8480	397.2	4.812	0.790	1.703
17.0	329.93	0.8546	394.7	4.785	0.791	1.712
18.0	346.05	0.8627	392.4	4.761	0.793	1.717
19.0	361.31	0.8722	390.3	4.737	0.794	1.718
20.0	375.74	0.8828	388.5	4.716	0.796	1.716
21.0	389.39	0.8945	386.9	4.696	0.797	1.712
22.0	402.29	0.9071	385.5	4.677	0.798	1.706
23.0	414.48	0.9204	384.2	4.659	0.799	1.698
24.0	426.03	0.9344	383.1	4.642	0.801	1.690
25.0	436.97	0.9489	382.1	4.627	0.802	1.682
26.0	447.35	0.9640	381.2	4.612	0.803	1.673
27.0	457.22	0.9795	380.5	4.598	0.804	1.664
28.0	466.60	0.9953	379.8	4.584	0.806	1.655
29.0	475.55	1.0115	379.3	4.571	0.807	1.646
30.0	484.08	1.0279	378.8	4.559	0.808	1.637
35.0	521.63	1.1129	377.5	4.506	0.814	1.598
40.0	552.62	1.2006	377.5	4.462	0.820	1.567
45.0	578.92	1.2893	378.4	4.424	0.826	1.541
50.0	601.76	1.3781	380.0	4.391	0.832	1.521
55.0	621.95	1.4668	381.9	4.361	0.837	1.504
60.0	640.04	1.5549	384.3	4.335	0.841	1.489
65.0	656.43	1.6424	386.8	4.310	0.846	1.477
70.0	671.41	1.7292	389.7	4.288	0.850	1.466
75.0	685.22	1.8154	392.7	4.267	0.854	1.456
80.0	698.01	1.9010	395.8	4.248	0.858	1.448
85.0	709.93	1.9859	399.1	4.229	0.863	1.440
90.0	721.08	2.0702	402.4	4.212	0.867	1.433
95.0	731.55	2.1539	405.9	4.196	0.871	1.426
100.0	741.41	2.2371	409.4	4.180	0.875	1.420

p	ρ	Z	h	s	c_v	c_p
			$T = 220$ K			
0.1	1.59	0.9983	473.2	6.558	0.716	1.006
0.5	7.98	0.9916	471.5	6.091	0.718	1.020
1.0	16.10	0.9834	469.4	5.884	0.721	1.038
1.5	24.34	0.9755	467.3	5.761	0.724	1.056
2.0	32.71	0.9679	465.2	5.671	0.726	1.075
2.5	41.21	0.9605	463.1	5.599	0.729	1.095
3.0	49.82	0.9533	460.9	5.540	0.732	1.115
3.5	58.56	0.9463	458.9	5.488	0.734	1.135
4.0	67.40	0.9396	456.8	5.442	0.737	1.156
4.5	76.36	0.9330	454.7	5.401	0.739	1.178
5.0	85.42	0.9267	452.6	5.364	0.742	1.200
6.0	103.84	0.9148	448.4	5.297	0.747	1.246
7.0	122.59	0.9041	444.3	5.238	0.752	1.292
8.0	141.58	0.8946	440.3	5.185	0.756	1.339
9.0	160.72	0.8866	436.4	5.137	0.761	1.385
10.0	179.87	0.8802	432.6	5.093	0.765	1.429
11.0	198.91	0.8756	428.9	5.052	0.768	1.469
12.0	217.69	0.8728	425.5	5.015	0.772	1.505
13.0	236.09	0.8718	422.2	4.980	0.774	1.537
14.0	254.02	0.8726	419.1	4.947	0.777	1.564
15.0	271.38	0.8751	416.3	4.917	0.779	1.586
16.0	288.13	0.8792	413.7	4.889	0.781	1.604
17.0	304.21	0.8847	411.3	4.863	0.783	1.617
18.0	319.62	0.8916	409.1	4.838	0.785	1.627
19.0	334.34	0.8997	407.1	4.815	0.787	1.633
20.0	348.39	0.9089	405.3	4.794	0.788	1.637
21.0	361.79	0.9190	403.6	4.774	0.790	1.638
22.0	374.55	0.9300	402.2	4.754	0.791	1.637
23.0	386.70	0.9417	400.8	4.737	0.792	1.635
24.0	398.28	0.9540	399.7	4.720	0.794	1.632
25.0	409.32	0.9670	398.6	4.704	0.795	1.628
26.0	419.84	0.9805	397.7	4.688	0.796	1.623
27.0	429.88	0.9944	396.9	4.674	0.798	1.617
28.0	439.48	1.0087	396.1	4.660	0.799	1.611
29.0	448.65	1.0234	395.5	4.647	0.800	1.605
30.0	457.43	1.0383	395.0	4.635	0.802	1.599
35.0	496.27	1.1166	393.4	4.580	0.808	1.569
40.0	528.50	1.1983	393.1	4.534	0.814	1.542
45.0	555.92	1.2816	393.7	4.495	0.819	1.520
50.0	579.74	1.3655	395.1	4.461	0.824	1.502
55.0	600.78	1.4494	396.9	4.431	0.829	1.486
60.0	619.62	1.5331	399.1	4.404	0.834	1.474
65.0	636.68	1.6163	401.5	4.379	0.838	1.462
70.0	652.28	1.6991	404.3	4.356	0.842	1.453
75.0	666.64	1.7812	407.2	4.335	0.846	1.444

p	ρ	Z	h	s	c_v	c_p
			$T=220$ K			
80.0	679.94	1.8628	410.2	4.315	0.850	1.436
85.0	692.33	1.9438	413.4	4.296	0.854	1.429
90.0	703.93	2.0242	416.7	4.278	0.858	1.422
95.0	714.82	2.1041	420.1	4.262	0.862	1.415
100.0	725.09	2.1835	423.6	4.246	0.865	1.410
			$T=230$ K			
0.1	1.52	0.9986	483.3	6.603	0.716	1.006
0.5	7.63	0.9930	481.7	6.136	0.718	1.018
1.0	15.36	0.9862	479.7	5.930	0.721	1.034
1.5	23.19	0.9797	477.8	5.807	0.723	1.051
2.0	31.11	0.9735	475.9	5.718	0.725	1.067
2.5	39.14	0.9674	473.9	5.648	0.728	1.084
3.0	47.25	0.9616	472.0	5.589	0.730	1.101
3.5	55.45	0.9559	470.1	5.538	0.732	1.119
4.0	63.73	0.9505	468.2	5.493	0.734	1.137
4.5	72.09	0.9453	466.3	5.453	0.737	1.155
5.0	80.53	0.9403	464.4	5.416	0.739	1.174
6.0	97.60	0.9310	460.7	5.351	0.743	1.212
7.0	114.90	0.9226	457.0	5.294	0.747	1.250
8.0	132.36	0.9153	453.4	5.243	0.751	1.289
9.0	149.89	0.9093	449.9	5.197	0.755	1.327
10.0	167.43	0.9045	446.5	5.155	0.759	1.363
11.0	184.86	0.9012	443.2	5.116	0.762	1.397
12.0	202.09	0.8992	440.1	5.080	0.765	1.429
13.0	219.04	0.8988	437.2	5.046	0.768	1.457
14.0	235.62	0.8998	434.4	5.015	0.770	1.482
15.0	251.78	0.9022	431.7	4.986	0.773	1.504
16.0	267.45	0.9060	429.3	4.959	0.775	1.522
17.0	282.62	0.9109	427.0	4.933	0.777	1.537
18.0	297.25	0.9171	425.0	4.909	0.778	1.549
19.0	311.35	0.9242	423.0	4.886	0.780	1.558
20.0	324.86	0.9323	421.3	4.865	0,782	1.565
21.0	337.84	0.9413	419.7	4.845	0.783	1.570
22.0	350.30	0.9511	418.2	4.826	0.785	1.573
23.0	362.24	0.9616	416.9	4.808	0.786	1.575
24.0	373.67	0.9727	415.7	4.791	0.788	1.575
25.0	384.63	0.9843	414.6	4.775	0.789	1.574
26.0	395.13	0.9965	413.7	4.759	0.791	1.572
27.0	405.20	1.0091	412.8	4.745	0.792	1.570
28.0	414.86	1.0221	412.0	4.731	0.793	1.567
29.0	424.12	1.0355	411.4	4.718	0.795	1.563
30.0	433.02	1.0492	410.8	4.705	0.796	1.560
35.0	472.68	1.1214	408.9	4.649	0.802	1.538
40.0	505.84	1.1975	408.4	4.602	0.808	1.517
45.0	534.16	1.2758	408.8	4.562	0.813	1.498
50.0	558.79	1.3551	410.0	4.528	0.818	1.482

Table II.11 (*Continued*)

p	ρ	Z	h	s	c_v	c_p
			$T=230$ K			
55.0	580.57	1.4347	411.7	4.497	0.823	1.469
60.0	600.07	1.5142	413.7	4.469	0.828	1.457
65.0	617.73	1.5935	416.1	4.443	0.832	1.447
70.0	633.86	1.6724	418.7	4.420	0.836	1.439
75.0	648.72	1.7508	421.5	4.398	0.839	1.431
80.0	662.48	1.8288	424.5	4.378	0.843	1.424
85.0	675.30	1.9062	427.6	4.359	0.847	1.417
90.0	687.30	1.9831	430.9	4.341	0.850	1.411
95.0	698.58	2.0594	434.2	4.325	0.854	1.405
100.0	709.20	2.1353	437.6	4.309	0.857	1.400
			$T=240$ K			
0.1	1.45	0.9988	493.3	6.646	0.716	1.006
0.5	7.30	0.9942	491.9	6.179	0.718	1.017
1.0	14.68	0.9886	490.1	5.974	0.720	1.031
1.5	22.14	0.9832	488.3	5.852	0.722	1.046
2.0	29.68	0.9781	486.5	5.764	0.724	1.060
2.5	37.28	0.9732	484.7	5.694	0.726	1.075
3.0	44.95	0.9685	483.0	5.636	0.729	1.091
3.5	52.69	0.9640	481.2	5.586	0.731	1.106
4.0	60.49	0.9597	479.5	5.541	0.733	1.122
4.5	68.35	0.9555	477.8	5.502	0.735	1.137
5.0	76.26	0.9516	476.1	5.466	0.736	1.153
6.0	92.21	0.9443	472.7	5.402	0.740	1.186
7.0	108.31	0.9379	469.4	5.347	0.744	1.218
8.0	124.52	0.9324	466.1	5.297	0.747	1.251
9.0	140.76	0.9280	463.0	5.253	0.751	1.282
10.0	156.97	0.9246	459.9	5.212	0.754	1.313
11.0	173.09	0.9223	456.9	5.174	0.757	1.342
12.0	189.04	0.9213	454.1	5.139	0.760	1.370
13.0	204.76	0.9214	451.4	5.107	0.762	1.395
14.0	220.19	0.9227	448.8	5.077	0.765	1.418
15.0	235.28	0.9252	446.4	5,048	0.767	1.438
16.0	249.99	0.9289	444.2	5,022	0.769	1.455
17.0	264.29	0.9335	442.1	4,997	0.771	1.471
18.0	278.15	0.9392	440.1	4,973	0.773	1.483
19.0	291.57	0.9457	438.3	4,951	0.775	1.494
20.0	304.53	0.9531	436.6	4.930	0.776	1.503
21.0	317.04	0.9613	435.1	4.910	0.778	1.510
22.0	329.10	0.9702	433.7	4.892	0.780	1.515
23.0	340.72	0.9797	432.4	4.874	0.781	1.519
24.0	351.90	0.9898	431.2	4.857	0.783	1.522
25.0	362.67	1.0004	430.1	4.841	0.784	1.524
26.0	373.04	1.0115	429.1	4.825	0.785	1.524
27.0	383.02	1.0231	428.3	4.811	0.787	1.524
28.0	392.62	1.0350	427.5	4.797	0.788	1.523
29.0	401.88	1.0473	426.8	4.783	0.789	1.522

p	ρ	Z	h	s	c_v	c_p
			$T=240$ K			
30.0	410.79	1.0599	426.2	4.770	0.791	1.520
35.0	450.83	1.1267	424.1	4.714	0.797	1.507
40.0	484.63	1.1979	423.4	4.666	0.803	1.491
45.0	513.64	1.2715	423.7	4.626	0.808	1.476
50.0	538.94	1.3464	424.7	4.590	0.813	1.462
55.0	561.34	1.4220	426.3	4.559	0.818	1.451
60.0	581.40	1.4977	428.2	4.530	0.822	1.441
65.0	599.58	1.5733	430.5	4.505	0.826	1.432
70.0	616.19	1.6487	433.0	4.481	0.830	1.424
75.0	631.48	1.7237	435.8	4.459	0.834	1.418
80.0	645.65	1.7983	438.7	4.439	0.837	1.412
85.0	658.85	1.8724	441.7	4.419	0.840	1.406
90.0	671.21	1.9460	444.9	4.401	0.844	1.401
95.0	682.82	2.0192	448.2	4.384	0.847	1.396
100.0	693.78	2.0919	451.6	4.368	0.850	1.391
			$T=250$ K			
0.1	1.39	0.9990	503.4	6.687	0.717	1.006
0.5	7.00	0.9952	502.0	6.221	0.718	1.016
1.0	14.06	0.9906	500.4	6.016	0.720	1.029
1.5	21.19	0.9862	498.7	5.895	0.722	1.042
2.0	28.37	0.9821	497.1	5.807	0.724	1.055
2.5	35.61	0.9781	495.5	5.737	0.726	1.068
3.0	42.90	0.9743	493.8	5.680	0.727	1.082
3.5	50.23	0.9708	492.2	5.630	0.729	1.095
4.0	57.61	0.9673	490.7	5.587	0.731	1.109
4.5	65.03	0.9641	489.1	5.548	0.733	1.123
5.0	72.49	0.9611	487.5	5.513	0.735	1.136
6.0	87.49	0.9555	484.4	5.450	0.738	1.164
7.0	102.59	0.9507	481.4	5.396	0.741	1.192
8.0	117.74	0.9466	478.5	5.348	0.744	1.220
9.0	132.90	0.9435	475.6	5.304	0.747	1.248
10.0	148.02	0.9413	472.8	5.265	0.750	1.274
11.0	163.04	0.9400	470.1	5.228	0.753	1.299
12.0	177.92	0.9397	467.6	5.194	0.756	1.323
13.0	192.60	0.9404	465.1	5.163	0.758	1.346
14.0	207.04	0.9421	462.8	5.134	0.760	1.366
15.0	221.20	0.9448	460.5	5.106	0.762	1.384
16.0	235.05	0.9484	458.4	5.080	0.765	1.401
17.0	248.56	0.9529	456.5	5.056	0.766	1.416
18.0	261.70	0.9583	454.7	5.033	0.768	1.429
19.0	274.48	0.9644	453.0	5.011	0.770	1.440
20.0	286.87	0.9713	451.4	4.990	0.772	1.449
21.0	298.88	0.9789	449.9	4.971	0.773	1.457
22.0	310.51	0.9871	448.5	4.952	0.775	1.464
23.0	321.75	0.9959	447.3	4.935	0.776	1.470
24.0	332.63	1.0053	446.2	4.918	0.778	1.474

Table II.11 (*Continued*)

p	ρ	Z	h	s	c_v	c_p
			$T=250$ K			
25.0	343.14	1.0151	445.1	4.902	0.779	1.477
26.0	353.29	1.0253	444.2	4.887	0.781	1.480
27.0	363.11	1.0360	443.3	4.872	0.782	1.481
28.0	372.59	1.0470	442.5	4.858	0.784	1.482
29.0	381.75	1.0584	441.8	4.845	0.785	1.483
30.0	390.60	1.0701	441.2	4.832	0.786	1.482
35.0	430.67	1.1323	439.0	4.774	0.793	1.476
40.0	464.84	1.1989	438.2	4.726	0.798	1.465
45.0	494.34	1.2683	438.4	4.685	0.804	1.453
50.0	520.17	1.3392	439.2	4.649	0.809	1.442
55.0	543.07	1.4110	440.7	4.618	0.813	1.432
60.0	563.62	1.4832	442.5	4.589	0.817	1.424
65.0	582.24	1.5554	444.7	4.563	0.821	1.416
70.0	599.26	1.6275	447.2	4.539	0.825	1.410
75.0	614.93	1.6993	449.9	4.517	0.828	1.404
80.0	629.45	1.7707	452.7	4.496	0.832	1.399
85.0	642.99	1.8418	455.7	4.477	0.835	1.394
90.0	655.66	1.9124	458.9	4.458	0.838	1.390
95.0	667.58	1.9827	462.1	4.441	0.841	1.385
100.0	678.83	2.0524	465.4	4.425	0.844	1.381
			$T=260$ K			
0.1	1.34	0.9992	513.5	6.726	0.717	1.006
0.5	6.72	0.9960	512.2	6.260	0.718	1.015
1.0	13.50	0.9923	510.7	6.057	0.720	1.027
1.5	20.32	0.9888	509.1	5.935	0.722	1.038
2.0	27.19	0.9854	507.6	5.848	0.723	1.050
2.5	34.10	0.9823	506.1	5.779	0.725	1.062
3.0	41.04	0.9793	504.6	5.722	0.727	1.074
3.5	48.02	0.9765	503.2	5.673	0.728	1.086
4.0	55.03	0.9738	501.7	5.630	0.730	1.098
4.5	62.06	0.9714	500.2	5.592	0.732	1.110
5.0	69.12	0.9690	498.8	5.557	0.733	1.123
6.0	83.31	0.9649	496.0	5.496	0.736	1.147
7.0	97.54	0.9614	493.2	5.442	0.739	1.172
8.0	111.80	0.9586	490.6	5.395	0.742	1.196
9.0	126.05	0.9565	487.9	5.353	0.745	1.220
10.0	140.24	0.9553	485.4	5.314	0.747	1.243
11.0	154.34	0.9548	482.9	5.278	0.750	1.265
12.0	168.30	0.9552	480.6	5.246	0.752	1.286
13.0	182.09	0.9564	478.3	5.215	0.755	1.306
14.0	195.68	0.9585	476.2	5.186	0.757	1.324
15.0	209.02	0.9614	474.2	5.159	0.759	1.341
16.0	222.10	0.9651	472.2	5.134	0.761	1.356
17.0	234.89	0.9696	470.4	5.110	0.763	1.370
18.0	247.38	0.9748	468.7	5.088	0.764	1.383
19.0	259.55	0.9807	467.1	5.067	0.766	1.394

Table II.11 (*Continued*)

p	ρ	Z	h	s	c_v	c_p
			$T=260$ K			
20.0	271.40	0.9872	465.6	5.046	0.768	1.404
21.0	282.92	0.9944	464.2	5.027	0.769	1.412
22.0	294.11	1.0021	463.0	5.009	0.771	1.419
23.0	304.97	1.0103	461.8	4.991	0.773	1.426
24.0	315.50	1.0191	460.7	4.975	0.774	1.431
25.0	325.72	1.0282	459.7	4.959	0.775	1.435
26.0	335.62	1.0378	458.7	4.944	0.777	1.439
27.0	345.22	1.0478	457.9	4.929	0.778	1.442
28.0	354.52	1.0580	457.1	4.915	0.780	1.444
29.0	363.54	1.0687	456.4	4.902	0.781	1.446
30.0	372.27	1.0796	455.8	4.889	0.782	1.447
35.0	412.11	1.1377	453.6	4.832	0.788	1.446
40.0	446.41	1.2004	452.7	4.783	0.794	1.440
45.0	476.23	1.2659	452.8	4.742	0.800	1.431
50.0	502.45	1.3331	453.6	4.706	0.804	1.423
55.0	525.76	1.4014	454.9	4.673	0.809	1.415
60.0	546.70	1.4703	456.7	4.644	0.813	1.407
65.0	565.70	1.5393	458.8	4.618	0.817	1.401
70.0	583.07	1.6083	461.2	4.594	0.820	1.395
75.0	599.07	1.6772	463.9	4.571	0.824	1.391
80.0	613.90	1.7458	466.7	4.551	0.827	1.386
85.0	627.73	1.8140	469.6	4.531	0.830	1.382
90.0	640.68	1.8819	472.7	4.513	0.833	1.379
95.0	652.86	1.9494	475.9	4.495	0.836	1.375
100.0	664.36	2.0165	479.2	4.479	0.839	1.372
			$T=270$ K			
0.1	1.29	0.9993	523.5	6.764	0.717	1.006
0.5	6.47	0.9968	522.3	6.299	0.718	1.014
1.0	12.98	0.9938	520.9	6.095	0.720	1.025
1.5	19.53	0.9910	519.5	5.975	0.721	1.036
2.0	26.11	0.9883	518.1	5.888	0.723	1.046
2.5	32.71	0.9858	516.7	5.819	0.725	1.057
3.0	39.35	0.9835	515.3	5.763	0.726	1.068
3.5	46.01	0.9814	514.0	5.714	0.728	1.079
4.0	52.69	0.9794	512.6	5.672	0.729	1.089
4.5	59.39	0.9775	511.3	5.634	0.731	1.100
5.0	66.10	0.9758	510.0	5.599	0.732	1.111
6.0	79.57	0.9728	507.4	5.539	0.735	1.133
7.0	93.05	0.9704	504.9	5.486	0.738	1.154
8.0	106.54	0.9687	502.4	5.440	0.740	1.176
9.0	120.00	0.9676	500.0	5.398	0.743	1.197
10.0	133.39	0.9671	497.7	5.360	0.745	1.217
11.0	146.69	0.9674	495.5	5.326	0.747	1.237
12.0	159.87	0.9683	493.3	5.294	0.750	1.255
13.0	172.89	0.9700	491.2	5.264	0.752	1.273
14.0	185.72	0.9724	489.3	5.236	0.754	1.290

p	ρ	Z	h	s	c_v	c_p
			$T=270$ K			
15.0	198.35	0.9756	487.4	5.209	0.756	1.305
16.0	210.75	0.9794	485.6	5.185	0.758	1.319
17.0	222.90	0.9839	483.9	5.161	0.759	1.332
18.0	234.79	0.9890	482.3	5.139	0.761	1.344
19.0	246.41	0.9947	480.9	5.118	0.763	1.355
20.0	257.74	1.0010	479.5	5.099	0.764	1.365
21.0	268.79	1.0079	478.2	5.080	0.766	1.373
22.0	279.56	1.0152	477.0	5.062	0.768	1.381
23.0	290.03	1.0230	475.8	5.045	0.769	1.387
24.0	300.22	1.0313	474.8	5.028	0.771	1.393
25.0	310.13	1.0399	473.8	5.012	0.772	1.398
26.0	319.76	1.0489	472.9	4.997	0.773	1.403
27.0	329.12	1.0583	472.1	4.983	0.775	1.406
28.0	338.21	1.0680	471.4	4.969	0.776	1.409
29.0	347.05	1.0780	470.7	4.956	0.777	1.412
30.0	355.63	1.0882	470.1	4.943	0.779	1.414
35.0	395.03	1.1430	468.0	4.886	0.785	1.418
40.0	429.27	1.2021	467.0	4.837	0.791	1.415
45.0	459.25	1.2640	467.0	4.796	0.796	1.410
50.0	485.74	1.3279	467.7	4.759	0.801	1.403
55.0	509.36	1.3930	469.0	4.726	0.805	1.397
60.0	530.62	1.4587	470.7	4.697	0.809	1.391
65.0	549.93	1.5248	472.8	4.671	0.813	1.386
70.0	567.60	1.5909	475.1	4.646	0.816	1.381
75.0	583.89	1.6570	477.7	4.624	0.820	1.377
80.0	598.99	1.7230	480.5	4.603	0.823	1.373
85.0	613.07	1.7886	483.4	4.583	0.826	1.370
90.0	626.26	1.8539	486.4	4.564	0.829	1.367
95.0	638.67	1.9189	489.6	4.547	0.831	1.364
100.0	650.38	1.9835	492.9	4.530	0.834	1.362
			$T=280$ K			
0.1	1.24	0.9995	533.6	6.801	0.717	1.006
0.5	6.24	0.9974	532.5	6.336	0.718	1.014
1.0	12.50	0.9950	531.2	6.133	0.720	1.024
1.5	18.79	0.9928	529.8	6.012	0.721	1.033
2.0	25.11	0.9908	528.5	5.926	0.723	1.043
2.5	31.45	0.9889	527.3	5.858	0.724	1.053
3.0	37.80	0.9872	526.0	5.801	0.726	1.062
3.5	44.18	0.9856	524.7	5.753	0.727	1.072
4.0	50.56	0.9841	523.5	5.711	0.729	1.082
4.5	56.96	0.9828	522.3	5.673	0.730	1.092
5.0	63.37	0.9816	521.0	5.639	0.731	1.101
6.0	76.19	0.9796	518.7	5.580	0.734	1.121
7.0	89.02	0.9782	516.3	5.528	0.736	1.140
8.0	101.83	0.9773	514.1	5.482	0.739	1.159
9.0	114.60	0.9769	511.9	5.441	0.741	1.178

p	ρ	Z	h	s	c_v	c_p
			$T=280$ K			
10.0	127.30	0.9772	509.8	5.404	0.743	1.196
11.0	139.91	0.9781	507.7	5.370	0.745	1.213
12.0	152.40	0.9795	505.7	5.339	0.747	1.230
13.0	164.74	0.9816	503.8	5.309	0.749	1.246
14.0	176.92	0.9844	502.0	5.282	0.751	1.261
15.0	188.92	0.9877	500.3	5.256	0.753	1.275
16.0	200.71	0.9916	498.6	5.232	0.755	1.288
17.0	212.29	0.9962	497.1	5.209	0.757	1.300
18.0	223.63	1.0013	495.6	5.188	0.758	1.311
19.0	234.74	1.0069	494.2	5.167	0.760	1.322
20.0	245.60	1.0130	492.9	5.148	0.762	1.331
21.0	256.21	1.0196	491.7	5.129	0.763	1.339
22.0	266.56	1.0267	490.6	5.111	0.765	1.347
23.0	276.67	1.0341	489.5	5.094	0.766	1.354
24.0	286.52	1.0420	488.6	5.078	0.768	1.360
25.0	296.11	1.0502	487.6	5.063	0.769	1.365
26.0	305.47	1.0588	486.8	5.048	0.770	1.370
27.0	314.57	1.0677	486.0	5.034	0.772	1.374
28.0	323.44	1.0769	485.3	5.020	0.773	1.378
29.0	332.08	1.0863	484.7	5.007	0.774	1.381
30.0	340.49	1.0961	484.1	4.994	0.776	1.384
35.0	379.31	1.1478	482.0	4.937	0.782	1.391
40.0	413.35	1.2038	481.0	4.888	0.787	1.392
45.0	443.36	1.2626	481.0	4.846	0.792	1.389
50.0	470.00	1.3234	481.6	4.810	0.797	1.385
55.0	493.84	1.3854	482.8	4.777	0.801	1.380
60.0	515.35	1.4483	484.5	4.748	0.805	1.375
65.0	534.92	1.5116	486.5	4.721	0.809	1.371
70.0	552.84	1.5751	488.9	4.696	0.813	1.367
75.0	569.36	1.6386	491.4	4.673	0.816	1.364
80.0	584.69	1.7020	494.1	4.652	0.819	1.361
85.0	598.99	1.7653	497.0	4.633	0.822	1.358
90.0	612.39	1.8282	500.1	4.614	0.825	1.356
95.0	625.00	1.8908	503.2	4.596	0.827	1.354
100.0	636.90	1.9531	506.4	4.579	0.830	1.352
			$T=290$ K			
0.1	1.20	0.9996	543.6	6.836	0.718	1.006
0.5	6.02	0.9980	542.6	6.371	0.719	1.013
1.0	12.06	0.9961	541.4	6.168	0.720	1.022
1.5	18.12	0.9945	540.2	6.048	0.722	1.031
2.0	24.19	0.9929	539.0	5.962	0.723	1.040
2.5	30.28	0.9915	537.8	5.894	0.724	1.049
3.0	36.39	0.9903	536.6	5.839	0.726	1.058
3.5	42.50	0.9892	535.4	5.791	0.727	1.067
4.0	48.62	0.9882	534.3	5.749	0.728	1.076
4.5	54.74	0.9873	533.1	5.712	0.729	1.084

Table II.11 (*Continued*)

p	ρ	Z	h	s	c_v	c_p
			$T=290$ K			
5.0	60.87	0.9866	532.0	5.678	0.731	1.093
6.0	73.13	0.9854	529.8	5.619	0.733	1.111
7.0	85.38	0.9848	527.7	5.568	0.735	1.128
8.0	97.59	0.9846	525.6	5.523	0.737	1.145
9.0	109.75	0.9849	523.6	5.482	0.740	1.162
10.0	121.84	0.9858	521.6	5.446	0.742	1.178
11.0	133.83	0.9872	519.7	5.412	0.744	1.194
12.0	145.72	0.9891	517.9	5.381	0.746	1.209
13.0	157.47	0.9916	516.2	5.353	0.748	1.223
14.0	169.07	0.9946	514.5	5.326	0.749	1.237
15.0	180.50	0.9981	512.9	5.301	0.751	1.250
16.0	191.75	1.0022	511.4	5.277	0.753	1.262
17.0	202.81	1.0068	509.9	5.254	0.755	1.273
18.0	213.66	1.0119	508.6	5.233	0.756	1.284
19.0	224.30	1.0174	507.3	5.213	0.758	1.293
20.0	234.72	1.0234	506.1	5.194	0.759	1.302
21.0	244.92	1.0298	505.0	5.175	0.761	1.310
22.0	254.89	1.0367	503.9	5.158	0.762	1.318
23.0	264.64	1.0439	502.9	5.141	0.764	1.325
24.0	274.16	1.0514	502.0	5.125	0.765	1.331
25.0	283.46	1.0593	501.2	5.110	0.766	1.336
26.0	292.53	1.0675	500.4	5.095	0.768	1.341
27.0	301.39	1.0760	499.6	5.081	0.769	1.346
28.0	310.02	1.0848	499.0	5.068	0.770	1.350
29.0	318.45	1.0938	498.4	5.055	0.772	1.353
30.0	326.67	1.1030	497.8	5.042	0.773	1.356
35.0	364.83	1.1523	495.8	4.985	0.779	1.367
40.0	498.55	1.2054	494.8	4.937	0.784	1.370
45.0	428.47	1.2614	494.8	4.895	0.789	1.370
50.0	455.17	1.3194	495.4	4.858	0.794	1.367
55.0	479.16	1.3787	496.6	4.825	0.798	1.364
60.0	500.85	1.4388	498.2	4.796	0.802	1.360
65.0	520.62	1.4996	500.2	4.769	0.806	1.356
70.0	538.75	1.5606	502.5	4.744	0.809	1.353
75.0	555.48	1.6217	505.0	4.721	0.812	1.351
80.0	571.00	1.6828	507.7	4.700	0.815	1.348
85.0	585.49	1.7437	510.6	4.680	0.818	1.346
90.0	599.06	1.8044	513.6	4.661	0.821	1.344
95.0	611.84	1.8649	516.7	4.644	0.823	1.343
100.0	623.92	1.9251	519.9	4.627	0.826	1.341
			$T=300$ K			
0.1	1.16	0.9997	553.7	6.870	0.718	1.007
0.5	5.81	0.9985	552.8	6.406	0.719	1.013
1.0	11.64	0.9971	551.6	6.203	0.720	1.021
1.5	17.49	0.9959	550.5	6.083	0.722	1.030
2.0	23.34	0.9948	549.3	5.997	0.723	1.038

Table II.11 (*Continued*)

p	ρ	Z	h	s	c_v	c_p
			$T=300$ K			
2.5	29.21	0.9938	548.2	5.930	0.724	1.046
3.0	35.08	0.9930	547.1	5.874	0.725	1.054
3.5	40.95	0.9923	546.1	5.827	0.727	1.062
4.0	46.83	0.9917	545.0	5.785	0.728	1.070
4.5	52.71	0.9912	543.9	5.748	0.729	1.078
5.0	58.59	0.9908	542.9	5.715	0.730	1.086
6.0	70.34	0.9904	540.9	5.656	0.732	1.102
7.0	82.06	0.9905	538.9	5.606	0.734	1.118
8.0	93.73	0.9909	537.0	5.561	0.737	1.133
9.0	105.35	0.9918	535.1	5.522	0.739	1.148
10.0	116.90	0.9932	533.3	5.486	0.741	1.163
11.0	128.35	0.9950	531.6	5.453	0.742	1.177
12.0	139.70	0.9973	529.9	5.422	0.744	1.191
13.0	150.92	1.0001	528.3	5.394	0.746	1.204
14.0	162.00	1.0034	526.8	5.367	0.748	1.216
15.0	172.93	1.0071	525.3	5.343	0.749	1.228
16.0	183.69	1.0113	523.9	5.319	0.751	1.239
17.0	194.28	1.0159	522.6	5.297	0.753	1.250
18.0	204.68	1.0210	521.3	5.276	0.754	1.260
19.0	214.90	1.0265	520.1	5.256	0.756	1.269
20.0	224.92	1.0324	519.0	5.237	0.757	1.277
21.0	234.73	1.0387	517.9	5.219	0.759	1.285
22.0	244.35	1.0453	517.0	5.202	0.760	1.292
23.0	253.76	1.0523	516.0	5.186	0.761	1.299
24.0	262.97	1.0596	515.2	5.170	0.763	1.305
25.0	271.97	1.0672	514.4	5.155	0.764	1.311
26.0	280.77	1.0751	513.6	5.140	0.765	1.316
27.0	289.38	1.0833	513.0	5.126	0.767	1.320
28.0	297.79	1.0917	512.3	5.113	0.768	1.324
29.0	306.00	1.1003	511.8	5.100	0.769	1.328
30.0	314.03	1.1092	511.2	5.088	0.770	1.332
35.0	351.46	1.1562	509.3	5.031	0.776	1.344
40.0	384.78	1.2069	508.4	4.983	0.781	1.349
45.0	414.53	1.2604	508.4	4.941	0.786	1.351
50.0	441.21	1.3157	509.0	4.904	0.791	1.350
55.0	465.26	1.3725	510.1	4.871	0.795	1.348
60.0	487.09	1.4302	511.7	4.841	0.799	1.345
65.0	507.01	1.4885	513.7	4.814	0.803	1.343
70.0	525.30	1.5472	515.9	4.790	0.806	1.340
75.0	542.20	1.6060	518.4	4.767	0.809	1.338
80.0	557.89	1.6649	521.1	4.745	0.812	1.336
85.0	572.54	1.7237	524.0	4.725	0.815	1.334
90.0	586.27	1.7824	527.0	4.707	0.817	1.333
95.0	599.19	1.8408	530.1	4.689	0.820	1.332
100.0	611.41	1.8990	533.3	4.672	0.822	1.331

Table II.11 (*Continued*)

p	ρ	Z	h	s	c_v	c_p
			$T = 350$ K			
0.1	1.00	1.0000	604.1	7.026	0.721	1.009
0.5	4.98	1.0001	603.4	6.562	0.722	1.014
1.0	9.95	1.0003	602.6	6.360	0.723	1.020
1.5	14.92	1.0006	601.8	6.242	0.724	1.025
2.0	19.88	1.0011	601.0	6.157	0.725	1.031
2.5	24.84	1.0015	600.3	6.090	0.726	1.036
3.0	29.79	1.0021	599.5	6.036	0.727	1.042
3.5	34.73	1.0028	598.8	5.989	0.728	1.047
4.0	39.67	1.0035	598.0	5.949	0.728	1.053
4.5	44.59	1.0043	597.3	5.913	0.729	1.058
5.0	49.50	1.0052	596.6	5.880	0.730	1.063
6.0	59.29	1.0071	595.2	5.824	0.732	1.074
7.0	69.02	1.0094	593.9	5.775	0.733	1.084
8.0	78.68	1.0119	592.6	5.733	0.735	1.094
9.0	88.27	1.0146	591.3	5.695	0.737	1.104
10.0	97.79	1.0177	590.1	5.661	0.738	1.114
11.0	107.22	1.0210	588.9	5.629	0.739	1.123
12.0	116.56	1.0245	587.8	5.601	0.741	1.132
13.0	125.81	1.0283	586.7	5.574	0.742	1.141
14.0	134.95	1.0325	585.7	5.549	0.744	1.149
15.0	143.98	1.0368	584.7	5.526	0.745	1.157
16.0	152.89	1.0414	583.8	5.504	0.746	1.165
17.0	161.69	1.0463	582.9	5.483	0.748	1.172
18.0	170.36	1.0515	582.1	5.464	0.749	1.179
19.0	178.91	1.0569	581.3	5.445	0.750	1.186
20.0	187.33	1.0625	580.5	5.427	0.751	1.193
21.0	195.62	1.0684	579.8	5.410	0.752	1.199
22.0	203.77	1.0744	579.1	5.394	0.753	1.204
23.0	211.79	1.0807	578.5	5.379	0.755	1.210
24.0	219.68	1.0872	578.0	5.364	0.756	1.215
25.0	227.44	1.0939	577.4	5.349	0.757	1.220
26.0	235.06	1.1007	576.9	5.336	0.758	1.224
27.0	242.55	1.1078	576.5	5.322	0.759	1.229
28.0	249.91	1.1150	576.1	5.310	0.760	1.233
29.0	257.15	1.1223	575.7	5.297	0.761	1.237
30.0	264.25	1.1298	575.3	5.285	0.762	1.240
35.0	297.90	1.1692	574.1	5.231	0.767	1.255
40.0	328.65	1.2112	573.7	5.184	0.772	1.266
45.0	356.77	1.2552	573.9	5.143	0.776	1.273
50.0	382.53	1.3008	574.6	5.106	0.780	1.278
55.0	406.17	1.3476	575.8	5.074	0.784	1.280
60.0	427.94	1.3953	577.3	5.044	0.787	1.282
65.0	448.04	1.4438	579.2	5.017	0.791	1.282
70.0	466.66	1.4928	581.4	4.992	0.794	1.282
75.0	483.97	1.5422	583.9	4.969	0.797	1.282

p	ρ	Z	h	s	c_v	c_p
			$T=350$ K			
80.0	500.12	1.5919	586.5	4.947	0.799	1.282
85.0	515.25	1.6417	589.3	4.927	0.802	1.282
90.0	529.46	1.6916	592.3	4.908	0.804	1.282
95.0	542.86	1.7415	595.4	4.890	0.807	1.282
100.0	555.54	1.7914	598.6	4.873	0.809	1.282
			$T=400$ K			
0.1	0.87	1.0002	654.7	7.161	0.727	1.014
0.5	4.35	1.0010	654.2	6.697	0.727	1.018
1.0	8.69	1.0020	653.6	6.497	0.728	1.022
1.5	13.02	1.0032	653.1	6.379	0.729	1.026
2.0	17.34	1.0043	652.5	6.294	0.729	1.030
2.5	21.65	1.0056	652.0	6.229	0.730	1.034
3.0	25.95	1.0069	651.5	6.175	0.731	1.038
3.5	30.23	1.0082	651.0	6.129	0.731	1.042
4.0	34.50	1.0096	650.5	6.089	0.732	1.046
4.5	38.76	1.0111	650.0	6.053	0.733	1.050
5.0	43.00	1.0125	649.5	6.022	0.733	1.054
6.0	51.44	1.0157	648.5	5.966	0.735	1.061
7.0	59.82	1.0190	647.6	5.919	0.736	1.068
8.0	68.13	1.0225	646.7	5.878	0.737	1.076
9.0	76.38	1.0261	645.9	5.841	0.738	1.083
10.0	84.55	1.0299	645.1	5.808	0.740	1.089
11.0	92.64	1.0339	644.3	5.777	0.741	1.096
12.0	100.66	1.0381	643.6	5.750	0.742	1.102
13.0	108.60	1.0424	642.9	5.724	0.743	1.109
14.0	116.45	1.0469	642.2	5.700	0.744	1.115
15.0	124.22	1.0515	641.6	5.678	0.745	1.121
16.0	131.89	1.0563	641.0	5.657	0.746	1.126
17.0	139.48	1.0613	640.4	5.637	0.747	1.132
18.0	146.97	1.0664	639.9	5.618	0.748	1.137
19.0	154.37	1.0717	639.4	5.600	0.749	1.142
20.0	161.68	1.0772	638.9	5.583	0.750	1.147
21.0	168.89	1.0828	638.5	5.567	0.751	1.152
22.0	176.00	1.0885	638.0	5.552	0.752	1.156
23.0	183.01	1.0944	637.7	5.537	0.753	1.161
24.0	189.92	1.1004	637.3	5.522	0.754	1.165
25.0	196.74	1.1065	637.0	5.509	0.755	1.169
26.0	203.46	1.1127	636.7	5.496	0.756	1.172
27.0	210.09	1.1191	636.5	5.483	0.757	1.176
28.0	216.61	1.1256	636.3	5.471	0.758	1.179
29.0	223.04	1.1322	636.1	5.459	0.758	1.183
30.0	229.38	1.1389	635.9	5.447	0.759	1.186
35.0	259.68	1.1736	635.4	5.395	0.763	1.200
40.0	287.80	1.2103	635.5	5.349	0.767	1.211
45.0	313.89	1.2483	636.1	5.309	0.771	1.219
50.0	338.15	1.2876	637.1	5.273	0.774	1.226

p	ρ	Z	h	s	c_v	c_p
			$T=400$ K			
55.0	360.72	1.3277	638.5	5.241	0.778	1.231
60.0	381.76	1.3686	640.2	5.212	0.781	1.234
65.0	401.39	1.4101	642.2	5.185	0.784	1.237
70.0	419.76	1.4521	644.4	5.160	0.787	1.239
75.0	436.97	1.4946	646.9	5.137	0.789	1.240
80.0	453.13	1.5374	649.5	5.115	0.792	1.241
85.0	468.33	1.5804	652.4	5.095	0.794	1.242
90.0	482.68	1.6236	655.3	5.077	0.796	1.242
95.0	496.25	1.6670	658.5	5.059	0.798	1.243
100.0	509.11	1.7104	661.7	5.042	0.800	1.244
			$T=450$ K			
0.1	0.77	1.0003	705.6	7.281	0.734	1.022
0.5	3.86	1.0015	705.2	6.818	0.734	1.024
1.0	7.72	1.0029	704.9	6.617	0.735	1.027
1.5	11.56	1.0045	704.5	6.500	0.736	1.030
2.0	15.39	1.0061	704.1	6.416	0.736	1.033
2.5	19.20	1.0077	703.7	6.350	0.737	1.036
3.0	23.00	1.0094	703.4	6.297	0.737	1.039
3.5	26.79	1.0111	703.0	6.251	0.738	1.042
4.0	30.57	1.0128	702.7	6.212	0.738	1.045
4.5	34.33	1.0146	702.4	6.177	0.739	1.048
5.0	38.08	1.0164	702.1	6.145	0.739	1.051
6.0	45.53	1.0201	701.4	6.091	0.740	1.057
7.0	52.91	1.0239	700.8	6.044	0.741	1.062
8.0	60.24	1.0279	700.3	6.004	0.742	1.067
9.0	67.50	1.0320	699.7	5.968	0.743	1.073
10.0	74.70	1.0361	699.2	5.935	0.744	1.078
11.0	81.83	1.0404	698.8	5.906	0.745	1.083
12.0	88.90	1.0448	698.3	5.879	0.746	1.088
13.0	95.89	1.0493	697.9	5.854	0.747	1.092
14.0	102.82	1.0540	697.5	5.830	0.748	1.097
15.0	109.67	1.0587	697.1	5.809	0.749	1.102
16.0	116.45	1.0635	696.7	5.788	0.750	1.106
17.0	123.15	1.0685	696.4	5.769	0.750	1.110
18.0	129.78	1.0735	696.1	5.751	0.751	1.114
19.0	136.34	1.0787	695.8	5.733	0.752	1.118
20.0	142.82	1.0839	695.6	5.717	0.753	1.122
21.0	149.23	1.0893	695.3	5.701	0.754	1.126
22.0	155.55	1.0947	695.1	5.686	0.754	1.129
23.0	161.81	1.1002	694.9	5.672	0.755	1.133
24.0	167.98	1.1059	694.8	5.658	0.756	1.136
25.0	174.08	1.1116	694.6	5.644	0.757	1.139
26.0	180.10	1.1174	694.5	5.632	0.757	1.142
27.0	186.05	1.1233	694.4	5.619	0.758	1.145
28.0	191.92	1.1292	694.4	5.607	0.759	1.148
29.0	197.72	1.1353	694.3	5.596	0.760	1.151

Table II.11 (*Continued*)

p	ρ	Z	h	s	c_v	c_p
			$T=450$ K			
30.0	203.44	1.1414	694.3	5.585	0.760	1.154
35.0	230.97	1.1729	694.5	5.534	0.764	1.166
40.0	256.76	1.2058	695.1	5.490	0.767	1.176
45.0	280.94	1.2398	696.1	5.451	0.770	1.184
50.0	303.64	1.2746	697.4	5.416	0.773	1.191
55.0	324.96	1.3101	699.1	5.384	0.776	1.196
60.0	345.01	1.3461	701.0	5.355	0.779	1.201
65.0	363.89	1.3826	703.1	5.328	0.781	1.204
70.0	381.69	1.4195	705.5	5.304	0.784	1.207
75.0	398.49	1.4568	708.1	5.281	0.786	1.209
80.0	414.37	1.4943	710.8	5.260	0.788	1.211
85.0	429.40	1.5322	713.7	5.240	0.790	1.213
90.0	443.65	1.5702	716.7	5.221	0.792	1.214
95.0	457.18	1.6084	719.9	5.204	0.794	1.215
100.0	470.04	1.6467	723.1	5.187	0.796	1.216
			$T=500$ K			
0.1	0.70	1.0003	756.9	7.389	0.743	1.030
0.5	3.48	1.0017	756.6	6.926	0.743	1.032
1.0	6.94	1.0034	756.4	6.726	0.744	1.035
1.5	10.40	1.0052	756.2	6.609	0.744	1.037
2.0	13.84	1.0070	755.9	6.525	0.745	1.039
2.5	17.26	1.0088	755.7	6.460	0.745	1.042
3.0	20.68	1.0107	755.5	6.407	0.745	1.044
3.5	24.08	1.0126	755.3	6.361	0.746	1.046
4.0	27.47	1.0145	755.0	6.322	0.746	1.049
4.5	30.84	1.0164	754.8	6.287	0.747	1.051
5.0	34.20	1.0184	754.6	6.256	0.747	1.053
6.0	40.88	1.0224	754.3	6.202	0.748	1.058
7.0	47.51	1.0264	753.9	6.156	0.749	1.062
8.0	54.08	1.0306	753.6	6.116	0.750	1.066
9.0	60.59	1.0348	753.3	6.080	0.750	1.070
10.0	67.04	1.0391	753.0	6.049	0.751	1.074
11.0	73.43	1.0435	752.7	6.020	0.752	1.078
12.0	79.77	1.0480	752.5	5.993	0.753	1.082
13.0	86.04	1.0525	752.3	5.968	0.753	1.086
14.0	92.26	1.0571	752.1	5.945	0.754	1.089
15.0	98.41	1.0618	751.9	5.924	0.755	1.093
16.0	104.50	1.0666	751.7	5.904	0.756	1.096
17.0	110.53	1.0714	751.6	5.885	0.756	1.099
18.0	116.50	1.0763	751.5	5.867	0.757	1.103
19.0	122.41	1.0813	751.4	5.850	0.758	1.106
20.0	128.25	1.0863	751.3	5.834	0.758	1.109
21.0	134.03	1.0915	751.2	5.819	0.759	1.112
22.0	139.75	1.0966	751.2	5.804	0.760	1.115
23.0	145.41	1.1019	751.2	5.790	0.760	1.118
24.0	151.00	1.1072	751.1	5.777	0.761	1.120

Table II.11 (*Continued*)

p	ρ	Z	h	s	c_v	c_p
			$T = 500$ K			
25.0	156.53	1.1126	751.2	5.764	0.762	1.123
26.0	162.00	1.1180	751.2	5.751	0.762	1.125
27.0	167.41	1.1235	751.2	5.739	0.763	1.128
28.0	172.75	1.1291	751.3	5.727	0.763	1.130
29.0	178.04	1.1347	751.4	5.716	0.764	1.133
30.0	183.26	1.1404	751.5	5.705	0.765	1.135
35.0	208.50	1.1694	752.2	5.656	0.767	1.145
40.0	232.31	1.1995	753.3	5.612	0.770	1.154
45.0	254.79	1.2303	754.7	5.574	0.773	1.161
50.0	276.04	1.2618	756.4	5.540	0.775	1.168
55.0	296.13	1.2938	758.3	5.509	0.778	1.173
60.0	315.15	1.3263	760.4	5.480	0.780	1.178
65.0	333.18	1.3590	762.8	5.454	0.782	1.182
70.0	350.29	1.3921	765.3	5.430	0.784	1.185
75.0	366.53	1.4254	768.0	5.407	0.786	1.186
80.0	381.98	1.4590	770.8	5.386	0.788	1.190
85.0	396.67	1.4928	773.8	5.367	0.790	1.192
90.0	410.66	1.5267	776.9	5.348	0.792	1.194
95.0	424.00	1.5608	780.1	5.330	0.794	1.195
100.0	436.73	1.5951	783.4	5.314	0.795	1.196
			$T = 550$ K			
0.1	0.63	1.0004	808.6	7.487	0.753	1.040
0.5	3.16	1.0018	808.5	7.025	0.753	1.042
1.0	6.31	1.0037	808.4	6.825	0.754	1.044
1.5	9.45	1.0056	808.2	6.708	0.754	1.046
2.0	12.57	1.0075	808.1	6.624	0.754	1.048
2.5	15.69	1.0094	808.0	6.560	0.755	1.050
3.0	18.79	1.0113	807.9	6.506	0.755	1.051
3.5	21.87	1.0133	807.7	6.461	0.756	1.053
4.0	24.95	1.0153	807.6	6.422	0.756	1.055
4.5	28.01	1.0173	807.5	6.388	0.756	1.057
5.0	31.07	1.0193	807.4	6.357	0.757	1.059
6.0	37.13	1.0234	807.3	6.303	0.757	1.062
7.0	43.14	1.0275	807.1	6.257	0.758	1.066
8.0	49.11	1.0317	807.0	6.218	0.759	1.069
9.0	55.02	1.0360	806.8	6.183	0.759	1.072
10.0	60.88	1.0403	806.7	6.151	0.760	1.075
11.0	66.68	1.0447	806.6	6.122	0.761	1.078
12.0	72.44	1.0491	806.6	6.096	0.761	1.082
13.0	78.14	1.0536	806.5	6.072	0.762	1.084
14.0	83.79	1.0581	806.5	6.049	0.762	1.087
15.0	89.39	1.0627	806.5	6.028	0.763	1.090
16.0	94.93	1.0673	806.4	6.008	0.764	1.093
17.0	100.43	1.0720	806.5	5.990	0.764	1.096
18.0	105.87	1.0768	806.5	5.972	0.765	1.098
19.0	111.25	1.0816	806.5	5.956	0.765	1.101

p	ρ	Z	h	s	c_v	c_p
			$T=550$ K			
20.0	116.59	1.0864	806.6	5.940	0.766	1.103
21.0	121.87	1.0913	806.6	5.925	0.766	1.106
22.0	127.09	1.0962	806.7	5.910	0.767	1.108
23.0	132.27	1.1012	806.8	5.896	0.768	1.111
24.0	137.39	1.1063	806.9	5.883	0.768	1.113
25.0	142.46	1.1113	807.1	5.870	0.769	1.115
26.0	147.48	1.1165	807.2	5.858	0.769	1.117
27.0	152.45	1.1216	807.4	5.846	0.770	1.119
28.0	157.36	1.1268	807.6	5.835	0.770	1.121
29.0	162.23	1.1321	807.7	5.824	0.771	1.123
30.0	167.04	1.1374	807.9	5.813	0.771	1.125
35.0	190.36	1.1644	809.1	5.764	0.774	1.134
40.0	212.48	1.1922	810.6	5.722	0.776	1.141
45.0	233.48	1.2206	812.4	5.684	0.778	1.148
50.0	253.42	1.2495	814.4	5.650	0.780	1.154
55.0	272.37	1.2788	816.6	5.620	0.782	1.159
60.0	290.41	1.3084	818.9	5.592	0.784	1.164
65.0	307.59	1.3383	821.5	5.566	0.786	1.168
70.0	323.98	1.3683	824.2	5.542	0.788	1.171
75.0	339.61	1.3986	827.0	5.520	0.790	1.174
80.0	354.53	1.4290	830.0	5.499	0.792	1.177
85.0	368.80	1.4596	833.0	5.480	0.793	1.179
90.0	382.44	1.4903	836.2	5.461	0.795	1.181
95.0	395.50	1.5212	839.5	5.444	0.796	1.182
100.0	408.01	1.5522	842.9	5.427	0.798	1.184
			$T=600$ K			
0.1	0.58	1.0004	860.9	7.578	0.764	1.052
0.5	2.90	1.0019	860.9	7.116	0.764	1.053
1.0	5.78	1.0038	860.8	6.916	0.765	1.054
1.5	8.66	1.0057	860.8	6.799	0.765	1.056
2.0	11.52	1.0076	860.7	6.716	0.765	1.057
2.5	14.38	1.0096	860.7	6.651	0.766	1.059
3.0	17.22	1.0116	860.6	6.598	0.766	1.060
3.5	20.05	1.0135	860.6	6.553	0.766	1.062
4.0	22.87	1.0155	860.6	6.515	0.767	1.063
4.5	25.67	1.0175	860.6	6.480	0.767	1.065
5.0	28.47	1.0196	860.6	6.449	0.767	1.066
6.0	34.03	1.0236	860.5	6.396	0.768	1.069
7.0	39.54	1.0278	860.5	6.350	0.768	1.072
8.0	45.00	1.0319	860.5	6.311	0.769	1.075
9.0	50.42	1.0362	860.6	6.276	0.769	1.077
10.0	55.80	1.0404	860.6	6.245	0.770	1.080
11.0	61.12	1.0447	860.7	6.216	0.770	1.082
12.0	66.40	1.0491	860.7	6.190	0.771	1.085
13.0	71.64	1.0534	860.8	6.166	0.772	1.087
14.0	76.83	1.0579	860.9	6.144	0.772	1.090

p	ρ	Z	h	s	c_v	c_p
			$T=600$ K			
15.0	81.97	1.0623	861.0	6.123	0.773	1.092
16.0	87.07	1.0668	861.1	6.104	0.773	1.094
17.0	92.12	1.0713	861.2	6.085	0.774	1.097
18.0	97.12	1.0759	861.4	6.068	0.774	1.099
19.0	102.08	1.0805	861.5	6.051	0.775	1.101
20.0	106.99	1.0851	861.7	6.036	0.775	1.103
21.0	111.86	1.0898	861.9	6.021	0.775	1.105
22.0	116.68	1.0945	862.1	6.006	0.776	1.107
23.0	121.46	1.0993	862.3	5.993	0.776	1.109
24.0	126.19	1.1041	862.5	5.980	0.777	1.111
25.0	130.88	1.1089	862.8	5.967	0.777	1.113
26.0	135.52	1.1137	863.0	5.955	0.778	1.115
27.0	140.12	1.1186	863.2	5.943	0.778	1.116
28.0	144.68	1.1235	863.5	5.932	0.779	1.118
29.0	149.19	1.1285	863.8	5.921	0.779	1.120
30.0	153.65	1.1334	864.1	5.911	0.779	1.121
35.0	175.35	1.1587	865.7	5.863	0.782	1.129
40.0	196.02	1.1846	867.5	5.821	0.784	1.135
45.0	215.72	1.2110	869.6	5.784	0.785	1.141
50.0	234.50	1.2378	871.9	5.750	0.787	1.147
55.0	252.43	1.2648	874.3	5.720	0.789	1.151
60.0	269.55	1.2922	876.9	5.693	0.791	1.156
65.0	285.93	1.3197	879.6	5.667	0.792	1.159
70.0	301.60	1.3473	882.5	5.644	0.794	1.163
75.0	316.61	1.3751	885.5	5.622	0.795	1.166
80.0	331.00	1.4031	888.6	5.601	0.797	1.168
85.0	344.80	1.4311	891.7	5.582	0.798	1.171
90.0	358.05	1.4592	895.0	5.563	0.800	1.173
95.0	370.77	1.4874	898.4	5.546	0.801	1.174
100.0	382.99	1.5157	901.8	5.530	0.802	1.176
			$T=650$ K			
0.1	0.54	1.0004	913.8	7.663	0.776	1.063
0.5	2.67	1.0019	913.8	7.201	0.776	1.064
1.0	5.34	1.0038	913.8	7.001	0.776	1.066
1.5	7.99	1.0057	913.8	6.884	0.777	1.067
2.0	10.64	1.0076	913.9	6.801	0.777	1.068
2.5	13.27	1.0096	913.9	6.736	0.777	1.069
3.0	15.89	1.0115	913.9	6.684	0.777	1.071
3.5	18.51	1.0135	914.0	6.639	0.778	1.072
4.0	21.11	1.0155	914.0	6.600	0.778	1.073
4.5	23.70	1.0175	914.0	6.566	0.778	1.074
5.0	26.28	1.0195	914.1	6.535	0.778	1.075
6.0	31.41	1.0235	914.2	6.482	0.779	1.078
7.0	36.50	1.0276	914.3	6.437	0.779	1.080
8.0	41.55	1.0317	914.5	6.397	0.780	1.082
9.0	46.56	1.0358	914.6	6.363	0.780	1.084

p	ρ	Z	h	s	c_v	c_p
			$T = 650$ K			
10.0	51.53	1.0400	914.8	6.331	0.781	1.087
11.0	56.45	1.0441	914.9	6.303	0.781	1.089
12.0	61.34	1.0484	915.1	6.277	0.782	1.091
13.0	66.18	1.0526	915.3	6.253	0.782	1.093
14.0	70.98	1.0569	915.5	6.231	0.783	1.095
15.0	75.74	1.0612	915.7	6.211	0.783	1.097
16.0	80.47	1.0655	915.9	6.191	0.783	1.099
17.0	85.15	1.0699	916.2	6.173	0.784	1.101
18.0	89.79	1.0743	916.4	6.156	0.784	1.102
19.0	94.39	1.0787	916.7	6.140	0.785	1.104
20.0	98.95	1.0831	916.9	6.124	0.785	1.106
21.0	103.47	1.0876	917.2	6.109	0.785	1.108
22.0	107.95	1.0921	917.5	6.095	0.786	1.109
23.0	112.39	1.0966	917.8	6.082	0.786	1.111
24.0	116.79	1.1012	918.1	6.069	0.787	1.113
25.0	121.15	1.1058	918.4	6.056	0.787	1.114
26.0	125.48	1.1104	918.7	6.044	0.787	1.116
27.0	129.76	1.1150	919.1	6.033	0.788	1.117
28.0	134.01	1.1196	919.4	6.022	0.788	1.119
29.0	138.22	1.1243	919.8	6.011	0.789	1.120
30.0	142.39	1.1290	920.1	6.000	0.789	1.122
35.0	162.69	1.1528	922.1	5.953	0.791	1.128
40.0	182.10	1.1771	924.2	5.912	0.792	1.134
45.0	200.66	1.2017	926.6	5.875	0.794	1.139
50.0	218.41	1.2267	929.1	5.842	0.796	1.144
55.0	235.41	1.2519	931.8	5.812	0.797	1.148
60.0	251.70	1.2774	934.6	5.785	0.799	1.152
65.0	267.33	1.3029	937.5	5.760	0.800	1.155
70.0	282.33	1.3286	940.5	5.737	0.801	1.159
75.0	296.74	1.3544	943.6	5.715	0.803	1.162
80.0	310.59	1.3802	946.9	5.694	0.804	1.164
85.0	323.92	1.4061	950.2	5.675	0.805	1.166
90.0	336.75	1.4321	953.5	5.657	0.806	1.168
95.0	349.11	1.4582	957.0	5.640	0.808	1.170
100.0	361.02	1.4843	960.5	5.624	0.809	1.172
			$T = 700$ K			
0.1	0.50	1.0004	967.2	7.742	0.788	1.075
0.5	2.48	1.0019	967.3	7.280	0.788	1.076
1.0	4.96	1.0038	967.4	7.080	0.788	1.077
1.5	7.42	1.0057	967.5	6.964	0.789	1.078
2.0	9.88	1.0076	967.5	6.881	0.789	1.079
2.5	12.32	1.0095	967.6	6.816	0.789	1.080
3.0	14.76	1.0114	967.7	6.763	0.789	1.081
3.5	17.19	1.0133	967.8	6.719	0.789	1.082
4.0	19.60	1.0153	967.9	6.680	0.790	1.083
4.5	22.01	1.0172	968.0	6.646	0.790	1.084

p	ρ	Z	h	s	c_v	c_p
			$T=700$ K			
5.0	24.41	1.0192	968.1	6.615	0.790	1.085
6.0	29.18	1.0231	968.3	6.562	0.790	1.087
7.0	33.91	1.0271	968.5	6.517	0.791	1.089
8.0	38.61	1.0311	968.8	6.478	0.791	1.091
9.0	43.26	1.0351	969.0	6.443	0.792	1.093
10.0	47.88	1.0391	969.3	6.412	0.792	1.095
11.0	52.47	1.0432	969.6	6.384	0.792	1.097
12.0	57.01	1.0473	969.8	6.358	0.793	1.098
13.0	61.52	1.0514	970.1	6.335	0.793	1.100
14.0	66.00	1.0555	970.4	6.313	0.794	1.102
15.0	70.43	1.0597	970.7	6.292	0.794	1.103
16.0	74.83	1.0639	971.0	6.273	0.794	1.105
17.0	79.20	1.0681	971.3	6.255	0.795	1.107
18.0	83.53	1.0723	971.7	6.238	0.795	1.108
19.0	87.82	1.0765	972.0	6.222	0.795	1.110
20.0	92.08	1.0808	972.4	6.206	0.796	1.111
21.0	96.30	1.0850	972.7	6.191	0.796	1.113
22.0	100.49	1.0893	973.1	6.177	0.796	1.114
23.0	104.65	1.0936	973.4	6.164	0.797	1.116
24.0	108.76	1.0980	973.8	6.151	0.797	1.117
25.0	112.85	1.1023	974.2	6.139	0.797	1.118
26.0	116.90	1.1067	974.6	6.127	0.798	1.120
27.0	120.91	1.1111	975.0	6.116	0.798	1.121
28.0	124.90	1.1155	975.4	6.105	0.798	1.122
29.0	128.85	1.1199	975.9	6.094	0.799	1.123
30.0	132.76	1.1244	976.3	6.084	0.799	1.125
35.0	151.85	1.1469	978.5	6.037	0.801	1.130
40.0	170.15	1.1698	981.0	5.996	0.802	1.135
45.0	187.70	1.1929	983.6	5.959	0.803	1.140
50.0	204.53	1.2164	986.3	5.927	0.805	1.144
55.0	220.70	1.2400	989.2	5.897	0.806	1.148
60.0	236.23	1.2638	992.2	5.870	0.807	1.152
65.0	251.16	1.2877	995.2	5.845	0.809	1.155
70.0	265.53	1.3117	998.4	5.822	0.810	1.158
75.0	279.38	1.3358	1001.7	5.801	0.811	1.161
80.0	292.72	1.3599	1005.0	5.781	0.812	1.163
85.0	305.58	1.3841	1008.4	5.762	0.813	1.165
90.0	318.00	1.4083	1011.9	5.744	0.814	1.167
95.0	329.99	1.4325	1015.5	5.727	0.815	1.169
100.0	341.57	1.4568	1019.1	5.711	0.816	1.171
			$T=750$ K			
0.1	0.46	1.0004	1021.3	7.817	0.800	1.087
0.5	2.32	1.0018	1021.4	7.355	0.800	1.088
1.0	4.63	1.0037	1021.5	7.155	0.800	1.089
1.5	6.93	1.0055	1021.7	7.038	0.801	1.090
2.0	9.22	1.0074	1021.8	6.955	0.801	1.091

p	ρ	Z	h	s	c_v	c_p
			$T=750$ K			
2.5	11.50	1.0093	1021.9	6.891	0.801	1.092
3.0	13.78	1.0112	1022.1	6.838	0.801	1.092
3.5	16.04	1.0131	1022.2	6.794	0.801	1.093
4.0	18.30	1.0150	1022.3	6.755	0.801	1.094
4.5	20.55	1.0169	1022.5	6.721	0.802	1.095
5.0	22.79	1.0188	1022.6	6.690	0.802	1.096
6.0	27.25	1.0226	1022.9	6.637	0.802	1.097
7.0	31.67	1.0265	1023.3	6.592	0.802	1.099
8.0	36.06	1.0303	1023.6	6.553	0.803	1.101
9.0	40.41	1.0342	1023.9	6.519	0.803	1.102
10.0	44.74	1.0381	1024.2	6.488	0.804	1.104
11.0	49.02	1.0421	1024.6	6.460	0.804	1.105
12.0	53.28	1.0460	1025.0	6.434	0.804	1.107
13.0	57.50	1.0500	1025.3	6.411	0.805	1.108
14.0	61.69	1.0540	1025.7	6.389	0.805	1.110
15.0	65.85	1.0580	1026.1	6.369	0.805	1.111
16.0	69.97	1.0620	1026.5	6.349	0.805	1.113
17.0	74.06	1.0660	1026.8	6.331	0.806	1.114
18.0	78.12	1.0701	1027.3	6.314	0.806	1.115
19.0	82.15	1.0741	1027.7	6.298	0.806	1.117
20.0	86.15	1.0782	1028.1	6.283	0.807	1.118
21.0	90.11	1.0823	1028.5	6.268	0.807	1.119
22.0	94.05	1.0864	1028.9	6.255	0.807	1.120
23.0	97.95	1.0905	1029.4	6.241	0.808	1.122
24.0	101.82	1.0947	1029.8	6.228	0.808	1.123
25.0	105.66	1.0988	1030.3	6.216	0.808	1.124
26.0	109.47	1.1030	1030.7	6.204	0.808	1.125
27.0	113.25	1.1072	1031.2	6.193	0.809	1.126
28.0	117.00	1.1114	1031.7	6.182	0.809	1.127
29.0	120.73	1.1156	1032.1	6.172	0.809	1.128
30.0	124.42	1.1198	1032.6	6.161	0.810	1.129
35.0	142.44	1.1411	1035.1	6.115	0.811	1.134
40.0	159.76	1.1628	1037.8	6.074	0.812	1.139
45.0	176.41	1.1847	1040.6	6.038	0.813	1.143
50.0	192.42	1.2068	1043.6	6.006	0.815	1.147
55.0	207.83	1.2290	1046.6	5.977	0.816	1.150
60.0	222.67	1.2514	1049.8	5.950	0.817	1.153
65.0	236.96	1.2739	1053.0	5.925	0.818	1.156
70.0	250.75	1.2965	1056.3	5.902	0.819	1.159
75.0	264.06	1.3190	1059.7	5.881	0.820	1.162
80.0	276.92	1.3417	1063.2	5.861	0.821	1.164
85.0	289.34	1.3643	1066.7	5.842	0.822	1.166
90.0	301.35	1.3870	1070.3	5.824	0.823	1.168
95.0	312.97	1.4097	1073.9	5.807	0.824	1.170
100.0	324.22	1.4324	1077.6	5.791	0.825	1.172

p	ρ	Z	h	s	c_v	c_p
			$T=800$ K			
0.1	0.44	1.0004	1076.0	7.887	0.812	1.099
0.5	2.17	1.0018	1076.1	7.425	0.812	1.100
1.0	4.34	1.0036	1076.3	7.226	0.812	1.100
1.5	6.50	1.0054	1076.4	7.109	0.812	1.101
2.0	8.65	1.0072	1076.6	7.026	0.812	1.102
2.5	10.79	1.0091	1076.8	6.962	0.813	1.103
3.0	12.92	1.0109	1077.0	6.909	0.813	1.103
3.5	15.05	1.0127	1077.1	6.865	0.813	1.104
4.0	17.17	1.0146	1077.3	6.826	0.813	1.105
4.5	19.28	1.0164	1077.5	6.792	0.813	1.106
5.0	21.38	1.0183	1077.7	6.761	0.813	1.106
6.0	25.56	1.0220	1078.1	6.708	0.814	1.108
7.0	29.71	1.0257	1078.5	6.664	0.814	1.109
8.0	33.83	1.0295	1078.9	6.625	0.814	1.111
9.0	37.92	1.0333	1079.3	6.590	0.815	1.112
10.0	41.98	1.0370	1079.7	6.560	0.815	1.113
11.0	46.01	1.0408	1080.1	6.532	0.815	1.115
12.0	50.01	1.0446	1080.5	6.506	0.815	1.116
13.0	53.98	1.0485	1080.9	6.483	0.816	1.117
14.0	57.92	1.0523	1081.4	6.461	0.816	1.118
15.0	61.84	1.0562	1081.8	6.440	0.816	1.120
16.0	65.72	1.0600	1082.3	6.421	0.817	1.121
17.0	69.57	1.0639	1082.7	6.404	0.817	1.122
18.0	73.40	1.0678	1083.2	6.387	0.817	1.123
19.0	77.19	1.0717	1083.7	6.371	0.817	1.124
20.0	80.96	1.0756	1084.2	6.355	0.818	1.125
21.0	84.70	1.0795	1084.6	6.341	0.818	1.126
22.0	88.41	1.0834	1085.1	6.327	0.818	1.127
23.0	92.09	1.0874	1085.6	6.314	0.818	1.129
24.0	95.75	1.0914	1086.1	6.301	0.819	1.130
25.0	99.37	1.0953	1086.6	6.289	0.819	1.131
26.0	102.97	1.0993	1087.1	6.277	0.819	1.132
27.0	106.55	1.1033	1087.7	6.266	0.819	1.133
28.0	110.09	1.1073	1088.2	6.255	0.820	1.134
29.0	113.61	1.1113	1088.7	6.245	0.820	1.134
30.0	117.11	1.1153	1089.2	6.234	0.820	1.135
35.0	134.19	1.1356	1092.0	6.188	0.821	1.140
40.0	150.63	1.1562	1094.9	6.148	0.822	1.144
45.0	166.47	1.1769	1097.9	6.112	0.823	1.148
50.0	181.74	1.1978	1101.0	6.080	0.824	1.151
55.0	196.46	1.2189	1104.2	6.051	0.825	1.154
60.0	210.67	1.2400	1107.5	6.024	0.826	1.157
65.0	224.38	1.2613	1110.9	6.000	0.827	1.160
70.0	237.63	1.2825	1114.4	5.977	0.828	1.162
75.0	250.44	1.3039	1117.9	5.956	0.829	1.165

p	ρ	Z	h	s	c_v	c_p
			$T=800$ K			
80.0	262.84	1.3252	1121.5	5.936	0.830	1.167
85.0	274.84	1.3465	1125.1	5.917	0.831	1.169
90.0	286.46	1.3679	1128.8	5.900	0.832	1.171
95.0	297.73	1.3893	1132.5	5.883	0.833	1.173
100.0	308.65	1.4106	1136.3	5.867	0.833	1.174
			$T=850$ K			
0.1	0.41	1.0003	1131.2	7.954	0.823	1.110
0.5	2.05	1.0018	1131.4	7.492	0.823	1.111
1.0	4.08	1.0035	1131.6	7.293	0.823	1.112
1.5	6.11	1.0053	1131.8	7.176	0.824	1.112
2.0	8.14	1.0070	1132.0	7.093	0.824	1.113
2.5	10.15	1.0088	1132.2	7.029	0.824	1.114
3.0	12.16	1.0106	1132.4	6.976	0.824	1.114
3.5	14.17	1.0124	1132.6	6.932	0.824	1.115
4.0	16.16	1.0142	1132.8	6.893	0.824	1.116
4.5	18.15	1.0160	1133.1	6.859	0.824	1.116
5.0	20.13	1.0178	1133.3	6.829	0.825	1.117
6.0	24.07	1.0214	1133.7	6.776	0.825	1.118
7.0	27.99	1.0250	1134.2	6.731	0.825	1.119
8.0	31.87	1.0286	1134.6	6.692	0.825	1.120
9.0	35.73	1.0322	1135.1	6.658	0.826	1.122
10.0	39.56	1.0359	1135.6	6.627	0.826	1.123
11.0	43.36	1.0396	1136.0	6.599	0.826	1.124
12.0	47.14	1.0432	1136.5	6.574	0.826	1.125
13.0	50.88	1.0469	1137.0	6.551	0.827	1.126
14.0	54.60	1.0506	1137.5	6.529	0.827	1.127
15.0	58.30	1.0543	1138.0	6.509	0.827	1.128
16.0	61.97	1.0580	1138.5	6.490	0.827	1.129
17.0	65.61	1.0618	1139.0	6.472	0.828	1.130
18.0	69.23	1.0655	1139.6	6.455	0.828	1.131
19.0	72.82	1.0692	1140.1	6.439	0.828	1.132
20.0	76.38	1.0730	1140.6	6.424	0.828	1.133
21.0	79.92	1.0768	1141.1	6.409	0.828	1.134
22.0	83.43	1.0805	1141.7	6.396	0.829	1.135
23.0	86.92	1.0843	1142.2	6.382	0.829	1.136
24.0	90.38	1.0881	1142.8	6.370	0.829	1.137
25.0	93.82	1.0919	1143.3	6.358	0.829	1.138
26.0	97.23	1.0957	1143.9	6.346	0.830	1.139
27.0	100.62	1.0995	1144.5	6.335	0.830	1.140
28.0	103.99	1.1034	1145.0	6.324	0.830	1.140
29.0	107.33	1.1072	1145.6	6.314	0.830	1.141
30.0	110.65	1.1111	1146.2	6.303	0.830	1.142
35.0	126.88	1.1304	1149.1	6.257	0.831	1.146
40.0	142.54	1.1499	1152.2	6.217	0.832	1.150
45.0	157.65	1.1697	1155.4	6.182	0.833	1.153
50.0	172.25	1.1895	1158.7	6.150	0.834	1.156

Table II.11 (*Continued*)

p	ρ	Z	h	s	c_v	c_p
			$T=850$ K			
55.0	186.34	1.2095	1162.0	6.121	0.835	1.159
60.0	199.97	1.2295	1165.5	6.095	0.836	1.161
65.0	213.14	1.2497	1169.0	6.070	0.837	1.164
70.0	225.89	1.2698	1172.6	6.048	0.838	1.166
75.0	238.24	1.2900	1176.2	6.027	0.839	1.168
80.0	250.21	1.3102	1179.9	6.007	0.839	1.171
85.0	261.81	1.3304	1183.6	5.988	0.840	1.172
90.0	273.06	1.3506	1187.4	5.971	0.841	1.174
95.0	283.98	1.3708	1191.2	5.954	0.841	1.176
100.0	294.59	1.3910	1195.1	5.938	0.842	1.178
			$T=900$ K			
0.1	0.39	1.0003	1187.0	8.018	0.834	1.121
0.5	1.93	1.0017	1187.2	7.556	0.834	1.122
1.0	3.86	1.0034	1187.4	7.357	0.834	1.122
1.5	5.78	1.0051	1187.7	7.240	0.835	1.123
2.0	7.69	1.0068	1187.9	7.157	0.835	1.124
2.5	9.59	1.0086	1188.1	7.093	0.835	1.124
3.0	11.49	1.0103	1188.4	7.040	0.835	1.125
3.5	13.38	1.0120	1188.6	6.996	0.835	1.125
4.0	15.27	1.0137	1188.9	6.957	0.835	1.126
4.5	17.15	1.0155	1189.1	6.923	0.835	1.126
5.0	19.02	1.0172	1189.4	6.893	0.835	1.127
6.0	22.75	1.0207	1189.9	6.840	0.836	1.128
7.0	26.45	1.0242	1190.4	6.795	0.836	1.129
8.0	30.13	1.0277	1190.9	6.757	0.836	1.130
9.0	33.78	1.0312	1191.4	6.722	0.836	1.131
10.0	37.40	1.0348	1191.9	6.692	0.836	1.132
11.0	41.00	1.0383	1192.5	6.664	0.837	1.133
12.0	44.58	1.0418	1193.0	6.639	0.837	1.134
13.0	48.13	1.0454	1193.5	6.615	0.837	1.135
14.0	51.65	1.0490	1194.1	6.594	0.837	1.136
15.0	55.16	1.0525	1194.6	6.573	0.838	1.137
16.0	58.63	1.0561	1195.2	6.554	0.838	1.138
17.0	62.09	1.0597	1195.8	6.537	0.838	1.139
18.0	65.52	1.0633	1196.3	6.520	0.838	1.139
19.0	68.92	1.0669	1196.9	6.504	0.838	1.140
20.0	72.31	1.0705	1197.5	6.489	0.839	1.141
21.0	75.67	1.0741	1198.1	6.474	0.839	1.142
22.0	79.00	1.0777	1198.6	6.461	0.839	1.143
23.0	82.32	1.0813	1199.2	6.448	0.839	1.144
24.0	85.61	1.0850	1199.8	6.435	0.839	1.145
25.0	88.88	1.0886	1200.4	6.423	0.840	1.145
26.0	92.12	1.0923	1201.0	6.411	0.840	1.146
27.0	95.35	1.0959	1201.6	6.400	0.840	1.147
28.0	98.55	1.0996	1202.2	6.389	0.840	1.148
29.0	101.73	1.1033	1202.8	6.379	0.840	1.148

p	ρ	Z	h	s	c_v	c_p
			$T=900$ K			
30.0	104.89	1.1070	1203.5	6.369	0.840	1.149
35.0	120.36	1.1254	1206.6	6.323	0.841	1.153
40.0	135.31	1.1441	1209.9	6.283	0.842	1.156
45.0	149.76	1.1629	1213.2	6.248	0.843	1.159
50.0	163.74	1.1818	1216.6	6.216	0.844	1.162
55.0	177.26	1.2008	1220.1	6.187	0.845	1.164
60.0	190.35	1.2199	1223.7	6.161	0.845	1.167
65.0	203.03	1.2390	1227.3	6.137	0.846	1.169
70.0	215.32	1.2582	1231.0	6.115	0.847	1.171
75.0	227.24	1.2773	1234.7	6.094	0.848	1.173
80.0	238.80	1.2965	1238.5	6.074	0.848	1.175
85.0	250.02	1.3157	1242.4	6.055	0.849	1.177
90.0	260.93	1.3349	1246.2	6.038	0.850	1.178
95.0	271.53	1.3541	1250.1	6.021	0.850	1.180
100.0	281.83	1.3732	1254.0	6.006	0.851	1.182
			$T=950$ K			
0.1	0.37	1.0003	1243.3	8.079	0.845	1.132
0.5	1.83	1.0017	1243.5	7.617	0.845	1.132
1.0	3.65	1.0033	1243.8	7.418	0.845	1.133
1.5	5.47	1.0050	1244.1	7.301	0.845	1.133
2.0	7.28	1.0066	1244.3	7.218	0.845	1.134
2.5	9.09	1.0083	1244.6	7.154	0.845	1.134
3.0	10.89	1.0100	1244.9	7.101	0.845	1.135
3.5	12.68	1.0116	1245.1	7.057	0.845	1.135
4.0	14.47	1.0133	1245.4	7.018	0.845	1.136
4.5	16.26	1.0150	1245.7	6.984	0.845	1.136
5.0	18.03	1.0167	1245.9	6.954	0.846	1.137
6.0	21.57	1.0201	1246.5	6.901	0.846	1.137
7.0	25.08	1.0234	1247.1	6.857	0.846	1.138
8.0	28.57	1.0268	1247.6	6.818	0.846	1.139
9.0	32.03	1.0302	1248.2	6.784	0.846	1.140
10.0	35.47	1.0336	1248.8	6.753	0.847	1.141
11.0	38.89	1.0370	1249.3	6.725	0.847	1.142
12.0	42.29	1.0405	1249.9	6.700	0.847	1.143
13.0	45.66	1.0439	1250.5	6.677	0.847	1.144
14.0	49.01	1.0473	1251.1	6.655	0.847	1.144
15.0	52.34	1.0508	1251.7	6.635	0.848	1.145
16.0	55.65	1.0542	1252.3	6.616	0.848	1.146
17.0	58.93	1.0577	1252.9	6.598	0.848	1.147
18.0	62.19	1.0611	1253.5	6.582	0.848	1.148
19.0	65.44	1.0646	1254.1	6.566	0.848	1.148
20.0	68.66	1.0680	1254.7	6.551	0.848	1.149
21.0	71.86	1.0715	1255.4	6.536	0.849	1.150
22.0	75.03	1.0750	1256.0	6.523	0.849	1.151
23.0	78.19	1.0785	1256.6	6.510	0.849	1.151
24.0	81.33	1.0820	1257.2	6.497	0.849	1.152

p	ρ	Z	h	s	c_v	c_p
			$T = 950$ K			
25.0	84.44	1.0855	1257.9	6.485	0.849	1.153
26.0	87.54	1.0890	1258.5	6.474	0.849	1.154
27.0	90.61	1.0925	1259.1	6.462	0.850	1.154
28.0	93.67	1.0960	1259.8	6.452	0.850	1.155
29.0	96.70	1.0995	1260.4	6.441	0.850	1.156
30.0	99.72	1.1031	1261.1	6.431	0.850	1.156
35.0	114.50	1.1208	1264.4	6.386	0.851	1.159
40.0	128.80	1.1386	1267.8	6.346	0.852	1.162
45.0	142.66	1.1566	1271.3	6.311	0.852	1.165
50.0	156.07	1.1746	1274.8	6.279	0.853	1.167
55.0	169.07	1.1927	1278.5	6.251	0.854	1.170
60.0	181.67	1.2109	1282.1	6.224	0.855	1.172
65.0	193.89	1.2292	1285.9	6.200	0.855	1.174
70.0	205.74	1.2474	1289.7	6.178	0.856	1.176
75.0	217.26	1.2657	1293.5	6.157	0.856	1.178
80.0	228.44	1.2840	1297.4	6.138	0.857	1.180
85.0	239.31	1.3023	1301.3	6.119	0.858	1.181
90.0	249.88	1.3205	1305.2	6.102	0.858	1.183
95.0	260.17	1.3388	1309.2	6.085	0.859	1.185
100.0	270.19	1.3570	1313.2	6.070	0.859	1.186
			$T = 1000$ K			
0.1	0.35	1.0003	1300.2	8.137	0.854	1.141
0.5	1.74	1.0016	1300.4	7.675	0.854	1.142
1.0	3.47	1.0032	1300.7	7.476	0.854	1.142
1.5	5.20	1.0048	1301.0	7.359	0.855	1.143
2.0	6.92	1.0064	1301.2	7.277	0.855	1.143
2.5	8.64	1.0080	1301.5	7.212	0.855	1.144
3.0	10.35	1.0097	1301.8	7.160	0.855	1.144
3.5	12.05	1.0113	1302.1	7.115	0.855	1.144
4.0	13.75	1.0129	1302.4	7.077	0.855	1.145
4.5	15.45	1.0145	1302.7	7.043	0.855	1.145
5.0	17.14	1.0162	1303.0	7.013	0.855	1.146
6.0	20.50	1.0194	1303.6	6.960	0.855	1.147
7.0	23.84	1.0227	1304.2	6.915	0.856	1.147
8.0	27.16	1.0260	1304.8	6.877	0.856	1.148
9.0	30.46	1.0292	1305.4	6.843	0.856	1.149
10.0	33.73	1.0325	1306.0	6.812	0.856	1.150
11.0	36.99	1.0358	1306.7	6.784	0.856	1.150
12.0	40.22	1.0391	1307.3	6.759	0.856	1.151
13.0	43.44	1.0424	1307.9	6.736	0.857	1.152
14.0	46.63	1.0457	1308.5	6.714	0.857	1.153
15.0	49.80	1.0491	1309.2	6.694	0.857	1.153
16.0	52.96	1.0524	1309.8	6.675	0.857	1.154
17.0	56.09	1.0557	1310.4	6.657	0.857	1.155
18.0	59.20	1.0590	1311.1	6.641	0.857	1.156
19.0	62.29	1.0624	1311.7	6.625	0.858	1.156

p	ρ	Z	h	s	c_v	c_p
			$T=1000$ K			
20.0	65.37	1.0657	1312.4	6.610	0.858	1.157
21.0	68.42	1.0691	1313.0	6.596	0.858	1.158
22.0	71.45	1.0724	1313.7	6.582	0.858	1.158
23.0	74.47	1.0758	1314.4	6.569	0.858	1.159
24.0	77.46	1.0791	1315.0	6.556	0.858	1.160
25.0	80.44	1.0825	1315.7	6.544	0.859	1.160
26.0	83.40	1.0859	1316.4	6.533	0.859	1.161
27.0	86.34	1.0892	1317.0	6.522	0.859	1.161
28.0	89.26	1.0926	1317.7	6.511	0.859	1.162
29.0	92.16	1.0960	1318.4	6.501	0.859	1.163
30.0	95.05	1.0994	1319.1	6.491	0.859	1.163
35.0	109.20	1.1164	1322.5	6.445	0.860	1.166
40.0	122.92	1.1335	1326.1	6.406	0.861	1.169
45.0	136.22	1.1507	1329.7	6.371	0.861	1.171
50.0	149.12	1.1679	1333.4	6.339	0.862	1.173
55.0	161.63	1.1853	1337.1	6.311	0.863	1.176
60.0	173.77	1.2026	1340.9	6.285	0.863	1.178
65.0	185.57	1.2201	1344.7	6.261	0.864	1.179
70.0	197.02	1.2375	1348.6	6.238	0.864	1.181
75.0	208.16	1.2550	1352.5	6.218	0.865	1.183
80.0	218.99	1.2724	1356.5	6.198	0.866	1.185
85.0	229.53	1.2899	1360.5	6.180	0.866	1.186
90.0	239.79	1.3073	1364.5	6.163	0.867	1.188
95.0	249.78	1.3248	1368.6	6.146	0.867	1.189
100.0	259.52	1.3422	1372.6	6.131	0.868	1.191
			$T=1050$ K			
0.1	0.33	1.0003	1357.5	8.193	0.864	1.151
0.5	1.66	1.0016	1357.7	7.731	0.864	1.151
1.0	3.31	1.0031	1358.0	7.532	0.864	1.151
1.5	4.95	1.0047	1358.3	7.415	0.864	1.152
2.0	6.59	1.0062	1358.6	7.333	0.864	1.152
2.5	8.23	1.0078	1358.9	7.268	0.864	1.153
3.0	9.86	1.0094	1359.3	7.216	0.864	1.153
3.5	11.48	1.0109	1359.6	7.172	0.864	1.153
4.0	13.11	1.0125	1359.9	7.133	0.864	1.154
4.5	14.72	1.0141	1360.2	7.099	0.864	1.154
5.0	16.33	1.0156	1360.5	7.069	0.864	1.154
6.0	19.54	1.0188	1361.1	7.016	0.864	1.155
7.0	22.72	1.0220	1361.8	6.972	0.865	1.156
8.0	25.89	1.0251	1362.4	6.933	0.865	1.157
9.0	29.03	1.0283	1363.1	6.899	0.865	1.157
10.0	32.16	1.0315	1363.7	6.868	0.865	1.158
11.0	35.27	1.0347	1364.4	6.841	0.865	1.159
12.0	38.36	1.0379	1365.0	6.815	0.865	1.159
13.0	41.42	1.0410	1365.7	6.792	0.866	1.160
14.0	44.47	1.0442	1366.4	6.771	0.866	1.161

Table II.11 (*Continued*)

p	ρ	Z	h	s	c_v	c_p
			$T = 1050$ K			
15.0	47.50	1.0474	1367.0	6.750	0.866	1.161
16.0	50.52	1.0506	1367.7	6.732	0.866	1.162
17.0	53.51	1.0539	1368.4	6.714	0.866	1.163
18.0	56.49	1.0571	1369.1	6.697	0.866	1.163
19.0	59.44	1.0603	1369.7	6.682	0.866	1.164
20.0	62.38	1.0635	1370.4	6.667	0.867	1.164
21.0	65.30	1.0667	1371.1	6.652	0.867	1.165
22.0	68.21	1.0700	1371.8	6.639	0.867	1.166
23.0	71.09	1.0732	1372.5	6.626	0.867	1.166
24.0	73.96	1.0764	1373.2	6.613	0.867	1.167
25.0	76.81	1.0797	1373.9	6.601	0.867	1.167
26.0	79.65	1.0829	1374.6	6.590	0.867	1.168
27.0	82.46	1.0861	1375.3	6.579	0.868	1.168
28.0	85.26	1.0894	1376.0	6.568	0.868	1.169
29.0	88.04	1.0926	1376.7	6.558	0.868	1.170
30.0	90.81	1.0959	1377.4	6.548	0.868	1.170
35.0	104.39	1.1122	1381.0	6.502	0.869	1.173
40.0	117.57	1.1286	1384.7	6.463	0.869	1.175
45.0	130.36	1.1451	1388.4	6.428	0.870	1.177
50.0	142.78	1.1617	1392.2	6.397	0.870	1.179
55.0	154.84	1.1783	1396.0	6.368	0.871	1.181
60.0	166.56	1.1950	1399.9	6.342	0.871	1.183
65.0	177.96	1.2116	1403.8	6.318	0.872	1.185
70.0	189.04	1.2283	1407.8	6.296	0.872	1.187
75.0	199.82	1.2451	1411.8	6.276	0.873	1.188
80.0	210.32	1.2618	1415.9	6.256	0.874	1.190
85.0	220.55	1.2785	1419.9	6.238	0.874	1.191
90.0	230.51	1.2952	1424.0	6.221	0.874	1.193
95.0	240.22	1.3119	1428.1	6.204	0.875	1.194
100.0	249.70	1.3285	1432.3	6.189	0.875	1.195
			$T = 1100$ K			
0.1	0.32	1.0003	1415.2	8.247	0.872	1.159
0.5	1.58	1.0015	1415.5	7.785	0.872	1.160
1.0	3.16	1.0030	1415.8	7.586	0.872	1.160
1.5	4.73	1.0045	1416.1	7.469	0.872	1.160
2.0	6.29	1.0060	1416.5	7.386	0.872	1.161
2.5	7.86	1.0076	1416.8	7.322	0.872	1.161
3.0	9.41	1.0091	1417.1	7.270	0.873	1.161
3.5	10.97	1.0106	1417.4	7.225	0.873	1.162
4.0	12.51	1.0121	1417.8	7.187	0.873	1.162
4.5	14.06	1.0136	1418.1	7.153	0.873	1.162
5.0	15.60	1.0152	1418.4	7.123	0.873	1.163
6.0	18.66	1.0182	1419.1	7.070	0.873	1.163
7.0	21.70	1.0213	1419.8	7.025	0.873	1.164
8.0	24.73	1.0243	1420.5	6.987	0.873	1.164
9.0	27.74	1.0274	1421.1	6.953	0.873	1.165

Table II.11 (*Continued*)

p	ρ	Z	h	s	c_v	c_p
			$T = 1100$ K			
10.0	30.73	1.0305	1421.8	6.922	0.874	1.166
11.0	33.70	1.0335	1422.5	6.895	0.874	1.166
12.0	36.65	1.0366	1423.2	6.869	0.874	1.167
13.0	39.59	1.0397	1423.9	6.846	0.874	1.168
14.0	42.51	1.0428	1424.6	6.825	0.874	1.168
15.0	45.41	1.0459	1425.3	6.805	0.874	1.169
16.0	48.30	1.0490	1426.0	6.786	0.874	1.169
17.0	51.16	1.0521	1426.7	6.768	0.874	1.170
18.0	54.02	1.0552	1427.4	6.752	0.875	1.170
19.0	56.85	1.0583	1428.1	6.736	0.875	1.171
20.0	59.67	1.0614	1428.8	6.721	0.875	1.172
21.0	62.47	1.0645	1429.5	6.707	0.875	1.172
22.0	65.25	1.0676	1430.3	6.693	0.875	1.173
23.0	68.02	1.0707	1431.0	6.680	0.875	1.173
24.0	70.77	1.0739	1431.7	6.668	0.875	1.174
25.0	73.50	1.0770	1432.4	6.656	0.875	1.174
26.0	76.22	1.0801	1433.2	6.644	0.876	1.175
27.0	78.93	1.0832	1433.9	6.633	0.876	1.175
28.0	81.61	1.0864	1434.6	6.622	0.876	1.176
29.0	84.29	1.0895	1435.3	6.612	0.876	1.176
30.0	86.94	1.0926	1436.1	6.602	0.876	1.177
35.0	99.99	1.1083	1439.8	6.557	0.877	1.179
40.0	112.67	1.1241	1443.6	6.518	0.877	1.181
45.0	125.00	1.1400	1447.4	6.483	0.878	1.183
50.0	136.97	1.1559	1451.3	6.452	0.878	1.185
55.0	148.62	1.1718	1455.2	6.423	0.879	1.187
60.0	159.95	1.1878	1459.2	6.397	0.879	1.189
65.0	170.97	1.2038	1463.2	6.374	0.880	1.190
70.0	181.71	1.2198	1467.3	6.352	0.880	1.192
75.0	192.16	1.2359	1471.4	6.331	0.881	1.193
80.0	202.35	1.2519	1475.5	6.312	0.881	1.195
85.0	212.27	1.2679	1479.6	6.293	0.882	1.196
90.0	221.96	1.2839	1483.8	6.276	0.882	1.197
95.0	231.41	1.2999	1488.0	6.260	0.882	1.199
100.0	240.63	1.3159	1492.2	6.245	0.883	1.200
			$T = 1150$ K			
0.1	0.30	1.0003	1473.4	8.299	0.880	1.167
0.5	1.51	1.0015	1473.7	7.837	0.880	1.168
1.0	3.02	1.0029	1474.0	7.637	0.880	1.168
1.5	4.52	1.0044	1474.3	7.521	0.880	1.168
2.0	6.02	1.0059	1474.7	7.438	0.880	1.168
2.5	7.52	1.0073	1475.0	7.374	0.880	1.169
3.0	9.01	1.0088	1475.4	7.322	0.880	1.169
3.5	10.49	1.0103	1475.7	7.277	0.881	1.169
4.0	11.97	1.0117	1476.1	7.239	0.881	1.170
4.5	13.45	1.0132	1476.4	7.205	0.881	1.170

p	ρ	Z	h	s	c_v	c_p
			$T=1150$ K			
5.0	14.92	1.0147	1476.8	7.174	0.881	1.170
6.0	17.86	1.0176	1477.5	7.122	0.881	1.171
7.0	20.77	1.0206	1478.2	7.077	0.881	1.171
8.0	23.67	1.0236	1478.9	7.039	0.881	1.172
9.0	26.55	1.0265	1479.6	7.005	0.881	1.173
10.0	29.42	1.0295	1480.3	6.974	0.881	1.173
11.0	32.27	1.0325	1481.0	6.947	0.881	1.174
12.0	35.10	1.0355	1481.7	6.921	0.882	1.174
13.0	37.92	1.0384	1482.5	6.898	0.882	1.175
14.0	40.72	1.0414	1483.2	6.877	0.882	1.175
15.0	43.50	1.0444	1483.9	6.857	0.882	1.176
16.0	46.27	1.0474	1484.6	6.838	0.882	1.176
17.0	49.02	1.0504	1485.4	6.820	0.882	1.177
18.0	51.75	1.0534	1486.1	6.804	0.882	1.177
19.0	54.47	1.0564	1486.8	6.788	0.882	1.178
20.0	57.18	1.0594	1487.6	6.773	0.883	1.178
21.0	59.87	1.0624	1488.3	6.759	0.883	1.179
22.0	62.54	1.0654	1489.1	6.745	0.883	1.179
23.0	65.20	1.0684	1489.8	6.732	0.883	1.180
24.0	67.85	1.0714	1490.5	6.720	0.883	1.180
25.0	70.47	1.0744	1491.3	6.708	0.883	1.181
26.0	73.09	1.0774	1492.0	6.697	0.883	1.181
27.0	75.69	1.0805	1492.8	6.686	0.883	1.182
28.0	78.27	1.0835	1493.6	6.675	0.883	1.182
29.0	80.84	1.0865	1494.3	6.665	0.884	1.182
30.0	83.40	1.0895	1495.1	6.655	0.884	1.183
35.0	95.96	1.1047	1498.9	6.609	0.884	1.185
40.0	108.18	1.1199	1502.8	6.570	0.885	1.187
45.0	120.07	1.1351	1506.7	6.535	0.885	1.189
50.0	131.64	1.1504	1510.7	6.504	0.886	1.191
55.0	142.90	1.1658	1514.7	6.476	0.886	1.192
60.0	153.86	1.1811	1518.8	6.450	0.887	1.194
65.0	164.54	1.1965	1522.9	6.427	0.887	1.195
70.0	174.94	1.2119	1527.0	6.405	0.887	1.197
75.0	185.08	1.2273	1531.1	6.384	0.888	1.198
80.0	194.98	1.2427	1535.3	6.365	0.888	1.200
85.0	204.63	1.2581	1539.5	6.347	0.889	1.201
90.0	214.05	1.2735	1543.8	6.330	0.889	1.202
95.0	223.24	1.2889	1548.0	6.313	0.889	1.203
100.0	232.23	1.3042	1552.3	6.298	0.890	1.204
			$T=1200$ K			
0.1	0.29	1.0003	1531.9	8.349	0.888	1.175
0.5	1.45	1.0014	1532.2	7.886	0.888	1.175
1.0	2.89	1.0028	1532.6	7.687	0.888	1.175
1.5	4.34	1.0043	1532.9	7.571	0.888	1.176
2.0	5.77	1.0057	1533.3	7.488	0.888	1.176

p	ρ	Z	h	s	c_v	c_p
			$T = 1200$ K			
2.5	7.21	1.0071	1533.6	7.424	0.888	1.176
3.0	8.63	1.0085	1534.0	7.371	0.888	1.176
3.5	10.06	1.0099	1534.4	7.327	0.888	1.177
4.0	11.48	1.0114	1534.7	7.289	0.888	1.177
4.5	12.90	1.0128	1535.1	7.255	0.888	1.177
5.0	14.31	1.0142	1535.5	7.224	0.888	1.177
6.0	17.12	1.0171	1536.2	7.172	0.888	1.178
7.0	19.92	1.0200	1536.9	7.127	0.888	1.178
8.0	22.70	1.0228	1537.6	7.089	0.889	1.179
9.0	25.47	1.0257	1538.4	7.055	0.889	1.179
10.0	28.22	1.0286	1539.1	7.024	0.889	1.180
11.0	30.95	1.0315	1539.9	6.997	0.889	1.180
12.0	33.67	1.0344	1540.6	6.972	0.889	1.181
13.0	36.38	1.0372	1541.4	6.948	0.889	1.181
14.0	39.07	1.0401	1542.1	6.927	0.889	1.182
15.0	41.74	1.0430	1542.9	6.907	0.889	1.182
16.0	44.40	1.0459	1543.6	6.888	0.889	1.183
17.0	47.05	1.0488	1544.4	6.871	0.890	1.183
18.0	49.68	1.0517	1545.1	6.854	0.890	1.184
19.0	52.29	1.0546	1545.9	6.838	0.890	1.184
20.0	54.90	1.0575	1546.6	6.823	0.890	1.185
21.0	57.48	1.0604	1547.4	6.809	0.890	1.185
22.0	60.06	1.0633	1548.2	6.796	0.890	1.186
23.0	62.61	1.0662	1548.9	6.783	0.890	1.186
24.0	65.16	1.0691	1549.7	6.770	0.890	1.186
25.0	67.69	1.0720	1550.5	6.758	0.890	1.187
26.0	70.21	1.0749	1551.3	6.747	0.890	1.187
27.0	72.71	1.0779	1552.0	6.736	0.891	1.188
28.0	75.20	1.0808	1552.8	6.725	0.891	1.188
29.0	77.67	1.0837	1553.6	6.715	0.891	1.189
30.0	80.14	1.0866	1554.4	6.705	0.891	1.189
35.0	92.25	1.1012	1558.3	6.660	0.891	1.191
40.0	104.05	1.1159	1562.3	6.621	0.892	1.193
45.0	115.53	1.1306	1566.3	6.586	0.892	1.194
50.0	126.71	1.1453	1570.4	6.555	0.893	1.196
55.0	137.61	1.1601	1574.5	6.527	0.893	1.198
60.0	148.23	1.1749	1578.6	6.501	0.893	1.199
65.0	158.58	1.1897	1582.8	6.478	0.894	1.200
70.0	168.68	1.2045	1587.0	6.456	0.894	1.202
75.0	178.53	1.2194	1591.2	6.435	0.895	1.203
80.0	188.14	1.2342	1595.4	6.416	0.895	1.204
85.0	197.53	1.2490	1599.7	6.398	0.895	1.205
90.0	206.70	1.2638	1604.0	6.381	0.896	1.207
95.0	215.66	1.2786	1608.3	6.365	0.896	1.208
100.0	224.42	1.2934	1612.6	6.349	0.896	1.209

Table II.11 (*Continued*)

p	ρ	Z	h	s	c_v	c_p
			$T=1250$ K			
0.1	0.28	1.0003	1590.9	8.397	0.895	1.182
0.5	1.39	1.0014	1591.1	7.935	0.895	1.182
1.0	2.78	1.0027	1591.5	7.735	0.895	1.182
1.5	4.16	1.0041	1591.9	7.619	0.895	1.183
2.0	5.54	1.0055	1592.3	7.536	0.895	1.183
2.5	6.92	1.0069	1592.6	7.472	0.895	1.183
3.0	8.29	1.0083	1593.0	7.420	0.895	1.183
3.5	9.66	1.0097	1593.4	7.375	0.895	1.183
4.0	11.02	1.0110	1593.7	7.337	0.895	1.184
4.5	12.39	1.0124	1594.1	7.303	0.895	1.184
5.0	13.74	1.0138	1594.5	7.273	0.895	1.184
6.0	16.45	1.0166	1595.2	7.220	0.895	1.185
7.0	19.13	1.0194	1596.0	7.176	0.895	1.185
8.0	21.81	1.0221	1596.8	7.137	0.895	1.186
9.0	24.47	1.0249	1597.5	7.103	0.896	1.186
10.0	27.11	1.0277	1598.3	7.073	0.896	1.187
11.0	29.74	1.0305	1599.1	7.045	0.896	1.187
12.0	32.36	1.0333	1599.8	7.020	0.896	1.187
13.0	34.96	1.0361	1600.6	6.997	0.896	1.188
14.0	37.55	1.0389	1601.4	6.975	0.896	1.188
15.0	40.12	1.0417	1602.1	6.955	0.896	1.189
16.0	42.68	1.0445	1602.9	6.937	0.896	1.189
17.0	45.23	1.0473	1603.7	6.919	0.896	1.190
18.0	49.76	1.0501	1604.5	6.902	0.896	1.190
19.0	50.28	1.0529	1605.3	6.887	0.897	1.190
20.0	52.79	1.0557	1606.0	6.872	0.897	1.191
21.0	55.28	1.0585	1606.8	6.858	0.897	1.191
22.0	57.76	1.0613	1607.6	6.844	0.897	1.192
23.0	60.23	1.0641	1608.4	6.831	0.897	1.192
24.0	62.68	1.0669	1609.2	6.819	0.897	1.192
25.0	65.12	1.0698	1610.0	6.807	0.897	1.193
26.0	67.55	1.0726	1610.8	6.796	0.897	1.193
27.0	69.96	1.0754	1611.6	6.785	0.897	1.194
28.0	72.36	1.0782	1612.4	6.774	0.897	1.194
29.0	74.75	1.0810	1613.2	6.764	0.897	1.194
30.0	77.13	1.0838	1614.0	6.754	0.898	1.195
35.0	88.82	1.0980	1618.0	6.709	0.898	1.196
40.0	100.22	1.1121	1622.0	6.670	0.898	1.198
45.0	111.33	1.1263	1626.1	6.635	0.899	1.200
50.0	122.16	1.1406	1630.3	6.604	0.899	1.201
55.0	132.71	1.1548	1634.5	6.576	0.900	1.203
60.0	143.01	1.1691	1638.7	6.550	0.900	1.204
65.0	153.06	1.1834	1642.9	6.527	0.900	1.205
70.0	162.86	1.1976	1647.2	6.505	0.901	1.207
75.0	172.44	1.2119	1651.4	6.484	0.901	1.208

p	ρ	Z	h	s	c_v	c_p
			$T=1250$ K			
80.0	181.79	1.2262	1655.8	6.465	0.901	1.209
85.0	190.93	1.2405	1660.1	6.447	0.902	1.210
90.0	199.86	1.2548	1664.4	6.430	0.902	1.211
95.0	208.59	1.2690	1668.8	6.414	0.902	1.212
100.0	217.14	1.2833	1673.1	6.399	0.903	1.213
			$T=1300$ K			
0.1	0.27	1.0003	1650.1	8.443	0.901	1.188
0.5	1.34	1.0013	1650.4	7.981	0.901	1.189
1.0	2.67	1.0027	1650.8	7.782	0.901	1.189
1.5	4.00	1.0040	1651.2	7.665	0.901	1.189
2.0	5.33	1.0053	1651.6	7.583	0.901	1.189
2.5	6.65	1.0067	1651.9	7.519	0.902	1.189
3.0	7.97	1.0080	1652.3	7.466	0.902	1.190
3.5	9.29	1.0094	1652.7	7.422	0.902	1.190
4.0	10.60	1.0107	1653.1	7.383	0.902	1.190
4.5	11.91	1.0121	1653.5	7.350	0.902	1.190
5.0	13.22	1.0134	1653.9	7.319	0.902	1.191
6.0	15.82	1.0161	1654.6	7.267	0.902	1.191
7.0	18.41	1.0188	1655.4	7.222	0.902	1.191
8.0	20.98	1.0215	1656.2	7.184	0.902	1.192
9.0	23.54	1.0242	1657.0	7.150	0.902	1.192
10.0	26.09	1.0269	1657.8	7.119	0.902	1.193
11.0	28.63	1.0296	1658.6	7.092	0.902	1.193
12.0	31.15	1.0323	1659.3	7.067	0.902	1.194
13.0	33.65	1.0350	1660.1	7.043	0.902	1.194
14.0	36.15	1.0377	1660.9	7.022	0.903	1.194
15.0	38.63	1.0404	1661.7	7.002	0.903	1.195
16.0	41.10	1.0431	1662.5	6.983	0.903	1.195
17.0	43.55	1.0459	1663.3	6.966	0.903	1.195
18.0	45.99	1.0486	1664.1	6.949	0.903	1.196
19.0	48.42	1.0513	1664.9	6.934	0.903	1.196
20.0	50.84	1.0540	1665.7	6.919	0.903	1.197
21.0	53.25	1.0567	1666.5	6.905	0.903	1.197
22.0	55.64	1.0594	1667.3	6.891	0.903	1.197
23.0	58.02	1.0622	1668.1	6.878	0.903	1.198
24.0	60.39	1.0649	1668.9	6.866	0.903	1.198
25.0	62.74	1.0676	1669.8	6.854	0.903	1.198
26.0	65.08	1.0703	1670.6	6.842	0.904	1.199
27.0	67.42	1.0731	1671.4	6.831	0.904	1.199
28.0	69.74	1.0758	1672.2	6.821	0.904	1.199
29.0	72.04	1.0785	1673.0	6.811	0.904	1.200
30.0	74.34	1.0812	1673.8	6.801	0.904	1.200
35.0	85.65	1.0949	1677.9	6.756	0.904	1.202
40.0	96.67	1.1086	1682.1	6.717	0.905	1.203
45.0	107.43	1.1223	1686.3	6.682	0.905	1.205
50.0	117.92	1.1361	1690.5	6.651	0.905	1.206

p	ρ	Z	h	s	c_v	c_p
			$T=1300$ K			
55.0	128.16	1.1498	1694.7	6.623	0.906	1.207
60.0	138.16	1.1636	1699.0	6.598	0.906	1.209
65.0	147.92	1.1774	1703.3	6.574	0.906	1.210
70.0	157.45	1.1912	1707.6	6.552	0.907	1.211
75.0	166.76	1.2050	1711.9	6.532	0.907	1.212
80.0	175.87	1.2188	1716.3	6.513	0.907	1.213
85.0	184.77	1.2326	1720.7	6.495	0.908	1.214
90.0	193.48	1.2464	1725.1	6.478	0.908	1.215
95.0	201.99	1.2601	1729.5	6.462	0.908	1.216
100.0	210.33	1.2739	1733.9	6.447	0.908	1.217
			$T=1350$ K			
0.1	0.26	1.0003	1709.7	8.488	0.907	1.195
0.5	1.29	1.0013	1710.0	8.026	0.908	1.195
1.0	2.57	1.0026	1710.4	7.827	0.908	1.195
1.5	3.86	1.0039	1710.8	7.710	0.908	1.195
2.0	5.13	1.0052	1711.2	7.628	0.908	1.195
2.5	6.41	1.0065	1711.6	7.564	0.908	1.196
3.0	7.68	1.0078	1712.0	7.511	0.908	1.196
3.5	8.95	1.0091	1712.4	7.467	0.908	1.196
4.0	10.21	1.0104	1712.8	7.428	0.908	1.196
4.5	11.48	1.0117	1713.2	7.395	0.908	1.196
5.0	12.73	1.0130	1713.5	7.364	0.908	1.197
6.0	15.24	1.0156	1714.3	7.312	0.908	1.197
7.0	17.74	1.0182	1715.1	7.267	0.908	1.197
8.0	20.22	1.0209	1715.9	7.229	0.908	1.198
9.0	22.69	1.0235	1716.7	7.195	0.908	1.198
10.0	25.14	1.0261	1717.6	7.164	0.908	1.199
11.0	27.59	1.0287	1718.4	7.137	0.908	1.199
12.0	30.02	1.0313	1719.2	7.112	0.908	1.199
13.0	32.44	1.0340	1720.0	7.089	0.909	1.200
14.0	34.85	1.0366	1720.8	7.067	0.909	1.200
15.0	37.24	1.0392	1721.6	7.047	0.909	1.200
16.0	39.62	1.0419	1722.4	7.029	0.909	1.201
17.0	41.99	1.0445	1723.2	7.011	0.909	1.201
18.0	44.35	1.0471	1724.1	6.994	0.909	1.201
19.0	46.70	1.0498	1724.9	6.979	0.909	1.202
20.0	49.03	1.0524	1725.7	6.964	0.909	1.202
21.0	51.36	1.0550	1726.5	6.950	0.909	1.202
22.0	53.67	1.0577	1727.3	6.936	0.909	1.203
23.0	55.97	1.0603	1728.2	6.923	0.909	1.203
24.0	58.26	1.0629	1729.0	6.911	0.909	1.203
25.0	60.53	1.0656	1729.8	6.899	0.909	1.204
26.0	62.80	1.0682	1730.6	6.888	0.910	1.204
27.0	65.05	1.0709	1731.5	6.877	0.910	1.204
28.0	67.30	1.0735	1732.3	6.866	0.910	1.205
29.0	69.53	1.0761	1733.1	6.856	0.910	1.205

Table II.11 (*Continued*)

p	ρ	Z	h	s	c_v	c_p
			$T = 1350$ K			
30.0	71.75	1.0788	1734.0	6.846	0.910	1.205
35.0	82.69	1.0920	1738.2	6.801	0.910	1.207
40.0	93.37	1.1053	1742.4	6.762	0.911	1.208
45.0	103.80	1.1185	1746.6	6.728	0.911	1.210
50.0	113.98	1.1318	1750.9	6.697	0.911	1.211
55.0	123.92	1.1451	1755.2	6.669	0.912	1.212
60.0	133.63	1.1585	1759.5	6.643	0.912	1.213
65.0	143.12	1.1718	1763.9	6.620	0.912	1.214
70.0	152.39	1.1851	1768.3	6.598	0.912	1.216
75.0	161.46	1.1985	1772.7	6.578	0.913	1.217
80.0	170.33	1.2118	1777.1	6.559	0.913	1.218
85.0	179.01	1.2251	1781.5	6.541	0.913	1.219
90.0	187.50	1.2384	1786.0	6.524	0.914	1.220
95.0	195.81	1.2517	1790.4	6.508	0.914	1.220
100.0	203.95	1.2650	1794.9	6.493	0.914	1.221
			$T = 1400$ K			
0.1	0.25	1.0003	1769.6	8.532	0.913	1.201
0.5	1.24	1.0013	1769.9	8.070	0.913	1.201
1.0	2.48	1.0025	1770.3	7.870	0.913	1.201
1.5	3.72	1.0038	1770.7	7.754	0.913	1.201
2.0	4.95	1.0050	1771.1	7.671	0.914	1.201
2.5	6.18	1.0063	1771.5	7.607	0.914	1.201
3.0	7.41	1.0076	1771.9	7.555	0.914	1.202
3.5	8.63	1.0088	1772.3	7.510	0.914	1.202
4.0	9.85	1.0101	1772.7	7.472	0.914	1.202
4.5	11.07	1.0114	1773.1	7.438	0.914	1.202
5.0	12.28	1.0126	1773.5	7.408	0.914	1.202
6.0	14.70	1.0152	1774.3	7.355	0.914	1.203
7.0	17.11	1.0177	1775.2	7.311	0.914	1.203
8.0	19.51	1.0203	1776.0	7.272	0.914	1.203
9.0	21.89	1.0228	1776.8	7.239	0.914	1.204
10.0	24.26	1.0253	1777.6	7.208	0.914	1.204
11.0	26.62	1.0279	1778.4	7.181	0.914	1.204
12.0	28.97	1.0304	1779.3	7.156	0.914	1.205
13.0	31.31	1.0330	1780.1	7.132	0.914	1.205
14.0	33.64	1.0355	1780.9	7.111	0.914	1.205
15.0	35.95	1.0381	1781.8	7.091	0.915	1.206
16.0	38.25	1.0406	1782.6	7.072	0.915	1.206
17.0	40.54	1.0432	1783.4	7.055	0.915	1.206
18.0	42.82	1.0457	1784.3	7.038	0.915	1.207
19.0	45.09	1.0483	1785.1	7.023	0.915	1.207
20.0	47.35	1.0509	1785.9	7.008	0.915	1.207
21.0	49.60	1.0534	1786.8	6.994	0.915	1.208
22.0	51.83	1.0560	1787.6	6.980	0.915	1.208
23.0	54.06	1.0585	1788.4	6.967	0.915	1.208
24.0	56.27	1.0611	1789.3	6.955	0.915	1.209

p	ρ	Z	h	s	c_v	c_p
			$T=1400$ K			
25.0	58.48	1.0637	1790.1	6.943	0.915	1.209
26.0	60.67	1.0662	1791.0	6.932	0.915	1.209
27.0	62.85	1.0688	1791.8	6.921	0.915	1.209
28.0	65.02	1.0713	1792.7	6.910	0.915	1.210
29.0	67.19	1.0739	1793.5	6.900	0.915	1.210
30.0	69.34	1.0765	1794.4	6.890	0.916	1.210
35.0	79.94	1.0893	1798.6	6.845	0.916	1.212
40.0	90.30	1.1021	1802.9	6.806	0.916	1.213
45.0	100.41	1.1150	1807.2	6.772	0.916	1.214
50.0	110.30	1.1279	1811.6	6.741	0.917	1.216
55.0	119.95	1.1407	1815.9	6.713	0.917	1.217
60.0	129.40	1.1536	1820.3	6.688	0.917	1.218
65.0	138.63	1.1665	1824.7	6.664	0.918	1.219
70.0	147.66	1.1794	1829.1	6.642	0.918	1.220
75.0	156.49	1.1923	1833.6	6.622	0.918	1.221
80.0	165.14	1.2053	1838.1	6.603	0.918	1.222
85.0	173.60	1.2181	1842.5	6.585	0.919	1.223
90.0	181.89	1.2310	1847.0	6.568	0.919	1.224
95.0	190.01	1.2439	1851.5	6.552	0.919	1.224
100.0	197.97	1.2567	1856.0	6.537	0.919	1.225
			$T=1450$ K			
0.1	0.24	1.0002	1829.7	8.574	0.919	1.206
0.5	1.20	1.0012	1830.1	8.112	0.919	1.206
1.0	2.40	1.0024	1830.5	7.913	0.919	1.206
1.5	3.59	1.0037	1830.9	7.796	0.919	1.207
2.0	4.78	1.0049	1831.3	7.714	0.919	1.207
2.5	5.97	1.0061	1831.7	7.649	0.919	1.207
3.0	7.15	1.0074	1832.1	7.597	0.919	1.207
3.5	8.34	1.0086	1832.5	7.553	0.919	1.207
4.0	9.52	1.0098	1832.9	7.514	0.919	1.207
4.5	10.69	1.0111	1833.4	7.480	0.919	1.208
5.0	11.86	1.0123	1833.8	7.450	0.919	1.208
6.0	14.20	1.0147	1834.6	7.398	0.919	1.208
7.0	16.53	1.0172	1835.4	7.353	0.919	1.208
8.0	18.85	1.0197	1836.3	7.315	0.919	1.209
9.0	21.15	1.0222	1837.1	7.281	0.920	1.209
10.0	23.44	1.0246	1838.0	7.250	0.920	1.209
11.0	25.73	1.0271	1838.8	7.223	0.920	1.210
12.0	28.00	1.0296	1839.6	7.198	0.920	1.210
13.0	30.26	1.0321	1840.5	7.175	0.920	1.210
14.0	32.51	1.0345	1841.3	7.153	0.920	1.211
15.0	34.75	1.0370	1842.2	7.133	0.920	1.211
16.0	36.97	1.0395	1843.0	7.115	0.920	1.211
17.0	39.19	1.0420	1843.9	7.097	0.920	1.211
18.0	41.40	1.0444	1844.7	7.081	0.920	1.212
19.0	43.59	1.0469	1845.6	7.065	0.920	1.212

Table II.11 (*Continued*)

p	ρ	Z	h	s	c_v	c_p
			$T=1450$ K			
20.0	45.78	1.0494	1846.4	7.050	0.920	1.212
21.0	47.96	1.0519	1847.3	7.036	0.920	1.213
22.0	50.12	1.0544	1848.1	7.023	0.920	1.213
23.0	52.28	1.0569	1849.0	7.010	0.920	1.213
24.0	54.42	1.0593	1849.8	6.997	0.921	1.213
25.0	56.56	1.0618	1850.7	6.986	0.921	1.214
26.0	58.68	1.0643	1851.6	6.974	0.921	1.214
27.0	60.80	1.0668	1852.4	6.963	0.921	1.214
28.0	62.90	1.0693	1853.3	6.953	0.921	1.215
29.0	65.00	1.0718	1854.1	6.942	0.921	1.215
30.0	67.08	1.0743	1855.0	6.933	0.921	1.215
35.0	77.37	1.0867	1859.3	6.888	0.921	1.216
40.0	87.42	1.0992	1863.7	6.849	0.921	1.218
45.0	97.24	1.1116	1868.0	6.815	0.922	1.219
50.0	106.85	1.1241	1872.4	6.784	0.922	1.220
55.0	116.24	1.1366	1876.9	6.756	0.922	1.221
60.0	125.43	1.1491	1881.3	6.730	0.923	1.222
65.0	134.42	1.1616	1885.8	6.707	0.923	1.223
70.0	143.22	1.1741	1890.2	6.685	0.923	1.224
75.0	151.83	1.1866	1894.7	6.665	0.923	1.225
80.0	160.27	1.1991	1899.2	6.646	0.924	1.226
85.0	168.53	1.2116	1903.8	6.628	0.924	1.227
90.0	176.62	1.2241	1908.3	6.611	0.924	1.227
95.0	184.55	1.2365	1912.8	6.595	0.924	1.228
100.0	192.33	1.2490	1917.4	6.580	0.924	1.229
			$T=1500$ K			
0.1	0.23	1.0002	1890.2	8.615	0.924	1.211
0.5	1.16	1.0012	1890.5	8.153	0.924	1.211
1.0	2.32	1.0024	1890.9	7.954	0.924	1.212
1.5	3.47	1.0036	1891.3	7.837	0.924	1.212
2.0	4.62	1.0048	1891.8	7.755	0.924	1.212
2.5	5.77	1.0060	1892.2	7.690	0.924	1.212
3.0	6.92	1.0072	1892.6	7.638	0.924	1.212
3.5	8.06	1.0084	1893.0	7.594	0.924	1.212
4.0	9.20	1.0096	1893.4	7.555	0.924	1.213
4.5	10.34	1.0107	1893.9	7.521	0.924	1.213
5.0	11.47	1.0119	1894.3	7.491	0.925	1.213
6.0	13.74	1.0143	1895.1	7.439	0.925	1.213
7.0	15.99	1.0167	1896.0	7.394	0.925	1.213
8.0	18.23	1.0191	1896.8	7.356	0.925	1.214
9.0	20.46	1.0215	1897.7	7.322	0.925	1.214
10.0	22.68	1.0239	1898.5	7.292	0.925	1.214
11.0	24.89	1.0264	1899.4	7.264	0.925	1.215
12.0	27.09	1.0288	1900.3	7.239	0.925	1.215
13.0	29.27	1.0312	1901.1	7.216	0.925	1.215
14.0	31.45	1.0336	1902.0	7.194	0.925	1.215

Table II.11 (*Continued*)

p	ρ	Z	h	s	c_v	c_p
			$T=1500$ K			
15.0	33.62	1.0360	1902.8	7.175	0.925	1.216
16.0	35.78	1.0384	1903.7	7.156	0.925	1.216
17.0	37.93	1.0408	1904.6	7.138	0.925	1.216
18.0	40.07	1.0432	1905.4	7.122	0.925	1.217
19.0	42.19	1.0456	1906.3	7.106	0.925	1.217
20.0	44.31	1.0480	1907.2	7.091	0.925	1.217
21.0	46.42	1.0504	1908.0	7.077	0.925	1.217
22.0	48.52	1.0529	1908.9	7.064	0.926	1.218
23.0	50.61	1.0553	1909.8	7.051	0.926	1.218
24.0	52.69	1.0577	1910.6	7.039	0.926	1.218
25.0	54.76	1.0601	1911.5	7.027	0.926	1.218
26.0	56.82	1.0625	1912.4	7.015	0.926	1.219
27.0	58.87	1.0649	1913.2	7.004	0.926	1.219
28.0	60.92	1.0673	1914.1	6.994	0.926	1.219
29.0	62.95	1.0698	1915.0	6.984	0.926	1.219
30.0	64.97	1.0722	1915.9	6.974	0.926	1.220
35.0	74.96	1.0843	1920.3	6.929	0.926	1.221
40.0	84.72	1.0963	1924.7	6.890	0.927	1.222
45.0	94.27	1.1084	1929.1	6.856	0.927	1.223
50.0	103.61	1.1205	1933.5	6.825	0.927	1.224
55.0	112.76	1.1327	1938.0	6.797	0.927	1.225
60.0	121.70	1.1448	1942.5	6.772	0.928	1.226
65.0	130.46	1.1569	1947.0	6.748	0.928	1.227
70.0	139.04	1.1690	1951.5	6.727	0.928	1.228
75.0	147.45	1.1812	1956.1	6.706	0.928	1.229
80.0	155.68	1.1933	1960.6	6.688	0.928	1.230
85.0	163.75	1.2054	1965.2	6.670	0.929	1.230
90.0	171.65	1.2175	1969.8	6.653	0.929	1.231
95.0	179.41	1.2296	1974.4	6.637	0.929	1.232
100.0	187.02	1.2416	1978.9	6.622	0.929	1.233

Table II.12

p	w	μ	k	f	α/α_0	γ/γ_0
			$T=100$ K			
0.1	198.0	16.89	1.39	0.10	1.063	1.041
0.5	186.0	17.65	1.36	0.45	1.453	1.267
1.0	648.5	—0.14	323.61	0.55	0.766	114.161
1.5	658.1	—0.16	222.89	0.57	0.749	77.202
2.0	667.2	—0.17	172.40	0.58	0.732	58.691
2.5	676.0	—0.18	142.01	0.59	0.717	47.562
3.0	684.4	—0.19	121.67	0.60	0.703	40.125
3.5	692.5	—0.20	107.08	0.62	0.690	34.798
4.0	700.3	—0.20	96.08	0.63	0.677	30.792
4.5	707.8	—0.21	87.48	0.65	0.665	27.666

Table II.12 (*Continued*)

p	w	μ	k	f	α/α_0	γ/γ_0
			$T=100$ K			
5.0	715.1	—0.22	80.56	0.66	0.654	25.156
6.0	728.8	—0.23	70.08	0.69	0.633	21.373
7.0	741.5	—0.25	62.50	0.72	0.615	18.651
8.0	753.5	—0.26	56.73	0.75	0.597	16.593
9.0	764.8	—0.27	52.17	0.79	0.582	14.979
10.0	775.3	—0.28	48.47	0.82	0.567	13.677
11.0	785.3	—0.29	45.38	0.86	0.554	12.603
12.0	794.7	—0.30	42.77	0.90	0.541	11.700
13.0	803.6	—0.31	40.53	0.94	0.530	10.929
14.0	812.1	—0.31	38.57	0.98	0.519	10.236
15.0	820.0	—0.32	36.84	1.02	0.508	9.680
16.0	827.6	—0.33	35.30	1.06	0.499	9.166
17.0	834.9	—0.33	33.92	1.11	0.490	8.709
18.0	841.7	—0.34	32.67	1.16	0.481	8.299
19.0	848.3	—0.35	31.53	1.20	0.473	7.928
20.0	854.5	—0.35	30.50	1.26	0.465	7.593
21.0	860.5	—0.36	29.54	1.31	0.458	7.286
22.0	866.2	—0.36	28.66	1.37	0.451	7.005
23.0	871.6	—0.37	27.84	1.42	0.444	6.747
24.0	876.9	—0.37	27.07	1.48	0.437	6.508
25.0	881.8	—0.38	26.36	1.55	0.431	6.286
26.0	886.6	—0.38	25.69	1.61	0.425	6.080
27.0	891.2	—0.38	25.06	1.68	0.420	5.888
28.0	895.6	—0.39	24.47	1.75	0.414	5.708
29.0	899.8	—0.39	23.91	1.82	0.409	5.539
30.0	903.8	—0.39	23.38	1.90	0.404	5.381
35.0	921.7	—0.41	21.09	2.32	0.382	4.710
40.0	936.4	—0.42	19.26	2.83	0.363	4.189
45.0	948.4	—0.43	17.75	3.45	0.347	3.772
50.0	958.1	—0.44	16.47	4.19	0.334	3.428
55.0	965.8	—0.44	15.36	5.08	0.322	3.141
60.0	971.8	—0.45	14.38	6.15	0.313	2.896
65.0	976.1	—0.46	13.51	7.44	0.305	2.685
70.0	978.9	—0.46	12.73	8.98	0.300	2.502
75.0	980.3	—0.46	12.02	10.82	0.296	2.343
80.0	980.2	—0.46	11.36	13.02	0.294	2.203
85.0	978.7	—0.46	10.75	15.64	0.295	2.081
90.0	975.7	—0.46	10.17	18.76	0.299	1.975
95.0	971.4	—0.46	9.63	22.46	0.306	1.884
100.0	965.9	—0.45	9.13	26.86	0.320	1.808
			$T=105$ K			
0.1	203.3	15.12	1.40	0.10	1.053	1.034
0.5	193.6	15.84	1.37	0.46	1.355	1.212
1.0	584.2	—0.04	252.01	0.75	0.941	103.791
1.5	595.5	—0.05	175.30	0.76	0.911	70.471
2.0	606.1	—0.07	136.80	0.78	0.884	53.768

Table II.12 (*Continued*)

p	w	μ	k	f	α/α_0	γ/γ_0
			$T=105$ K			
2.5	616.3	—0.09	113.59	0.80	0.859	43.715
3.0	625.9	—0.10	98.03	0.82	0.836	36.990
3.5	635.2	—0.12	86.84	0.83	0.815	32.168
4.0	644.0	—0.13	78.39	0.85	0.796	28.536
4.5	652.5	—0.14	71.78	0.87	0.778	25.699
5.0	660.6	—0.15	66.44	0.89	0.761	23.420
6.0	676.0	—0.17	58.33	0.93	0.731	19.976
7.0	690.3	—0.19	52.44	0.97	0.704	17.493
8.0	703.6	—0.21	47.94	1.01	0.680	15.611
9.0	716.0	—0.22	44.37	1.06	0.658	14.132
10.0	727.7	—0.23	41.46	1.10	0.639	12.937
11.0	738.7	—0.25	39.03	1.15	0.621	11.949
12.0	749.1	—0.26	36.96	1.20	0.604	11.118
13.0	759.0	—0.27	35.18	1.25	0.589	10.407
14.0	768.3	—0.28	33.62	1.30	0.575	9.791
15.0	777.2	—0.29	32.24	1.36	0.562	9.253
16.0	785.7	—0.30	31.01	1.42	0.550	8.777
17.0	793.7	—0.31	29.90	1.48	0.538	8.353
18.0	801.4	—0.31	28.90	1.54	0.528	7.972
19.0	808.8	—0.32	27.98	1.60	0.517	7.629
20.0	815.9	—0.33	27.15	1.67	0.508	7.316
21.0	822.6	—0.33	26.37	1.74	0.499	7.032
22.0	829.1	—0.34	25.66	1.81	0.490	6.770
23.0	835.3	—0.34	25.00	1.89	0.482	6.530
24.0	841.3	—0.35	24.38	1.96	0.475	6.307
25.0	847.1	—0.35	23.80	2.04	0.467	6.101
26.0	852.7	—0.36	23.25	2.13	0.460	5.909
27.0	858.1	—0.36	22.74	2.21	0.454	5.730
28.0	863.2	—0.37	22.26	2.30	0.447	5.563
29.0	868.3	—0.37	21.81	2.40	0.441	5.405
30.0	873.1	—0.38	21.37	2.49	0.435	5.258
35.0	895.1	—0.39	19.51	3.03	0.410	4.633
40.0	914.1	—0.41	18.02	3.68	0.389	4.150
45.0	930.6	—0.42	16.79	4.45	0.371	3.764
50.0	945.1	—0.42	15.76	5.38	0.357	3.448
55.0	958.0	—0.43	14.87	6.49	0.345	3.185
60.0	969.6	—0.44	14.09	7.81	0.335	2.964
65.0	980.0	—0.44	13.41	9.38	0.327	2.775
70.0	989.5	—0.44	12.81	11.25	0.322	2.614
75.0	998.3	—0.44	12.28	13.47	0.318	2.475
80.0	1006.5	—0.44	11.80	16.11	0.316	2.356
85.0	1014.3	—0.44	11.37	19.23	0.316	2.254
90.0	1022.2	—0.44	11.00	22.93	0.319	2.168
95.0	1030.3	—0.43	10.67	27.29	0.324	2.097
100.0	1039.4	—0.42	10.40	32.45	0.333	2.040

Table II.12 (*Continued*)

p	w	μ	k	f	α/α_0	γ/γ_0
			$T=110$ K			
0.1	208.5	13.74	1.40	0.10	1.045	1.029
0.5	200.3	14.34	1.38	0.46	1.289	1.174
1.0	187.9	15.02	1.37	0.85	1.870	1.461
1.5	528.3	0.09	131.54	0.99	1.159	62.765
2.0	541.3	0.06	104.22	1.01	1.108	48.182
2.5	553.6	0.04	87.69	1.04	1.063	39.385
3.0	565.2	0.01	76.56	1.06	1.024	33.485
3.5	576.1	—0.01	68.53	1.08	0.988	29.244
4.0	586.5	—0.03	62.44	1.11	0.957	26.043
4.5	596.4	—0.05	57.64	1.13	0.928	23.536
5.0	605.8	—0.06	53.76	1.16	0.901	21.517
6.0	623.4	—0.09	47.83	1.21	0.855	18.457
7.0	639.7	—0.12	43.48	1.26	0.815	16.241
8.0	654.7	—0.14	40.13	1.32	0.781	14.555
9.0	668.7	—0.16	37.45	1.37	0.750	13.226
10.0	681.8	—0.18	35.26	1.43	0.723	12.148
11.0	694.1	—0.20	33.41	1.49	0.699	11.254
12.0	705.6	—0.21	31.83	1.55	0.677	10.499
13.0	716.6	—0.23	30.46	1.62	0.657	9.853
14.0	726.9	—0.24	29.26	1.69	0.639	9.292
15.0	736.8	—0.25	28.19	1.76	0.622	8.800
16.0	746.2	—0.26	27.23	1.83	0.607	8.364
17.0	755.1	—0.27	26.36	1.91	0.592	7.975
18.0	763.7	—0.28	25.57	1.98	0.579	7.626
19.0	771.9	—0.29	24.85	2.06	0.566	7.310
20.0	779.8	—0.30	24.19	2.15	0.555	7.023
21.0	787.3	—0.31	23.58	2.24	0.544	6.760
22.0	794.6	—0.31	23.01	2.33	0.533	6.519
23.0	801.6	—0.32	22.48	2.42	0.523	6.297
24.0	808.3	—0.33	21.99	2.52	0.514	6.091
25.0	814.9	—0.33	21.53	2.62	0.505	5.901
26.0	821.2	—0.34	21.09	2.72	0.497	5.723
27.0	827.3	—0.35	20.68	2.83	0.489	5.557
28.0	833.2	—0.35	20.30	2.94	0.482	5.402
29.0	838.9	—0.35	19.93	3.06	0.475	5.256
30.0	844.5	—0.36	19.58	3.18	0.468	5.119
35.0	870.2	—0.38	18.08	3.84	0.438	4.540
40.0	892.8	—0.39	16.88	4.64	0.414	4.091
45.0	913.2	—0.41	15.88	5.59	0.395	3.734
50.0	931.7	—0.42	15.05	6.71	0.378	3.443
55.0	948.8	—0.42	14.34	8.05	0.365	3.202
60.0	964.8	—0.43	13.73	9.64	0.354	2.999
65.0	980.0	—0.43	13.21	11.51	0.346	2.828
70.0	994.6	—0.43	12.75	13.73	0.339	2.683
75.0	1008.9	—0.43	12.35	16.36	0.334	2.559

p	w	μ	k	f	α/α_0	γ/γ_0
			$T=110$ K			
80.0	1023.0	—0.43	12.01	19.45	0.332	2.454
85.0	1037.2	—0.43	11.71	23.10	0.330	2.366
90.0	1051.9	—0.43	11.47	27.39	0.331	2.292
95.0	1067.4	—0.42	11.28	32.43	0.334	2.232
100.0	1084.1	—0.41	11.14	38.35	0.339	2.184
			$T=115$ K			
0.1	213.5	12.61	1.40	0.10	1.040	1.025
0.5	206.5	13.09	1.39	0.47	1.242	1.147
1.0	196.5	13.60	1.38	0.87	1.664	1.368
1.5	183.7	14.04	1.38	1.20	2.607	1.739
2.0	469.5	0.28	74.08	1.28	1.492	41.788
2.5	485.4	0.23	63.87	1.31	1.396	34.494
3.0	500.0	0.19	56.92	1.34	1.316	29.570
3.5	513.5	0.15	51.83	1.37	1.249	26.009
4.0	526.2	0.12	47.94	1.40	1.191	23.306
4.5	538.1	0.09	44.83	1.43	1.141	21.177
5.0	549.3	0.06	42.30	1.46	1.096	19.454
6.0	570.0	0.01	38.36	1.52	1.021	16.826
7.0	588.8	—0.03	35.43	1.59	0.960	14.906
8.0	606.0	—0.06	33.13	1.66	0.908	13.436
9.0	621.9	—0.09	31.26	1.73	0.864	12.269
10.0	636.7	—0.12	29.71	1.80	0.826	11.318
11.0	650.4	—0.14	28.39	1.88	0.793	10.525
12.0	663.4	—0.16	27.25	1.96	0.763	9.853
13.0	675.5	—0.18	26.25	2.04	0.736	9.274
14.0	687.0	—0.19	25.36	2.12	0.712	8.771
15.0	697.9	—0.21	24.56	2.21	0.691	8.327
16.0	708.3	—0.22	23.85	2.30	0.671	7.934
17.0	718.2	—0.24	23.19	2.39	0.652	7.582
18.0	727.6	—0.25	22.60	2.49	0.636	7.265
19.0	736.6	—0.26	22.05	2.59	0.620	6.977
20.0	745.3	—0.27	21.54	2.69	0.606	6.715
21.0	753.6	—0.28	21.07	2.80	0.592	6.475
22.0	761.7	—0.29	20.63	2.91	0.579	6.254
23.0	769.4	—0.30	20.22	3.02	0.568	6.051
24.0	776.9	—0.31	19.83	3.14	0.556	5.862
25.0	784.1	—0.31	19.47	3.26	0.546	5.687
26.0	791.1	—0.32	19.13	3.39	0.536	5.523
27.0	797.9	—0.33	18.81	3.52	0.527	5.370
28.0	804.5	—0.33	18.50	3.66	0.518	5.227
29.0	810.9	—0.34	18.21	3.80	0.509	5.093
30.0	817.1	—0.34	17.94	3.94	0.501	4.966
35.0	845.9	—0.37	16.75	4.75	0.467	4.430
40.0	871.7	—0.38	15.79	5.71	0.439	4.015
45.0	895.2	—0.40	14.99	6.84	0.417	3.684
50.0	917.0	—0.41	14.33	8.18	0.399	3.414

Table II.12 (*Continued*)

p	w	μ	k	f	α/α_0	γ/γ_0
			$T=115$ K			
55.0	937.5	—0.42	13.77	9.76	0.384	3.191
60.0	957.0	—0.42	13.29	11.63	0.372	3.005
65.0	975.7	—0.43	12.89	13.82	0.362	2.848
70.0	994.1	—0.43	12.54	16.40	0.354	2.715
75.0	1012.2	—0.43	12.24	19.44	0.348	2.602
80.0	1030.4	—0.43	12.00	23.00	0.343	2.507
85.0	1048.8	—0.43	11.80	27.17	0.341	2.427
90.0	1067.7	—0.42	11.64	32.06	0.340	2.360
95.0	1087.4	—0.42	11.53	37.78	0.340	2.306
100.0	1108.3	—0.41	11.47	44.46	0.342	2.263
			$T=120$ K			
0.1	218.3	11.67	1.40	0.10	1.035	1.022
0.5	212.3	12.03	1.40	0.47	1.206	1.127
1.0	204.0	12.41	1.39	0.88	1.532	1.304
1.5	194.2	12.70	1.40	1.24	2.123	1.568
2.0	181.1	12.88	1.41	1.53	3.613	2.027
2.5	406.4	0.58	41.68	1.60	2.072	28.775
3.0	426.6	0.48	38.79	1.64	1.858	25.096
3.5	444.7	0.40	36.54	1.67	1.698	22.377
4.0	461.0	0.34	34.72	1.71	1.573	20.275
4.5	476.0	0.28	33.21	1.75	1.472	18.594
5.0	489.9	0.24	31.92	1.79	1.388	17.217
6.0	514.9	0.16	29.83	1.87	1.255	15.082
7.0	537.1	0.10	28.19	1.95	1.154	13.495
8.0	557.1	0.05	26.84	2.03	1.074	12.262
9.0	575.3	0.01	25.71	2.12	1.008	11.272
10.0	592.0	—0.03	24.74	2.21	0.953	10.456
11.0	607.5	—0.06	23.89	2.30	0.906	9.771
12.0	621.9	—0.09	23.14	2.40	0.865	9.186
13.0	635.5	—0.12	22.47	2.50	0.829	8.679
14.0	648.2	—0.14	21.86	2.60	0.797	8.235
15.0	660.2	—0.16	21.31	2.70	0.769	7.842
16.0	671.6	—0.18	20.81	2.81	0.743	7.492
17.0	682.5	—0.19	20.34	2.92	0.720	7.177
18.0	692.8	—0.21	19.91	3.04	0.699	6.893
19.0	702.7	—0.22	19.52	3.16	0.679	6.634
20.0	712.2	—0.24	19.14	3.29	0.662	6.398
21.0	721.3	—0.25	18.80	3.41	0.645	6.181
22.0	730.0	—0.26	18.47	3.55	0.630	5.981
23.0	738.4	—0.27	18.16	3.68	0.615	5.796
24.0	746.6	—0.28	17.87	3.83	0.602	5.624
25.0	754.5	—0.29	17.60	3.97	0.589	5.464
26.0	762.1	—0.30	17.34	4.12	0.577	5.314
27.0	769.5	—0.31	17.09	4.28	0.566	5.174
28.0	776.7	—0.31	16.86	4.44	0.556	5.043
29.0	783.7	—0.32	16.64	4.61	0.546	4.920

Table II.12 (*Continued*)

p	w	μ	k	f	α/α_0	γ/γ_0
			$T=120$ K			
30.0	790.5	—0.33	16.42	4.78	0.537	4.803
35.0	822.1	—0.35	15.50	5.74	0.497	4.309
40.0	850.6	—0.38	14.74	6.87	0.465	3.925
45.0	876.7	—0.39	14.12	8.20	0.439	3.618
50.0	901.1	—0.40	13.60	9.76	0.419	3.367
55.0	924.2	—0.41	13.16	11.59	0.402	3.160
60.0	946.3	—0.42	12.79	13.75	0.388	2.987
65.0	967.8	—0.43	12.48	16.27	0.376	2.841
70.0	988.8	—0.43	12.22	19.22	0.367	2.718
75.0	1009.7	—0.43	12.00	22.67	0.359	2.614
80.0	1030.6	—0.43	11.83	26.70	0.353	2.525
85.0	1051.7	—0.43	11.69	31.40	0.349	2.451
90.0	1073.3	—0.43	11.60	36.88	0.346	2.389
95.0	1095.5	—0.42	11.54	43.26	0.345	2.338
100.0	1118.6	—0.42	11.51	50.68	0.344	2.297
			$T=125$ K			
0.1	223.0	10.87	1.40	0.10	1.031	1.020
0.5	217.8	11.13	1.40	0.47	1.179	1.111
1.0	210.7	11.40	1.40	0.90	1.440	1.258
1.5	202.9	11.59	1.41	1.27	1.856	1.460
2.0	193.5	11.69	1.42	1.58	2.636	1.759
2.5	180.8	11.60	1.46	1.84	4.791	2.291
3.0	335.5	1.10	21.71	1.95	3.384	19.570
3.5	363.7	0.88	22.42	2.00	2.729	18.097
4.0	387.2	0.73	22.66	2.05	2.338	16.813
4.5	407.5	0.61	22.68	2.09	2.074	15.710
5.0	425.7	0.52	22.57	2.14	1.880	14.759
6.0	457.1	0.38	22.19	2.24	1.611	13.211
7.0	483.9	0.27	21.71	2.34	1.429	12.004
8.0	507.5	0.19	21.23	2.44	1.296	11.035
9.0	528.5	0.13	20.75	2.54	1.194	10.238
10.0	547.6	0.08	20.29	2.65	1.112	9.569
11.0	565.1	0.03	19.87	2.76	1.044	8.998
12.0	581.3	—0.01	19.46	2.88	0.987	8.504
13.0	596.3	—0.04	19.08	2.99	0.939	8.072
14.0	610.3	—0.07	18.73	3.11	0.896	7.690
15.0	623.6	—0.10	18.39	3.24	0.859	7.350
16.0	636.0	—0.12	18.07	3.37	0.826	7.044
17.0	647.9	—0.14	17.78	3.50	0.796	6.768
18.0	659.1	—0.16	17.49	3.64	0.770	6.517
19.0	669.9	—0.18	17.23	3.78	0.745	6.287
20.0	680.1	—0.20	16.98	3.93	0.723	6.077
21.0	690.0	—0.21	16.74	4.08	0.703	5.883
22.0	699.5	—0.23	16.51	4.24	0.684	5.703
23.0	708.6	—0.24	16.29	4.40	0.667	5.536
24.0	717.4	—0.25	16.08	4.57	0.651	5.381

Table II.12 (*Continued*)

p	w	μ	k	f	α/α_0	γ/γ_0
			$T=125$ K			
25.0	725.9	—0.26	15.89	4.74	0.636	5.236
26.0	734.1	—0.27	15.70	4.92	0.622	5.101
27.0	742.1	—0.28	15.52	5.10	0.609	4.974
28.0	749.8	—0.29	15.35	5.29	0.596	4.854
29.0	757.3	—0.30	15.18	5.49	0.585	4.741
30.0	764.7	—0.31	15.03	5.69	0.574	4.635
35.0	798.8	—0.34	14.33	6.80	0.527	4.181
40.0	829.5	—0.37	13.75	8.11	0.491	3.826
45.0	857.8	—0.39	13.27	9.64	0.462	3.541
50.0	884.3	—0.40	12.87	11.43	0.438	3.308
55.0	909.4	—0.41	12.53	13.53	0.419	3.114
60.0	933.5	—0.42	12.25	15.97	0.403	2.953
65.0	957.0	—0.43	12.01	18.82	0.389	2.816
70.0	980.0	—0.43	11.82	22.15	0.378	2.701
75.0	1002.8	—0.43	11.66	26.01	0.369	2.603
80.0	1025.6	—0.43	11.54	30.51	0.362	2.520
85.0	1048.5	—0.43	11.45	35.73	0.356	2.450
90.0	1071.7	—0.43	11.40	41.79	0.352	2.391
95.0	1095.4	—0.43	11.37	48.81	0.348	2.342
100.0	1119.7	—0.43	11.38	56.95	0.346	2.302
			$T=130$ K			
0.1	227.6	10.17	1.40	0.10	1.028	1.018
0.5	223.0	10.35	1.40	0.48	1.157	1.099
1.0	217.0	10.53	1.40	0.91	1.373	1.223
1.5	210.5	10.65	1.41	1.29	1.686	1.385
2.0	203.2	10.70	1.43	1.63	2.186	1.603
2.5	194.6	10.62	1.45	1.92	3.127	1.923
3.0	183.3	10.28	1.52	2.17	5.705	2.480
3.5	246.0	2.43	8.54	2.32	10.114	11.881
4.0	294.4	1.63	11.52	2.38	4.944	12.462
4.5	327.3	1.25	13.17	2.44	3.585	12.313
5.0	353.5	1.01	14.22	2.50	2.914	11.976
6.0	395.3	0.72	15.42	2.62	2.218	11.183
7.0	428.8	0.53	16.00	2.74	1.846	10.427
8.0	457.0	0.40	16.27	2.86	1.609	9.757
9.0	481.6	0.29	16.36	2.99	1.441	9.172
10.0	503.4	0.21	16.36	3.11	1.315	8.661
11.0	523.2	0.15	16.29	3.25	1.216	8.211
12.0	541.3	0.10	16.19	3.38	1.135	7.814
13.0	558.0	0.05	16.07	3.52	1.068	7.459
14.0	573.5	0.01	15.93	3.66	1.012	7.141
15.0	588.0	—0.03	15.78	3.81	0.963	6.855
16.0	601.6	—0.06	15.62	3.96	0.920	6.594
17.0	614.4	—0.09	15.47	4.12	0.882	6.357
18.0	626.6	—0.11	15.32	4.28	0.849	6.140
19.0	638.2	—0.13	15.17	4.44	0.818	5.940

Table II.12 (*Continued*)

p	w	μ	k	f	α/α_0	γ/γ_0
			$T=130$ K			
20.0	649.3	—0.15	15.02	4.61	0.791	5.755
21.0	659.9	—0.17	14.87	4.79	0.766	5.584
22.0	670.0	—0.19	14.73	4.97	0.744	5.425
23.0	679.8	—0.21	14.59	5.16	0.723	5.277
24.0	689.2	—0.22	14.46	5.35	0.703	5.138
25.0	698.3	—0.23	14.33	5.55	0.686	5.008
26.0	707.1	—0.25	14.20	5.76	0.669	4.886
27.0	715.6	—0.26	14.08	5.97	0.654	4.771
28.0	723.8	—0.27	13.96	6.19	0.639	4.663
29.0	731.9	—0.28	13.85	6.41	0.626	4.561
30.0	739.7	—0.29	13.74	6.64	0.613	4.464
35.0	776.0	—0.33	13.23	7.92	0.559	4.048
40.0	808.6	—0.36	12.81	9.42	0.518	3.721
45.0	838.7	—0.38	12.45	11.15	0.485	3.457
50.0	866.9	—0.40	12.15	13.18	0.458	3.240
55.0	893.6	—0.41	11.90	15.54	0.436	3.059
60.0	919.3	—0.42	11.69	18.28	0.418	2.907
65.0	944.3	—0.43	11.52	21.46	0.403	2.779
70.0	968.7	—0.43	11.38	25.15	0.390	2.670
75.0	992.9	—0.44	11.27	29.43	0.379	2.577
80.0	1016.9	—0.44	11.19	34.38	0.370	2.498
85.0	1041.0	—0.44	11.13	40.11	0.363	2.431
90.0	1065.2	—0.44	11.11	46.73	0.357	2.375
95.0	1089.8	—0.44	11.10	54.37	0.352	2.327
100.0	1114.8	—0.44	11.12	63.18	0.349	2.288
			$T=131$ K			
0.1	228.5	10.04	1.40	0.10	1.027	1.018
0.5	224.0	10.21	1.40	0.48	1.153	1.097
1.0	218.2	10.37	1.40	0.91	1.361	1.217
1.5	211.9	10.48	1.41	1.30	1.659	1.372
2.0	204.9	10.52	1.43	1.64	2.123	1.579
2.5	196.8	10.44	1.45	1.94	2.957	1.874
3.0	186.7	10.14	1.51	2.19	4.976	2.360
3.5 *	192.4	4.18	4.41	2.38	104.912	8.704
4.0	270.9	2.00	9.34	2.45	6.564	11.340
4.5	308.9	1.47	11.38	2.51	4.223	11.526
5.0	337.7	1.16	12.67	2.58	3.276	11.360
6.0	382.4	0.81	14.17	2.70	2.394	10.754
7.0	417.5	0.59	14.94	2.82	1.956	10.101
8.0	446.8	0.44	15.35	2.95	1.687	9.496
9.0	472.2	0.33	15.55	3.08	1.500	8.956
10.0	494.7	0.25	15.63	3.21	1.362	8.477
11.0	514.9	0.18	15.63	3.34	1.255	8.053
12.0	533.4	0.12	15.59	3.48	1.169	7.875
13.0	550.4	0.07	15.51	3.63	1.097	7.336
14.0	566.2	0.03	15.40	3.77	1.037	7.032

*This point lies in the two-phase region.

p	w	μ	k	f	α/α_0	γ/γ_0
			$T=131$ K			
15.0	581.0	—0.01	15.29	3.92	0.985	6.756
16.0	594.8	—0.04	15.17	4.08	0.940	6.505
17.0	607.9	—0.07	15.04	4.24	0.901	6.275
18.0	620.3	—0.10	14.91	4.41	0.865	6.064
19.0	632.0	—0.12	14.78	4.58	0.834	5.870
20.0	643.3	—0.14	14.65	4.75	0.805	5.691
21.0	654.0	—0.16	14.52	4.93	0.780	5.524
22.0	664.3	—0.18	14.40	5.12	0.756	5.369
23.0	674.2	—0.20	14.27	5.31	0.734	5.225
24.0	683.7	—0.21	14.15	5.51	0.714	5.089
25.0	692.9	—0.23	14.03	5.72	0.696	4.962
26.0	701.8	—0.24	13.92	5.93	0.679	4.843
27.0	710.4	—0.25	13.81	6.15	0.663	4.731
28.0	718.8	—0.26	13.70	6.37	0.648	4.625
29.0	726.9	—0.27	13.59	6.60	0.634	4.524
30.0	734.8	—0.28	13.49	6.84	0.621	4.430
35.0	771.5	—0.32	13.03	8.15	0.566	4.022
40.0	804.5	—0.36	12.63	9.68	0.523	3.700
45.0	834.9	—0.38	12.29	11.46	0.489	3.439
50.0	863.4	—0.40	12.01	13.54	0.462	3.225
55.0	890.4	—0.41	11.77	15.95	0.440	3.047
60.0	916.3	—0.42	11.58	18.75	0.421	2.897
65.0	941.5	—0.43	11.41	21.99	0.405	2.770
70.0	966.2	—0.43	11.28	25.76	0.392	2.663
75.0	990.6	—0.44	11.18	30.12	0.381	2.571
80.0	1014.8	—0.44	11.11	35.16	0.372	2.493
85.0	1039.1	—0.44	11.06	40.99	0.364	2.426
90.0	1063.5	—0.44	11.04	47.71	0.358	2.370
95.0	1088.2	—0.44	11.04	55.48	0.353	2.323
100.0	1113.3	—0.44	11.06	64.42	0.349	2.284
			$T=132$ K			
0.1	229.4	9.91	1.40	0.10	1.027	1.017
0.5	225.0	10.07	1.40	0.48	1.150	1.094
1.0	219.3	10.22	1.41	0.91	1.350	1.212
1.5	213.3	10.32	1.41	1.30	1.634	1.360
2.0	206.7	10.34	1.43	1.65	2.066	1.556
2.5	199.0	10.27	1.45	1.95	2.813	1.831
3.0	189.7	10.00	1.51	2.21	4.453	2.261
3.5	177.0	9.09	1.67	2.41	12.038	3.198
4.0	243.8	2.55	7.11	2.51	10.053	10.007
4.5	289.3	1.74	9.63	2.58	5.153	10.679
5.0	321.4	1.34	11.17	2.65	3.741	10.718
6.0	369.3	0.91	12.97	2.78	2.599	10.317
7.0	406.1	0.66	13.93	2.90	2.078	9.771
8.0	436.5	0.50	14.47	3.04	1.771	9.233
9.0	462.8	0.38	14.77	3.17	1.563	8.739

Table II.12 (*Continued*)

p	w	μ	k	f	α/α_0	γ/γ_0
			$T=132$ K			
10.0	485.9	0.28	14.92	3.30	1.412	8.293
11.0	506.7	0.21	14.99	3.44	1.296	7.894
12.0	525.6	0.14	14.99	3.59	1.203	7.536
13.0	542.9	0.09	14.96	3.74	1.127	7.214
14.0	559.0	0.05	14.90	3.89	1.063	6.922
15.0	574.0	0.01	14.81	4.04	1.008	6.657
16.0	588.1	—0.03	14.72	4.20	0.961	6.415
17.0	601.4	—0.06	14.62	4.37	0.919	6.193
18.0	614.0	—0.09	14.51	4.54	0.883	5.989
19.0	625.9	—0.11	14.40	4.71	0.850	5.801
20.0	637.3	—0.13	14.29	4.89	0.820	5.627
21.0	648.2	—0.15	14.18	5.08	0.793	5.465
22.0	658.6	—0.17	14.07	5.27	0.769	5.314
23.0	668.6	—0.19	13.96	5.47	0.746	5.173
24.0	678.3	—0.21	13.85	5.68	0.726	5.041
25.0	687.6	—0.22	13.74	5.89	0.706	4.917
26.0	696.6	—0.23	13.64	6.10	0.689	4.800
27.0	705.3	—0.25	13.54	6.33	0.672	4.690
28.0	713.7	—0.26	13.44	6.56	0.657	4.587
29.0	721.9	—0.27	13.34	6.80	0.643	4.488
30.0	729.9	—0.28	13.25	7.04	0.629	4.395
35.0	767.0	—0.32	12.82	8.39	0.572	3.995
40.0	800.4	—0.35	12.45	9.95	0.529	3.679
45.0	831.1	—0.38	12.14	11.78	0.494	3.422
50.0	859.8	—0.40	11.87	13.90	0.466	3.211
55.0	887.1	—0.41	11.65	16.36	0.443	3.035
60.0	913.3	—0.42	11.46	19.22	0.424	2.887
65.0	938.8	—0.43	11.31	22.53	0.408	2.762
70.0	963.7	—0.43	11.19	26.37	0.395	2.655
75.0	988.3	—0.44	11.10	30.81	0.383	2.564
80.0	1012.7	—0.44	11.03	35.94	0.374	2.487
85.0	1037.1	—0.44	10.99	41.86	0.366	2.421
90.0	1061.7	—0.44	10.97	48.70	0.359	2.365
95.0	1086.5	—0.44	10.97	56.58	0.354	2.318
100.0	1111.6	—0.44	11.00	65.66	0.350	2.279
			$T=133$ K			
0.1	230.3	9.79	1.40	0.10	1.026	1.017
0.5	226.0	9.93	1.40	0.48	1.146	1.092
1.0	220.5	10.07	1.41	0.91	1.340	1.206
1.5	214.7	10.15	1.41	1.31	1.611	1.349
2.0	208.3	10.17	1.43	1.66	2.015	1.536
2.5	201.1	10.10	1.45	1.96	2.687	1.791
3.0	192.5	9.85	1.50	2.23	4.056	2.178
3.5	181.6	9.14	1.62	2.44	8.640	2.912
4.0	209.8	3.60	4.71	2.58	23.110	8.174
4.5	268.2	2.10	7.93	2.65	6.623	9.750

Table II.12 (*Continued*)

p	w	μ	k	f	α/α_0	γ/γ_0
			$T=133$ K			
5.0	304.3	1.56	9.72	2.72	4.357	10.044
6.0	355.9	1.03	11.80	2.85	2.838	9.870
7.0	394.6	0.74	12.94	2.99	2.214	9.436
8.0	426.3	0.55	13.61	3.12	1.862	8.968
9.0	453.4	0.42	14.00	3.26	1.630	8.520
10.0	477.2	0.32	14.24	3.40	1.465	8.109
11.0	498.4	0.24	14.36	3.54	1.339	7.736
12.0	517.8	0.17	14.42	3.69	1.239	7.398
13.0	535.5	0.11	14.43	3.84	1.158	7.091
14.0	551.9	0.06	14.40	4.00	1.090	6.812
15.0	567.2	0.02	14.35	4.16	1.032	6.558
16.0	581.5	—0.01	14.28	4.33	0.982	6.325
17.0	595.0	—0.05	14.21	4.50	0.939	6.112
18.0	607.7	—0.07	14.12	4.67	0.900	5.914
19.0	619.8	—0.10	14.03	4.85	0.866	5.732
20.0	631.4	—0.12	13.94	5.04	0.835	5.563
21.0	642.4	—0.14	13.84	5.23	0.807	5.406
22.0	652.9	—0.16	13.75	5.43	0.782	5.259
23.0	663.1	—0.18	13.65	5.63	0.758	5.121
24.0	672.9	—0.20	13.56	5.84	0.737	4.993
25.0	682.3	—0.21	13.46	6.06	0.717	4.871
26.0	691.4	—0.23	13.37	6.28	0.699	4.758
27.0	700.2	—0.24	13.28	6.51	0.682	4.650
28.0	708.7	—0.25	13.19	6.74	0.666	4.548
29.0	717.0	—0.26	13.10	6.99	0.651	4.452
30.0	725.1	—0.27	13.01	7.24	0.637	4.361
35.0	762.6	—0.32	12.62	8.62	0.579	3.969
40.0	796.3	—0.35	12.27	10.22	0.534	3.657
45.0	827.3	—0.38	11.98	12.09	0.499	3.405
50.0	856.3	—0.39	11.73	14.26	0.470	3.197
55.0	883.9	—0.41	11.53	16.77	0.447	3.023
60.0	910.3	—0.42	11.35	19.69	0.427	2.877
65.0	936.0	—0.43	11.21	23.07	0.411	2.753
70.0	961.1	—0.44	11.10	26.98	0.397	2.647
75.0	985.9	—0.44	11.01	31.50	0.385	2.557
80.0	1010.5	—0.44	10.95	36.72	0.376	2.480
85.0	1035.0	—0.44	10.92	42.74	0.367	2.415
90.0	1059.7	—0.44	10.90	49.68	0.360	2.359
95.0	1084.6	—0.44	10.91	57.68	0.355	2.313
100.0	1109.8	—0.44	10.94	66.89	0.350	2.273
			$T=134$ K			
0.1	231.2	9.67	1.40	0.10	1.026	1.017
0.5	227.0	9.80	1.40	0.48	1.143	1.090
1.0	221.7	9.92	1.41	0.91	1.330	1.201
1.5	216.0	10.00	1.42	1.31	1.589	1.339
2.0	209.9	10.01	1.43	1.66	1.967	1.517

Table II.12 (*Continued*)

p	w	μ	k	f	α/α_0	γ/γ_0
			$T=134$ K			
2.5	203.1	9.94	1.45	1.98	2.578	1.756
3.0	195.1	9.71	1.50	2.24	3.743	2.106
3.5	185.5	9.12	1.60	2.47	6.938	2.714
4.0	176.9	6.40	2.30	2.63	50.433	4.915
4.5	245.5	2.59	6.26	2.72	9.229	8.705
5.0	286.6	1.82	8.32	2.79	5.203	9.331
6.0	342.4	1.16	10.68	2.93	3.120	9.413
7.0	383.1	0.83	11.99	3.07	2.366	9.098
8.0	416.0	0.62	12.78	3.21	1.961	8.701
9.0	444.0	0.47	13.26	3.35	1.702	8.301
10.0	468.4	0.36	13.57	3.50	1.520	7.924
11.0	490.3	0.27	13.75	3.65	1.383	7.577
12.0	510.0	0.20	13.86	3.80	1.276	7.259
13.0	528.1	0.14	13.91	3.95	1.189	6.968
14.0	544.8	0.09	13.92	4.12	1.117	6.703
15.0	560.3	0.04	13.90	4.28	1.056	6.460
16.0	574.9	0.00	13.86	4.45	1.004	6.236
17.0	588.6	—0.03	13.81	4.63	0.958	6.030
18.0	601.5	—0.06	13.74	4.81	0.918	5.840
19.0	613.8	—0.09	13.67	4.99	0.882	5.663
20.0	625.5	—0.11	13.59	5.18	0.850	5.499
21.0	636.7	—0.13	13.51	5.38	0.821	5.346
22.0	647.4	—0.16	13.43	5.58	0.795	5.204
23.0	657.6	—0.17	13.35	5.79	0.770	5.070
24.0	667.5	—0.19	13.27	6.01	0.748	4.944
25.0	677.0	—0.21	13.18	6.23	0.728	4.826
26.0	686.2	—0.22	13.10	6.46	0.709	4.715
27.0	695.1	—0.23	13.02	6.69	0.691	4.610
28.0	703.8	—0.25	12.94	6.93	0.675	4.510
29.0	712.2	—0.26	12.86	7.18	0.660	4.416
30.0	720.3	—0.27	12.78	7.44	0.646	4.327
35.0	758.2	—0.31	12.42	8.85	0.586	3.942
40.0	792.2	—0.35	12.10	10.50	0.540	3.636
45.0	823.5	—0.37	11.83	12.41	0.504	3.387
50.0	852.8	—0.39	11.60	14.62	0.474	3.182
55.0	880.6	—0.41	11.40	17.19	0.450	3.011
60.0	907.3	—0.42	11.24	20.16	0.430	2.866
65.0	933.2	—0.43	11.11	23.61	0.413	2.744
70.0	958.5	—0.44	11.00	27.59	0.399	2.639
75.0	983.4	—0.44	10.93	32.19	0.387	2.550
80.0	1008.2	—0.44	10.87	37.50	0.377	2.474
85.0	1032.9	—0.45	10.84	43.62	0.369	2.409
90.0	1057.7	—0.45	10.83	50.67	0.362	2.354
95.0	1082.6	—0.44	10.84	58.78	0.356	2.307
100.0	1107.9	—0.44	10.87	68.11	0.351	2.268

Table II.12 (*Continued*)

p	w	μ	k	f	α/α_0	γ/γ_0
			$T=135$ K			
0.1	232.1	9.55	1.40	0.10	1.025	1.016
0.5	228.0	9.67	1.40	0.48	1.139	1.088
1.0	222.8	9.78	1.41	0.92	1.321	1.196
1.5	217.4	9.85	1.42	1.31	1.568	1.329
2.0	211.5	9.85	1.43	1.67	1.924	1.498
2.5	205.0	9.78	1.45	1.99	2.482	1.723
3.0	197.6	9.57	1.50	2.26	3.490	2.043
3.5	188.9	9.06	1.58	2.49	5.891	2.563
4.0	179.5	7.50	1.90	2.67	18.463	3.798
4.5	221.3	3.32	4.67	2.78	14.459	7.483
5.0	268.3	2.15	6.98	2.86	6.413	8.572
6.0	328.7	1.31	9.61	3.01	3.457	8.946
7.0	371.6	0.92	11.07	3.15	2.537	8.757
8.0	405.8	0.68	11.97	3.30	2.069	8.433
9.0	434.6	0.52	12.55	3.44	1.779	8.082
10.0	459.8	0.40	12.92	3.59	1.578	7.740
11.0	482.1	0.30	13.16	3.75	1.430	7.418
12.0	502.3	0.22	13.31	3.90	1.315	7.120
13.0	520.7	0.16	13.40	4.07	1.222	6.846
14.0	537.7	0.11	13.45	4.23	1.146	6.594
15.0	553.6	0.06	13.46	4.40	1.081	6.361
16.0	568.3	0.02	13.44	4.58	1.026	6.147
17.0	582.2	—0.02	13.41	4.76	0.978	5.949
18.0	595.4	—0.05	13.37	4.94	0.936	5.765
19.0	607.8	—0.08	13.32	5.13	0.899	5.595
20.0	619.7	—0.10	13.26	5.33	0.865	5.436
21.0	631.0	—0.12	13.19	5.53	0.835	5.287
22.0	641.8	—0.15	13.13	5.74	0.808	5.149
23.0	652.2	—0.17	13.05	5.95	0.783	5.018
24.0	662.2	—0.18	12.98	6.17	0.760	4.896
25.0	671.8	—0.20	12.91	6.40	0.739	4.781
26.0	681.1	—0.21	12.84	6.63	0.719	4.672
27.0	690.1	—0.23	12.77	6.88	0.701	4.570
28.0	698.9	—0.24	12.69	7.12	0.684	4.473
29.0	707.3	—0.25	12.62	7.38	0.669	4.381
30.0	715.6	—0.26	12.55	7.64	0.654	4.293
35.0	753.8	—0.31	12.22	9.09	0.592	3.915
40.0	788.2	—0.35	11.93	10.77	0.545	3.614
45.0	819.8	—0.37	11.68	12.72	0.508	3.370
50.0	849.3	—0.39	11.46	14.99	0.478	3.167
55.0	877.3	—0.41	11.28	17.61	0.454	2.998
60.0	904.2	—0.42	11.13	20.64	0.433	2.855
65.0	930.3	—0.43	11.01	24.15	0.416	2.734
70.0	955.8	—0.44	10.91	28.20	0.402	2.631
75.0	980.9	—0.44	10.84	32.88	0.389	2.543

Table II.12 (*Continued*)

p	w	μ	k	f	α/α_0	γ/γ_0
			$T=135$ K			
80.0	1005.8	—0.44	10.79	38.28	0.379	2.467
85.0	1030.6	—0.45	10.76	44.49	0.370	2.402
90.0	1055.5	—0.45	10.76	51.65	0.363	2.347
95.0	1080.6	—0.45	10.77	59.88	0.356	2.301
100.0	1105.9	—0.45	10.80	69.33	0.351	2.262
			$T=136$ K			
0.1	232.9	9.44	1.40	0.10	1.025	1.016
0.5	229.0	9.54	1.40	0.48	1.136	1.087
1.0	223.9	9.64	1.41	0.92	1.312	1.191
1.5	218.7	9.70	1.42	1.32	1.549	1.320
2.0	213.0	9.70	1.43	1.68	1.884	1.482
2.5	206.9	9.62	1.45	2.00	2.396	1.693
3.0	199.9	9.43	1.49	2.28	3.280	1.987
3.5	192.0	8.98	1.57	2.52	5.174	2.441
4.0	183.5	7.84	1.79	2.71	11.788	3.339
4.5	199.5	4.38	3.31	2.84	23.068	6.070
5.0	249.6	2.57	5.73	2.93	8.190	7.759
6.0	314.9	1.48	8.58	3.08	3.861	8.467
7.0	360.0	1.02	10.19	3.23	2.728	8.411
8.0	395.6	0.75	11.20	3.38	2.186	8.163
9.0	425.3	0.57	11.86	3.54	1.860	7.862
10.0	451.1	0.44	12.29	3.69	1.640	7.555
11.0	474.0	0.34	12.59	3.85	1.479	7.259
12.0	494.6	0.25	12.78	4.01	1.355	6.982
13.0	513.4	0.19	12.91	4.18	1.256	6.724
14.0	530.8	0.13	12.99	4.35	1.175	6.484
15.0	546.8	0.08	13.03	4.52	1.107	6.263
16.0	561.9	0.04	13.04	4.70	1.049	6.058
17.0	576.0	0.00	13.03	4.89	0.999	5.868
18.0	589.3	—0.03	13.01	5.08	0.955	5.691
19.0	601.9	—0.06	12.97	5.27	0.916	5.526
20.0	613.9	—0.09	12.93	5.47	0.881	5.372
21.0	625.3	—0.11	12.88	5.68	0.849	5.229
22.0	636.3	—0.14	12.82	5.89	0.821	5.094
23.0	646.8	—0.16	12.77	6.11	0.795	4.967
24.0	656.9	—0.17	12.71	6.34	0.772	4.848
25.0	666.7	—0.19	12.64	6.57	0.750	4.736
26.0	676.1	—0.21	12.58	6.81	0.730	4.630
27.0	685.2	—0.22	12.52	7.06	0.711	4.530
28.0	694.0	—0.24	12.45	7.32	0.694	4.435
29.0	702.6	—0.25	12.39	7.58	0.678	4.345
30.0	710.9	—0.26	12.33	7.85	0.662	4.259
35.0	749.5	—0.31	12.03	9.33	0.599	3.889
40.0	784.2	—0.34	11.76	11.05	0.551	3.593
45.0	816.0	—0.37	11.53	13.04	0.513	3.352
50.0	845.8	—0.39	11.33	15.35	0.483	3.153

Table II.12 (*Continued*)

p	w	μ	k	f	α/α_0	γ/γ_0
			$T=136$ K			
55.0	874.0	—0.41	11.16	18.03	0.457	2.986
60.0	901.1	—0.42	11.02	21.12	0.436	2.845
65.0	927.4	—0.43	10.90	24.69	0.419	2.725
70.0	953.1	—0.44	10.82	28.82	0.404	2.622
75.0	978.3	—0.44	10.75	33.57	0.391	2.535
80.0	1003.4	—0.45	10.71	39.06	0.381	2.460
85.0	1028.3	—0.45	10.69	45.37	0.372	2.396
90.0	1053.3	—0.45	10.68	52.63	0.364	2.341
95.0	1078.4	—0.45	10.70	60.97	0.357	2.295
100.0	1103.8	—0.45	10.73	70.55	0.352	2.255
			$T=137$ K			
0.1	233.8	9.32	1.40	0.10	1.024	1.016
0.5	230.0	9.42	1.40	0.48	1.133	1.085
1.0	225.0	9.51	1.41	0.92	1.304	1.187
1.5	219.9	9.55	1.42	1.32	1.530	1.311
2.0	214.5	9.55	1.43	1.68	1.847	1.466
2.5	208.7	9.47	1.45	2.01	2.319	1.666
3.0	202.2	9.29	1.49	2.30	3.102	1.937
3.5	194.9	8.89	1.56	2.54	4.647	2.340
4.0	187.2	7.99	1.73	2.74	8.909	3.051
4.5	189.1	5.51	2.49	2.89	22.743	4.829
5.0	231.5	3.10	4.60	2.99	10.721	6.892
6.0	301.0	1.67	7.61	3.16	4.349	7.978
7.0	348.4	1.13	9.35	3.31	2.944	8.063
8.0	385.3	0.83	10.46	3.47	2.313	7.892
9.0	416.1	0.63	11.19	3.63	1.947	7.641
10.0	442.6	0.48	11.69	3.79	1.704	7.370
11.0	466.0	0.37	12.03	3.95	1.529	7.101
12.0	487.0	0.28	12.27	4.12	1.396	6.844
13.0	506.2	0.21	12.43	4.29	1.291	6.602
14.0	523.8	0.15	12.54	4.46	1.205	6.376
15.0	540.2	0.10	12.61	4.64	1.133	6.165
16.0	555.4	0.05	12.64	4.83	1.072	5.969
17.0	569.7	0.01	12.66	5.02	1.019	5.787
18.0	583.2	—0.02	12.65	5.21	0.973	5.617
19.0	596.0	—0.05	12.63	5.41	0.933	5.458
20.0	608.2	—0.08	12.61	5.62	0.897	5.309
21.0	619.8	—0.10	12.57	5.83	0.864	5.170
22.0	630.9	—0.13	12.53	6.05	0.835	5.039
23.0	641.5	—0.15	12.48	6.28	0.808	4.916
24.0	651.7	—0.17	12.43	6.51	0.783	4.800
25.0	661.6	—0.18	12.38	6.75	0.761	4.691
26.0	671.1	—0.20	12.33	6.99	0.740	4.588
27.0	680.3	—0.22	12.27	7.25	0.721	4.490
28.0	689.2	—0.23	12.22	7.51	0.703	4.397
29.0	697.8	—0.24	12.16	7.78	0.687	4.309

Table II.12 (*Continued*)

p	ω	μ	k	f	α/α_0	γ/γ_0
			T=137 K			
30.0	706.2	—0.25	12.11	8.05	0.671	4.225
35.0	745.2	—0.30	11.84	9.57	0.606	3.862
40.0	780.1	—0.34	11.59	11.33	0.557	3.572
45.0	812.3	—0.37	11.38	13.36	0.518	3.334
50.0	842.3	—0.39	11.19	15.72	0.487	3.138
55.0	870.7	—0.41	11.04	18.45	0.461	2.973
60.0	898.0	—0.42	10.91	21.60	0.440	2.834
65.0	924.5	—0.43	10.80	25.24	0.422	2.715
70.0	950.3	—0.44	10.72	29.43	0.406	2.614
75.0	975.7	—0.44	10.66	34.27	0.393	2.527
80.0	1000.9	—0.45	10.63	39.84	0.382	2.453
85.0	1026.0	—0.45	10.61	46.24	0.373	2.389
90.0	1051.0	—0.45	10.61	53.60	0.365	2.334
95.0	1076.2	—0.45	10.63	62.06	0.358	2.288
100.0	1101.6	—0.45	10.66	71.76	0.353	2.249
			T=138 K			
0.1	234.7	9.21	1.40	0.10	1.024	1.016
0.5	230.9	9.30	1.40	0.48	1.130	1.083
1.0	226.1	9.37	1.41	0.92	1.296	1.182
1.5	221.2	9.41	1.42	1.33	1.513	1.302
2.0	216.0	9.40	1.43	1.69	1.813	1.451
2.5	210.4	9.33	1.45	2.02	2.250	1.640
3.0	204.3	9.15	1.49	2.31	2.950	1.892
3.5	197.6	8.79	1.55	2.57	4.241	2.254
4.0	190.6	8.04	1.69	2.78	7.283	2.845
4.5	188.2	6.26	2.14	2.94	16.370	4.068
5.0	215.9	3.74	3.66	3.05	13.587	6.006
6.0	287.3	1.90	6.70	3.23	4.934	7.480
7.0	336.9	1.26	8.55	3.39	3.186	7.711
8.0	375.2	0.91	9.74	3.56	2.452	7.620
9.0	406.9	0.69	10.54	3.72	2.040	7.421
10.0	434.1	0.53	11.10	3.88	1.773	7.185
11.0	458.0	0.41	11.49	4.05	1.583	6.942
12.0	479.5	0.32	11.77	4.22	1.439	6.706
13.0	499.0	0.24	11.97	4.40	1.327	6.480
14.0	517.0	0.17	12.11	4.58	1.236	6.267
15.0	533.6	0.12	12.20	4.76	1.160	6.068
16.0	549.1	0.07	12.26	4.95	1.096	5.881
17.0	563.6	0.03	12.29	5.15	1.041	5.706
18.0	577.3	—0.01	12.31	5.35	0.993	5.543
19.0	590.2	—0.04	12.30	5.55	0.950	5.390
20.0	602.5	—0.07	12.29	5.77	0.913	5.246
21.0	614.2	—0.09	12.27	5.98	0.879	5.112
22.0	625.5	—0.12	12.24	6.21	0.848	4.985
23.0	636.2	—0.14	12.21	6.44	0.821	4.865
24.0	646.5	—0.16	12.17	6.68	0.795	4.753

Table II.12 (*Continued*)

p	w	μ	k	f	α/α_0	γ/γ_0
			$T=138$ K			
25.0	656.5	—0.18	12.13	6.92	0.772	4.646
26.0	666.1	—0.19	12.08	7.17	0.751	4.546
27.0	675.4	—0.21	12.03	7.43	0.731	4.450
28.0	684.4	—0.22	11.99	7.70	0.713	4.360
29.0	693.1	—0.24	11.94	7.98	0.696	4.274
30.0	701.6	—0.25	11.89	8.26	0.680	4.192
35.0	740.9	—0.30	11.65	9.81	0.613	3.836
40.0	776.2	—0.34	11.43	11.61	0.563	3.550
45.0	808.5	—0.37	11.23	13.68	0.523	3.317
50.0	838.8	—0.39	11.06	16.09	0.491	3.123
55.0	867.4	—0.41	10.92	18.87	0.465	2.960
60.0	894.9	—0.42	10.80	22.08	0.443	2.822
65.0	921.5	—0.43	10.70	25.78	0.424	2.705
70.0	947.5	—0.44	10.63	30.05	0.409	2.605
75.0	973.1	—0.44	10.58	34.96	0.395	2.519
80.0	998.4	—0.45	10.54	40.61	0.384	2.445
85.0	1023.5	—0.45	10.53	47.11	0.375	2.382
90.0	1048.7	—0.45	10.53	54.58	0.366	2.328
95.0	1073.9	—0.45	10.55	63.14	0.359	2.281
100.0	1099.3	—0.45	10.59	72.96	0.353	2.242
			$T=139$ K			
0.1	235.6	9.10	1.40	0.10	1.023	1.015
0.5	231.9	9.18	1.40	0.48	1.127	1.081
1.0	227.2	9.25	1.41	0.92	1.288	1.178
1.5	222.5	9.28	1.42	1.33	1.497	1.294
2.0	217.4	9.26	1.43	1.70	1.781	1.437
2.5	212.1	9.19	1.45	2.03	2.187	1.616
3.0	206.3	9.02	1.49	2.33	2.818	1.852
3.5	200.1	8.69	1.55	2.59	3.919	2.179
4.0	193.7	8.05	1.66	2.81	6.228	2.685
4.5	190.1	6.68	1.97	2.98	12.104	3.604
5.0	205.2	4.42	2.98	3.11	15.033	5.189
6.0	273.9	2.15	5.86	3.30	5.625	6.975
7.0	325.5	1.39	7.79	3.47	3.458	7.358
8.0	365.1	1.00	9.06	3.64	2.603	7.348
9.0	397.7	0.76	9.93	3.81	2.139	7.200
10.0	425.6	0.58	10.54	3.98	1.844	7.000
11.0	450.1	0.45	10.97	4.15	1.638	6.785
12.0	472.0	0.35	11.29	4.33	1.484	6.568
13.0	491.9	0.27	11.52	4.51	1.364	6.359
14.0	510.2	0.20	11.68	4.70	1.267	6.159
15.0	527.0	0.14	11.80	4.89	1.187	5.971
16.0	542.7	0.09	11.88	5.08	1.120	5.793
17.0	557.5	0.05	11.94	5.28	1.062	5.626
18.0	571.3	0.01	11.97	5.49	1.012	5.469
19.0	584.5	—0.02	11.98	5.70	0.968	5.322

Table II.12 (*Continued*)

p	w	μ	k	f	α/α_0	γ/γ_0
			$T=139$ K			
20.0	596.9	—0.05	11.98	5.91	0.929	5.184
21.0	608.8	—0.08	11.98	6.14	0.894	5.053
22.0	620.1	—0.11	11.96	6.37	0.862	4.931
23.0	631.0	—0.13	11.93	6.60	0.834	4.815
24.0	641.4	—0.15	11.91	6.85	0.808	4.705
25.0	651.5	—0.17	11.87	7.10	0.784	4.602
26.0	661.2	—0.19	11.84	7.36	0.762	4.504
27.0	670.6	—0.20	11.80	7.62	0.741	4.411
28.0	679.7	—0.22	11.76	7.90	0.722	4.322
29.0	688.5	—0.23	11.72	8.18	0.705	4.238
30.0	697.0	—0.24	11.68	8.47	0.688	4.158
35.0	736.7	—0.30	11.47	10.05	0.620	3.809
40.0	772.2	—0.34	11.27	11.89	0.568	3.529
45.0	804.8	—0.37	11.09	14.01	0.528	3.299
50.0	835.3	—0.39	10.93	16.46	0.495	3.108
55.0	864.2	—0.41	10.80	19.29	0.468	2.947
60.0	891.8	—0.42	10.69	22.56	0.446	2.811
65.0	918.6	—0.43	10.60	26.32	0.427	2.695
70.0	944.7	—0.44	10.53	30.66	0.411	2.596
75.0	970.4	—0.44	10.49	35.65	0.398	2.511
80.0	995.8	—0.45	10.46	41.39	0.386	2.438
85.0	1021.0	—0.45	10.45	47.98	0.376	2.375
90.0	1046.2	—0.45	10.46	55.55	0.367	2.321
95.0	1071.5	—0.45	10.48	64.22	0.360	2.275
100.0	1096.9	—0.45	10.52	74.16	0.354	2.235
			$T=140$ K			
0.1	236.4	9.00	1.40	0.10	1.023	1.015
0.5	232.8	9.06	1.40	0.48	1.124	1.080
1.0	228.3	9.12	1.41	0.93	1.280	1.174
1.5	223.7	9.15	1.42	1.33	1.482	1.287
2.0	218.9	9.13	1.43	1.70	1.751	1.424
2.5	213.7	9.05	1.45	2.04	2.130	1.594
3.0	208.3	8.89	1.49	2.34	2.703	1.815
3.5	202.4	8.58	1.54	2.61	3.655	2.114
4.0	196.6	8.02	1.64	2.84	5.485	2.555
4.5	192.6	6.92	1.88	3.02	9.551	3.287
5.0	200.2	5.02	2.55	3.17	14.134	4.530
6.0	261.2	2.44	5.11	3.37	6.404	6.469
7.0	314.3	1.54	7.08	3.55	3.760	7.003
8.0	355.2	1.10	8.41	3.73	2.766	7.076
9.0	388.7	0.82	9.33	3.90	2.245	6.980
10.0	417.3	0.63	9.99	4.08	1.920	6.816
11.0	442.3	0.49	10.47	4.26	1.696	6.627
12.0	464.6	0.38	10.82	4.44	1.530	6.431
13.0	484.9	0.30	11.08	4.62	1.402	6.238
14.0	503.4	0.22	11.27	4.81	1.300	6.052

Table II.12 (*Continued*)

d	w	μ	k	f	α/α_0	γ/γ_0
			$T=140$ K			
15.0	520.6	0.16	11.41	5.01	1.215	5.874
16.0	536.5	0.11	11.52	5.21	1.145	5.705
17.0	551.4	0.06	11.59	5.41	1.084	5.546
18.0	565.5	0.02	11.64	5.62	1.032	5.396
19.0	578.8	—0.01	11.67	5.84	0.986	5.255
20.0	591.4	—0.04	11.68	6.06	0.945	5.121
21.0	603.4	—0.07	11.69	6.29	0.909	4.995
22.0	614.8	—0.10	11.68	6.53	0.876	4.877
23.0	625.8	—0.12	11.67	6.77	0.847	4.764
24.0	636.4	—0.14	11.65	7.02	0.820	4.658
25.0	646.5	—0.16	11.63	7.28	0.795	4.557
26.0	656.3	—0.18	11.60	7.54	0.772	4.462
27.0	665.8	—0.20	11.57	7.81	0.751	4.371
28.0	675.0	—0.21	11.54	8.09	0.732	4.285
29.0	683.9	—0.22	11.50	8.38	0.714	4.203
30.0	692.5	—0.24	11.47	8.68	0.697	4.125
35.0	732.5	—0.29	11.29	10.30	0.627	3.783
40.0	768.3	—0.33	11.11	12.17	0.574	3.508
45.0	801.2	—0.36	10.94	14.33	0.533	3.281
50.0	831.8	—0.39	10.80	16.83	0.499	3.093
55.0	860.9	—0.41	10.68	19.71	0.472	2.935
60.0	888.7	—0.42	10.58	23.04	0.449	2.800
65.0	915.6	—0.43	10.50	26.87	0.430	2.685
70.0	941.9	—0.44	10.44	31.28	0.414	2.587
75.0	967.7	—0.45	10.40	36.34	0.400	2.502
80.0	993.2	—0.45	10.38	42.17	0.388	2.430
85.0	1018.5	—0.45	10.37	48.85	0.378	2.367
90.0	1043.8	—0.46	10.38	56.51	0.369	2.313
95.0	1069.1	—0.46	10.41	65.30	0.361	2.268
100.0	1094.5	—0.46	10.45	75.35	0.355	2.228
			$T=145$ K			
0.1	240.7	8.50	1.40	0.10	1.021	1.014
0.5	237.5	8.53	1.40	0.48	1.112	1.072
1.0	233.6	8.54	1.41	0.93	1.248	1.156
1.5	229.6	8.53	1.42	1.35	1.415	1.253
2.0	225.6	8.49	1.43	1.73	1.628	1.367
2.5	221.4	8.41	1.45	2.09	1.908	1.503
3.0	217.2	8.27	1.48	2.41	2.287	1.669
3.5	212.8	8.05	1.52	2.70	2.827	1.876
4.0	208.7	7.70	1.58	2.96	3.635	2.142
4.5	205.3	7.16	1.69	3.19	4.889	2.497
5.0	204.3	6.33	1.88	3.39	6.730	2.982
6.0	222.4	4.06	2.80	3.69	8.434	4.295
7.0	266.0	2.49	4.31	3.93	5.512	5.292
8.0	309.2	1.69	5.69	4.15	3.747	5.741
9.0	345.9	1.24	6.76	4.35	2.862	5.897

Table II.12 (*Continued*)

p	w	μ	k	f	α/α_0	γ/γ_0
			$T=145$ K			
10.0	377.4	0.94	7.58	4.56	2.350	5.911
11.0	404.8	0.73	8.22	4.76	2.017	5.853
12.0	429.1	0.58	8.71	4.97	1.784	5.758
13.0	451.0	0.46	9.09	5.19	1.609	5.644
14.0	470.9	0.36	9.40	5.40	1.474	5.522
15.0	489.3	0.28	9.64	5.62	1.365	5.398
16.0	506.3	0.21	9.83	5.85	1.275	5.274
17.0	522.2	0.16	9.99	6.08	1.200	5.153
18.0	537.1	0.11	10.11	6.32	1.135	5.035
19.0	551.2	0.06	10.21	6.56	1.079	4.922
20.0	564.5	0.02	10.28	6.81	1.031	4.814
21.0	577.2	−0.01	10.34	7.07	0.987	4.710
22.0	589.2	−0.04	10.39	7.33	0.949	4.611
23.0	600.8	−0.07	10.43	7.61	0.914	4.516
24.0	611.8	−0.09	10.45	7.88	0.883	4.425
25.0	622.5	−0.12	10.47	8.17	0.854	4.339
26.0	632.7	−0.14	10.48	8.47	0.828	4.256
27.0	642.6	−0.16	10.49	8.77	0.804	4.177
28.0	652.2	−0.18	10.49	9.08	0.781	4.101
29.0	661.5	−0.19	10.49	9.40	0.761	4.029
30.0	670.5	−0.21	10.48	9.73	0.742	3.959
35.0	712.0	−0.27	10.43	11.52	0.663	3.653
40.0	749.1	−0.32	10.34	13.59	0.604	3.402
45.0	783.1	−0.35	10.26	15.96	0.558	3.194
50.0	814.7	−0.38	10.18	18.70	0.521	3.019
55.0	844.6	−0.40	10.11	21.84	0.491	2.870
60.0	873.1	−0.42	10.05	25.45	0.465	2.744
65.0	900.7	−0.43	10.00	29.60	0.444	2.635
70.0	927.5	−0.44	9.97	34.35	0.426	2.541
75.0	953.8	−0.45	9.96	39.79	0.411	2.460
80.0	979.6	−0.46	9.96	46.02	0.397	2.390
85.0	1005.3	−0.46	9.97	53.15	0.385	2.329
90.0	1030.7	−0.46	9.99	61.29	0.375	2.276
95.0	1056.1	−0.46	10.03	70.59	0.367	2.231
100.0	1081.6	−0.47	10.07	81.21	0.359	2.192
			$T=150$ K			
0.1	244.9	8.04	1.40	0.10	1.019	1.013
0.5	242.1	8.04	1.41	0.48	1.101	1.066
1.0	238.6	8.03	1.41	0.94	1.220	1.141
1.5	235.2	7.99	1.42	1.36	1.363	1.226
2.0	231.8	7.94	1.44	1.76	1.537	1.323
2.5	228.4	7.85	1.45	2.13	1.753	1.435
3.0	224.9	7.73	1.48	2.47	2.028	1.566
3.5	221.6	7.55	1.51	2.78	2.385	1.722
4.0	218.5	7.29	1.56	3.07	2.859	1.909
4.5	215.9	6.94	1.63	3.33	3.494	2.138

Table II.12 (*Continued*)

p	w	μ	k	f	α/α_0	γ/γ_0
			$T=150$ K			
5.0	214.5	6.44	1.73	3.56	4.323	2.419
6.0	219.4	5.02	2.13	3.95	6.111	3.157
7.0	241.4	3.49	2.90	4.27	6.024	3.963
8.0	275.3	2.41	3.91	4,53	4.632	4.557
9.0	310.4	1.75	4.88	4,78	3.531	4.894
10.0	342.4	1.32	5.72	5,02	2.833	5.058
11.0	371.0	1.03	6.41	5,26	2.379	5.117
12.0	396.6	0.81	6.97	5,50	2.066	5.114
13.0	419.8	0.65	7.43	5,74	1.837	5.074
14.0	440.9	0.52	7.81	5.99	1.663	5.013
15.0	460.2	0.42	8.12	6.24	1.526	4.939
16.0	478.2	0.33	8.38	6.49	1.414	4.857
17.0	494.9	0.26	8.59	6.75	1.322	4.772
18.0	510.6	0.20	8.77	7.02	1.244	4.686
19.0	525.3	0.15	8.92	7.29	1.178	4.600
20.0	539.3	0.10	9.05	7.57	1.120	4.516
21.0	552.5	0.06	9.15	7.85	1.069	4.433
22.0	565.1	0.02	9.24	8.15	1.024	4.352
23.0	577.1	—0.01	9.32	8.45	0.984	4.274
24.0	588.7	—0.04	9.38	8.76	0.947	4.199
25.0	599.7	—0.07	9.43	9.07	0.914	4.126
26.0	610.4	—0.10	9.48	9.40	0.885	4.055
27.0	620.7	—0.12	9.51	9.73	0.857	3.987
28.0	630.6	—0.14	9.54	10.08	0.832	3.922
29.0	640.2	—0.16	9.57	10.43	0.809	3.859
30.0	649.6	—0.18	9.59	10.79	0.787	3.798
35.0	692.5	—0.25	9.64	12.76	0.699	3.526
40.0	730.8	—0.30	9.64	15.02	0.634	3.299
45.0	765.6	—0.34	9.62	17.61	0.583	3.108
50.0	798.0	—0.37	9.59	20.58	0.543	2.945
55.0	828.6	—0.40	9.56	23.97	0.510	2.807
60.0	857.7	—0.42	9.54	27.86	0.482	2.687
65.0	885.7	—0.43	9.53	32.31	0.459	2.584
70.0	913.0	—0.44	9.52	37.39	0.439	2.494
75.0	939.6	—0.45	9.53	43.20	0.422	2.416
80.0	965.7	—0.46	9.54	49.82	0.407	2.349
85.0	991.5	—0.47	9.57	57.37	0.394	2.289
90.0	1017.0	—0.47	9.60	65.96	0.383	2.238
95.0	1042.4	—0.47	9.65	75.75	0.373	2.193
100.0	1067.7	—0.47	9.70	86.88	0.364	2.153
			$T=155$ K			
0.1	249.1	7.62	1.40	0.10	1.017	1.012
0.5	246.5	7.60	1.41	0.49	1.092	1.061
1.0	243.5	7.56	1.41	0.94	1.198	1.128
1.5	240.5	7.51	1.42	1.38	1.321	1.203
2.0	237.6	7.45	1.44	1.78	1.466	1.287

Table II.12 (*Continued*)

p	ω	μ	k	j	α/α_0	γ/γ_0
			$T=155$ K			
2.5	234.8	7.36	1.45	2.16	1.640	1.382
3.0	232.0	7.24	1.48	2.52	1.850	1.490
3.5	229.3	7.09	1.50	2.85	2.108	1.614
4.0	226.9	6.88	1.54	3.16	2.427	1.756
4.5	225.0	6.62	1.59	3.45	2.821	1.922
5.0	223.7	6.28	1.66	3.71	3.299	2.114
6.0	225.4	5.32	1.90	4.17	4.406	2.590
7.0	236.6	4.14	2.32	4.55	5.068	3.146
8.0	258.3	3.07	2.94	4.88	4.731	3.663
9.0	286.2	2.29	3.67	5.18	3.950	4.050
10.0	315.2	1.74	4.39	5.46	3.247	4.300
11.0	343.0	1.36	5.04	5.74	2.726	4.445
12.0	368.6	1.08	5.60	6.01	2.348	4.517
13.0	392.2	0.87	6.08	6.29	2.070	4.541
14.0	413.9	0.70	6.50	6.56	1.858	4.533
15.0	433.9	0.57	6.85	6.84	1.692	4.504
16.0	452.5	0.47	7.14	7.13	1.558	4.461
17.0	469.8	0.38	7.40	7.41	1.448	4.410
18.0	486.0	0.30	7.62	7.71	1.356	4.352
19.0	501.3	0.24	7.81	8.01	1.278	4.292
20.0	515.8	0.18	7.97	8.32	1.211	4.229
21.0	529.5	0.13	8.11	8.64	1.152	4.166
22.0	542.6	0.09	8.23	8.96	1.100	4.103
23.0	555.0	0.05	8.34	9.29	1.054	4.041
24.0	567.0	0.01	8.43	9.63	1.013	3.980
25.0	578.4	—0.02	8.51	9.98	0.976	3.919
26.0	589.4	—0.05	8.58	10.34	0.942	3.861
27.0	600.1	—0.08	8.64	10.70	0.911	3.803
28.0	610.3	—0.10	8.69	11.08	0.883	3.748
29.0	620.2	—0.12	8.74	11.46	0.857	3.694
30.0	629.8	—0.14	8.78	11.86	0.833	3.641
35.0	674.0	—0.23	8.92	14.01	0.736	3.402
40.0	713.3	—0.28	8.99	16.46	0.664	3.198
45.0	749.0	—0.33	9.02	19.25	0.609	3.023
50.0	782.0	—0.36	9.04	22.45	0.565	2.873
55.0	813.1	—0.39	9.05	26.10	0.529	2.744
60.0	842.7	—0.41	9.06	30.26	0.500	2.632
65.0	871.1	—0.43	9.08	35.00	0.474	2.534
70.0	898.6	—0.44	9.09	40.40	0.453	2.448
75.0	925.4	—0.45	9.11	46.55	0.434	2.373
80.0	951.6	—0.46	9.14	53.54	0.418	2.308
85.0	977.5	—0.47	9.18	61.48	0.403	2.250
90.0	1003.0	—0.48	9.22	70.50	0.391	2.199
95.0	1028.3	—0.48	9.27	80.73	0.380	2.155
100.0	1053.5	—0.48	9.33	92.34	0.370	2.115

Table II.12 (*Continued*)

p	w	μ	k	f	α/α_0	γ/γ_0
			$T=160$ K			
0.1	253.1	7.24	1.40	0.10	1.016	1.011
0.5	250.9	7.20	1.41	0.49	1.084	1.056
1.0	248.2	7.15	1.41	0.95	1.178	1.117
1.5	245.6	7.08	1.42	1.39	1.286	1.184
2.0	243.1	7.01	1.44	1.80	1.410	1.258
2.5	240.7	6.91	1.45	2.20	1.553	1.340
3.0	238.5	6.80	1.47	2.57	1.721	1.431
3.5	236.3	6.67	1.50	2.91	1.918	1.532
4.0	234.5	6.50	1.53	3.24	2.150	1.646
4.5	232.9	6.29	1.57	3.55	2.423	1.774
5.0	231.9	6.02	1.62	3.84	2.738	1.918
6.0	232.7	5.33	1.79	4.35	3.460	2.258
7.0	239.3	4.44	2.05	4.80	4.078	2.653
8.0	253.2	3.53	2.45	5.19	4.235	3.056
9.0	273.5	2.75	2.95	5.54	3.929	3.406
10.0	297.3	2.15	3.51	5.87	3.439	3.676
11.0	322.0	1.69	4.05	6.19	2.968	3.864
12.0	346.2	1.36	4.57	6.50	2.582	3.986
13.0	369.1	1.10	5.03	6.81	2.279	4.057
14.0	390.7	0.90	5.44	7.12	2.040	4.091
15.0	410.8	0.74	5.80	7.43	1.851	4.100
16.0	429.6	0.61	6.12	7.74	1.698	4.091
17.0	447.2	0.50	6.39	8.06	1.572	4.068
18.0	463.8	0.41	6.63	8.39	1.467	4.037
19.0	479.4	0.33	6.84	8.72	1.378	3.999
20.0	494.3	0.27	7.03	9.06	1.301	3.957
21.0	508.3	0.21	7.20	9.41	1.235	3.912
22.0	521.7	0.16	7.34	9.76	1.176	3.865
23.0	534.5	0.11	7.47	10.13	1.125	3.818
24.0	546.8	0.07	7.58	10.50	1.078	3.769
25.0	558.6	0.03	7.68	10.88	1.037	3.721
26.0	569.9	0.00	7.77	11.27	0.999	3.674
27.0	580.8	—0.03	7.85	11.66	0.965	3.626
28.0	591.3	—0.06	7.93	12.07	0.934	3.580
29.0	601.5	—0.08	7.99	12.49	0.905	3.534
30.0	611.3	—0.11	8.05	12.92	0.879	3.490
35.0	656.6	—0.20	8.26	15.24	0.773	3.282
40.0	696.7	—0.27	8.39	17.88	0.695	3.100
45.0	733.1	—0.31	8.48	20.89	0.635	2.941
50.0	766.7	—0.35	8.53	24.31	0.588	2.803
55.0	798.2	—0.38	8.58	28.20	0.549	2.683
60.0	828.1	—0.41	8.61	32.62	0.517	2.578
65.0	856.8	—0.43	8.65	37.65	0.490	2.485
70.0	884.5	—0.44	8.68	43.35	0.467	2.404
75.0	911.4	—0.45	8.72	49.83	0.446	2.332

Table II.12 (*Continued*)

p	w	μ	k	f	α/α_0	γ/γ_0
			$T=160$ K			
80.0	937.7	—0.47	8.76	57.17	0.429	2.268
85.0	963.6	—0.47	8.81	65.48	0.413	2.212
90.0	989.0	—0.48	8.86	74.89	0.399	2.162
95.0	1014.2	—0.49	8.91	85.54	0.387	2.118
100.0	1039.2	—0.49	8.97	97.58	0.376	2.079
			$T=165$ K			
0.1	257.1	6.89	1.40	0.10	1.015	1.010
0.5	255.1	6.84	1.41	0.49	1.077	1.052
1.0	252.8	6.77	1.41	0.95	1.162	1.107
1.5	250.5	6.69	1.42	1.40	1.257	1.168
2.0	248.4	6.61	1.44	1.82	1.364	1.233
2.5	246.4	6.52	1.45	2.22	1.484	1.305
3.0	244.5	6.41	1.47	2.61	1.622	1.383
3.5	242.8	6.29	1.49	2.97	1.778	1.469
4.0	241.3	6.14	1.52	3.31	1.957	1.563
4.5	240.1	5.96	1.56	3.64	2.159	1.666
5.0	239.4	5.75	1.60	3.95	2.386	1.780
6.0	239.9	5.21	1.72	4.51	2.892	2.039
7.0	244.4	4.53	1.91	5.01	3.374	2.335
8.0	254.0	3.78	2.18	5.46	3.650	2.647
9.0	268.8	3.07	2.54	5.86	3.627	2.940
10.0	287.4	2.48	2.95	6.24	3.383	3.190
11.0	308.3	2.00	3.38	6.60	3.050	3.386
12.0	329.8	1.63	3.82	6.96	2.720	3.531
13.0	351.1	1.33	4.23	7.30	2.431	3.631
14.0	371.7	1.10	4.61	7.65	2.189	3.696
15.0	391.3	0.91	4.96	7.99	1.989	3.733
16.0	409.8	0.76	5.27	8.34	1.824	3.750
17.0	427.4	0.63	5.55	8.70	1.687	3.752
18.0	444.1	0.52	5.80	9.05	1.571	3.743
19.0	459.9	0.43	6.03	9.42	1.472	3.725
20.0	474.8	0.36	6.23	9.79	1.388	3.701
21.0	489.1	0.29	6.41	10.17	1.314	3.672
22.0	502.7	0.23	6.57	10.55	1.250	3.640
23.0	515.8	0.18	6.71	10.95	1.193	3.605
24.0	528.2	0.13	6.84	11.35	1.142	3.569
25.0	540.2	0.09	6.96	11.76	1.096	3.532
26.0	551.8	0.05	7.06	12.18	1.055	3.495
27.0	562.9	0.02	7.16	12.61	1.018	3.457
28.0	573.6	—0.01	7.24	13.06	0.984	3.419
29.0	584.0	—0.04	7.32	13.51	0.953	3.381
30.0	594.0	—0.07	7.39	13.97	0.924	3.344
35.0	640.2	—0.17	7.67	16.47	0.809	3.166
40.0	681.0	—0.25	7.85	19.30	0.725	3.005
45.0	718.0	—0.30	7.97	22.50	0.661	2.862
50.0	752.1	—0.34	8.06	26.14	0.611	2.735

Table II.12 (*Continued*)

p	w	μ	k	f	α/α_0	γ/γ_0
			$T=165$ K			
55.0	784.0	—0.37	8.13	30.27	0.570	2.624
60.0	814.2	—0.40	8.19	34.95	0.535	2.525
65.0	843.0	—0.42	8.24	40.24	0.506	2.438
70.0	870.9	—0.44	8.29	46.24	0.481	2.360
75.0	897.8	—0.45	8.34	53.02	0.459	2.291
80.0	924.2	—0.47	8.40	60.69	0.440	2.230
85.0	949.9	—0.48	8.45	69.35	0.424	2.175
90.0	975.3	—0.48	8.50	79.13	0.409	2.126
95.0	1000.3	—0.49	8.56	90.16	0.395	2.082
100.0	1025.0	—0.50	8.62	102.59	0.383	2.043
			$T=170$ K			
0.1	261.1	6.56	1.40	0.10	1.014	1.009
0.5	259.3	6.50	1.41	0.49	1.070	1.048
1.0	257.2	6.42	1.42	0.96	1.147	1.099
1.5	255.3	6.34	1.43	1.41	1.232	1.154
2.0	253.4	6.26	1.44	1.84	1.325	1.213
2.5	251.8	6.16	1.45	2.25	1.429	1.276
3.0	250.2	6.06	1.47	2.64	1.544	1.344
3.5	248.8	5.94	1.49	3.02	1.672	1.418
4.0	247.7	5.81	1.52	3.38	1.815	1.497
4.5	246.8	5.65	1.55	3.72	1.972	1.583
5.0	246.3	5.48	1.58	4.04	2.144	1.676
6.0	246.7	5.04	1.68	4.65	2.520	1.883
7.0	250.1	4.50	1.83	5.19	2.891	2.116
8.0	257.3	3.89	2.03	5.69	3.162	2.362
9.0	268.5	3.27	2.29	6.15	3.259	2.602
10.0	283.2	2.72	2.59	6.58	3.180	2.821
11.0	300.4	2.25	2.94	6.99	2.987	3.006
12.0	318.9	1.86	3.29	7.38	2.747	3.154
13.0	338.0	1.54	3.64	7.77	2.506	3.267
14.0	357.0	1.29	3.98	8.15	2.285	3.350
15.0	375.6	1.08	4.30	8.53	2.092	3.407
16.0	393.5	0.91	4.60	8.92	1.926	3.444
17.0	410.6	0.76	4.87	9.30	1.784	3.464
18.0	427.0	0.64	5.11	9.69	1.662	3.472
19.0	442.7	0.54	5.34	10.09	1.558	3.471
20.0	457.6	0.45	5.54	10.49	1.467	3.462
21.0	471.9	0.37	5.73	10.90	1.388	3.447
22.0	485.6	0.30	5.90	11.32	1.319	3.428
23.0	498.7	0.25	6.05	11.75	1.257	3.405
24.0	511.3	0.19	6.19	12.18	1.202	3.380
25.0	523.4	0.15	6.31	12.63	1.153	3.353
26.0	535.1	0.10	6.43	13.08	1.109	3.324
27.0	546.4	0.06	6.54	13.54	1.069	3.295
28.0	557.3	0.03	6.63	14.02	1.032	3.265
29.0	567.8	0.00	6.72	14.50	0.998	3.235

Table II.12 (*Continued*)

p	w	μ	k	f	α/α_0	γ/γ_0
			$T=170$ K			
30.0	578.0	—0.03	6.80	15.00	0.968	3.204
35.0	624.9	—0.14	7.13	17.67	0.844	3.053
40.0	666.3	—0.23	7.35	20.69	0.755	2.913
45.0	703.8	—0.28	7.51	24.09	0.687	2.785
50.0	738.2	—0.33	7.62	27.94	0.633	2.670
55.0	770.4	—0.37	7.72	32.30	0.590	2.567
60.0	800.9	—0.39	7.80	37.22	0.553	2.475
65.0	829.9	—0.42	7.87	42.77	0.522	2.392
70.0	857.8	—0.44	7.93	49.05	0.496	2.319
75.0	884.8	—0.45	7.99	56.12	0.473	2.253
80.0	911.0	—0.47	8.05	64.10	0.452	2.193
85.0	936.7	—0.48	8.11	73.09	0.435	2.140
90.0	961.9	—0.49	8.17	83.20	0.419	2.092
95.0	986.7	—0.49	8.23	94.58	0.404	2.049
100.0	1011.2	—0.50	8.30	107.38	0.391	2.010
			$T=175$ K			
0.1	264.9	6.25	1.40	0.10	1.013	1.009
0.5	263.4	6.19	1.41	0.49	1.065	1.045
1.0	261.5	6.11	1.42	0.96	1.135	1.092
1.5	259.8	6.02	1.43	1.42	1.211	1.142
2.0	258.3	5.93	1.44	1.85	1.293	1.195
2.5	256.9	5.84	1.45	2.27	1.383	1.251
3.0	255.6	5.74	1.47	2.67	1.481	1.312
3.5	254.5	5.63	1.49	3.06	1.588	1.376
4.0	253.6	5.50	1.51	3.43	1.705	1.444
4.5	253.0	5.37	1.54	3.79	1.832	1.518
5.0	252.6	5.21	1.57	4.13	1.968	1.596
6.0	253.2	4.85	1.65	4.77	2.260	1.767
7.0	256.0	4.40	1.77	5.36	2.550	1.955
8.0	261.7	3.90	1.92	5.90	2.788	2.154
9.0	270.6	3.37	2.12	6.41	2.921	2.353
10.0	282.4	2.87	2.36	6.88	2.932	2.541
11.0	296.6	2.43	2.63	7.34	2.842	2.707
12.0	312.4	2.05	2.92	7.77	2.688	2.848
13.0	329.2	1.73	3.21	8.20	2.507	2.963
14.0	346.4	1.46	3.51	8.62	2.324	3.054
15.0	363.5	1.23	3.79	9.04	2.152	3.122
16.0	380.4	1.05	4.06	9.46	1.996	3.172
17.0	396.8	0.89	4.31	9.88	1.857	3.206
18.0	412.7	0.75	4.55	10.31	1.736	3.227
19.0	428.0	0.64	4.76	10.74	1.630	3.239
20.0	442.7	0.54	4.96	11.17	1.536	3.242
21.0	456.8	0.46	5.15	11.61	1.454	3.239
22.0	470.4	0.38	5.32	12.06	1.381	3.230
23.0	483.5	0.31	5.48	12.52	1.316	3.218
24.0	496.1	0.26	5.62	12.99	1.258	3.202

Table II.12 (*Continued*)

p	w	μ	k	f	α/α_0	γ/γ_0
			$T=175$ K			
25.0	508.3	0.20	5.75	13.47	1.206	3.184
26.0	520.0	0.16	5.88	13.95	1.159	3.164
27.0	531.3	0.11	5.99	14.45	1.116	3.142
28.0	542.3	0.08	6.09	14.96	1.077	3.119
29.0	552.9	0.04	6.19	15.48	1.042	3.095
30.0	563.2	0.01	6.28	16.01	1.009	3.071
35.0	610.6	—0.12	6.64	18.85	0.878	2.946
40.0	652.5	—0.20	6.89	22.05	0.784	2.824
45.0	690.4	—0.27	7.08	25.65	0.712	2.710
50.0	725.2	—0.32	7.22	29.70	0.656	2.606
55.0	757.6	—0.36	7.34	34.28	0.610	2.511
60.0	788.2	—0.39	7.43	39.43	0.571	2.426
65.0	817.3	—0.41	7.51	45.24	0.539	2.348
70.0	845.3	—0.43	7.59	51.77	0.511	2.279
75.0	872.2	—0.45	7.66	59.12	0.486	2.215
80.0	898.4	—0.47	7.73	67.39	0.465	2.158
85.0	924.0	—0.48	7.79	76.68	0.446	2.107
90.0	949.0	—0.49	7.86	87.11	0.429	2.060
95.0	973.6	—0.50	7.92	98.81	0.414	2.018
100.0	997.8	—0.50	7.99	111.93	0.400	1.979
			$T=180$ K			
0.1	268.7	5.97	1.40	0.10	1.012	1.008
0.5	267.3	5.90	1.41	0.49	1.060	1.041
1.0	265.8	5.81	1.42	0.97	1.124	1.085
1.5	264.3	5.72	1.43	1.42	1.192	1.131
2.0	263.0	5.63	1.44	1.87	1.266	1.179
2.5	261.8	5.54	1.45	2.29	1.344	1.230
3.0	260.8	5.44	1.47	2.70	1.429	1.284
3.5	259.9	5.34	1.49	3.10	1.521	1.341
4.0	259.2	5.22	1.51	3.48	1.619	1.401
4.5	258.8	5.10	1.53	3.85	1.723	1.464
5.0	258.6	4.96	1.56	4.21	1.834	1.531
6.0	259.3	4.65	1.63	4.88	2.068	1.676
7.0	261.8	4.27	1.73	5.51	2.301	1.833
8.0	266.6	3.85	1.85	6.09	2.502	1.998
9.0	273.9	3.40	2.01	6.64	2.639	2.164
10.0	283.7	2.96	2.20	7.16	2.692	2.325
11.0	295.6	2.55	2.42	7.66	2.665	2.472
12.0	309.1	2.19	2.65	8.14	2.578	2.602
13.0	323.8	1.87	2.90	8.60	2.454	2.712
14.0	339.1	1.60	3.15	9.06	2.313	2.804
15.0	354.8	1.37	3.39	9.52	2.169	2.877
16.0	370.4	1.17	3.63	9.98	2.032	2.934
17.0	385.8	1.01	3.86	10.43	1.904	2.977
18.0	401.0	0.86	4.08	10.89	1.789	3.008
19.0	415.7	0.74	4.29	11.35	1.685	3.029

Table II.12 (*Continued*)

p	w	μ	k	f	α/α_0	γ/γ_0
			$T=180$ K			
20.0	430.0	0.63	4.48	11.82	1.592	3.042
21.0	443.0	0.54	4.66	12.30	1.509	3.048
22.0	457.2	0.46	4.83	12.78	1.435	3.048
23.0	470.1	0.38	4.98	13.27	1.368	3.044
24.0	482.6	0.32	5.13	13.77	1.308	3.036
25.0	494.7	0.26	5.26	14.28	1.254	3.026
26.0	506.4	0.21	5.39	14.80	1.205	3.013
27.0	517.7	0.16	5.51	15.33	1.160	2.998
28.0	528.7	0.12	5.61	15.87	1.119	2.981
29.0	539.3	0.08	5.71	16.42	1.082	2.963
30.0	549.7	0.05	5.81	16.98	1.048	2.944
35.0	597.3	−0.09	6.19	20.00	0.910	2.843
40.0	639.6	−0.18	6.47	23.37	0.811	2.738
45.0	677.8	−0.25	6.68	27.16	0.736	2.637
50.0	712.9	−0.30	6.85	31.42	0.677	2.544
55.0	745.5	−0.35	6.98	36.20	0.629	2.457
60.0	776.2	−0.38	7.09	41.58	0.589	2.378
65.0	805.4	−0.41	7.19	47.62	0.555	2.306
70.0	833.4	−0.43	7.27	54.41	0.526	2.240
75.0	860.3	−0.45	7.35	62.02	0.500	2.180
80.0	886.4	−0.46	7.42	70.56	0.478	2.125
85.0	911.8	−0.48	7.49	80.12	0.458	2.075
90.0	936.6	−0.49	7.56	90.84	0.440	2.030
95.0	961.0	−0.50	7.63	102.84	0.423	1.988
100.0	984.9	−0.51	7.69	116.25	0.409	1.950
			$T=185$ K			
0.1	272.5	5.70	1.40	0.10	1.011	1.008
0.5	271.3	5.63	1.41	0.49	1.055	1.039
1.0	269.9	5.54	1.42	0.97	1.114	1.079
1.5	268.6	5.45	1.43	1.43	1.176	1.121
2.0	267.5	5.36	1.44	1.88	1.242	1.166
2.5	266.6	5.27	1.45	2.31	1.312	1.212
3.0	265.8	5.17	1.47	2.73	1.386	1.260
3.5	265.1	5.07	1.48	3.14	1.465	1.311
4.0	264.6	4.97	1.50	3.53	1.549	1.364
4.5	264.3	4.85	1.53	3.91	1.637	1.420
5.0	264.3	4.73	1.55	4.28	1.729	1.478
6.0	265.1	4.45	1.61	4.98	1.921	1.603
7.0	267.4	4.13	1.70	5.64	2.112	1.736
8.0	271.6	3.77	1.80	6.26	2.283	1.876
9.0	277.8	3.38	1.93	6.85	2.410	2.017
10.0	286.2	2.99	2.09	7.41	2.480	2.155
11.0	296.3	2.62	2.26	7.95	2.490	2.285
12.0	308.0	2.28	2.46	8.47	2.447	2.402
13.0	320.9	1.98	2.66	8.97	2.368	2.505
14.0	334.6	1.71	2.88	9.47	2.266	2.594

Table II.12 (*Continued*)

p	w	μ	k	f	α/α_0	γ/γ_0
			$T=185$ K			
15.0	348.7	1.48	3.09	9.97	2.152	2.667
16.0	363.1	1.28	3.30	10.46	2.037	2.727
17.0	377.4	1.11	3.51	10.95	1.925	2.775
18.0	391.7	0.96	3.71	11.44	1.820	2.813
19.0	405.7	0.83	3.90	11.94	1.722	2.840
20.0	419.4	0.72	4.08	12.44	1.633	2.860
21.0	432.8	0.62	4.25	12.95	1.552	2.874
22.0	445.8	0.53	4.41	13.47	1.478	2.881
23.0	458.5	0.45	4.56	13.99	1.411	2.884
24.0	470.7	0.38	4.71	14.52	1.351	2.883
25.0	482.7	0.32	4.84	15.06	1.296	2.879
26.0	494.2	0.26	4.96	15.61	1.245	2.872
27.0	505.5	0.21	5.08	16.18	1.200	2.862
28.0	516.4	0.17	5.19	16.75	1.158	2.851
29.0	527.0	0.12	5.29	17.33	1.119	2.839
30.0	537.3	0.09	5.39	17.93	1.083	2.825
35.0	585.1	—0.06	5.79	21.11	0.940	2.744
40.0	627.6	—0.16	6.09	24.66	0.837	2.656
45.0	666.0	—0.24	6.32	28.63	0.759	2.568
50.0	701.3	—0.29	6.50	33.08	0.698	2.484
55.0	734.1	—0.34	6.65	38.07	0.648	2.405
60.0	764.9	—0.37	6.77	43.66	0.606	2.332
65.0	794.2	—0.40	6.88	49.93	0.571	2.265
70.0	822.1	—0.42	6.97	56.95	0.541	2.203
75.0	849.0	—0.44	7.06	64.80	0.514	2.146
80.0	875.0	—0.46	7.14	73.59	0.490	2.094
85.0	900.2	—0.48	7.21	83.42	0.469	2.046
90.0	924.9	—0.49	7.28	94.40	0.451	2.001
95.0	949.0	—0.50	7.35	106.67	0.434	1.961
100.0	972.6	—0.51	7.42	120.35	0.418	1.923
			$T=190$ K			
0.1	276.2	5,45	1.40	0.10	1.010	1.007
0.5	275.1	5,38	1.41	0.49	1.051	1.036
1.0	273.9	5,29	1.42	0.97	1.105	1.074
1.5	272.9	5,20	1.43	1.44	1.162	1.113
2.0	272.0	5,11	1.44	1.89	1.221	1.154
2.5	271.2	5,02	1.45	2.33	1.283	1.196
3.0	270.5	4,93	1.46	2.75	1.349	1.240
3.5	270.0	4,83	1.48	3.17	1.418	1.286
4.0	269.7	4,73	1.50	3.57	1.491	1.333
4.5	269.6	4,63	1.52	3.96	1.566	1.383
5.0	269.7	4,51	1.54	4.34	1.644	1.434
6.0	270.6	4,26	1.60	5.07	1.805	1.543
7.0	272.8	3,98	1.67	5.76	1.965	1.658
8.0	276.6	3,66	1.76	6.42	2.110	1.778
9.0	282.1	3,33	1.87	7.04	2.225	1.900

Table II.12 (*Continued*)

p	w	μ	k	f	α/α_0	γ/γ_0
			$T=190$ K			
10.0	289.3	2.98	2.00	7.64	2.299	2.019
11.0	298.2	2.65	2.15	8.21	2.328	2.134
12.0	308.4	2.34	2.31	8.77	2.315	2.239
13.0	319.8	2.05	2.49	9.32	2.268	2.334
14.0	332.0	1.80	2.67	9.85	2.197	2.417
15.0	344.8	1.57	2.85	10.38	2.111	2.489
16.0	357.9	1.37	3.04	10.91	2.018	2.550
17.0	371.2	1.20	3.22	11.44	1.923	2.600
18.0	384.5	1.05	3.40	11.96	1.831	2.641
19.0	397.8	0.91	3.58	12.50	1.742	2.673
20.0	410.8	0.79	3.75	13.03	1.659	2.698
21.0	423.7	0.69	3.91	13.57	1.582	2.716
22.0	436.2	0.59	4.06	14.12	1.511	2.729
23.0	448.5	0.51	4.20	14.68	1.446	2.737
24.0	460.5	0.44	4.34	15.24	1.386	2.742
25.0	472.2	0.37	4.47	15.81	1.331	2.743
26.0	483.6	0.31	4.59	16.40	1.280	2.741
27.0	494.6	0.26	4.71	16.99	1.234	2.736
28.0	505.4	0.21	4.82	17.59	1.191	2.730
29.0	516.0	0.16	4.92	18.21	1.152	2.722
30.0	526.2	0.12	5.02	18.84	1.115	2.712
35.0	573.9	—0.03	5.43	22.18	0.968	2.650
40.0	616.4	—0.14	5.75	25.90	0.862	2.577
45.0	655.0	—0.22	5.99	30.05	0.781	2.500
50.0	690.4	—0.28	6.19	34.69	0.718	2.426
55.0	723.4	—0.33	6.35	39.87	0.666	2.355
60.0	754.3	—0.36	6.48	45.67	0.623	2.288
65.0	783.6	—0.39	6.59	52.15	0.587	2.225
70.0	811.5	—0.42	6.70	59.39	0.555	2.167
75.0	838.3	—0.44	6.79	67.47	0.528	2.113
80.0	864.2	—0.46	6.87	76.50	0.503	2.064
85.0	889.3	—0.47	6.95	86.57	0.481	2.017
90.0	913.7	—0.49	7.02	97.79	0.462	1.975
95.0	937.6	—0.50	7.09	110.30	0.444	1.935
100.0	960.9	—0.51	7.16	124.22	0.428	1.898
			$T=195$ K			
0.1	279.8	5.22	1.40	0.10	1.009	1.007
0.5	278.9	5.15	1.41	0.49	1.048	1.034
1.0	277.9	5.06	1.42	0.97	1.097	1.069
1.5	277.0	4.97	1.43	1.44	1.149	1.105
2.0	276.3	4.88	1.44	1.90	1.203	1.143
2.5	275.6	4.79	1.45	2.34	1.259	1.182
3.0	275.2	4.70	1.46	2.77	1.318	1.222
3.5	274.8	4.61	1.48	3.20	1.378	1.263
4.0	274.6	4.51	1.50	3.61	1.442	1.306
4.5	274.6	4.41	1.52	4.01	1.507	1.351

Table II.12 (*Continued*)

p	w	μ	k	f	α/α_0	γ/γ_0
			$T=195$ K			
5.0	274.8	4.31	1.54	4.40	1.574	1.397
6.0	275.9	4.08	1.59	5.15	1.712	1.493
7.0	278.0	3.83	1.65	5.87	1.847	1.594
8.0	281.5	3.55	1.73	6.56	1.972	1.698
9.0	286.4	3.26	1.83	7.21	2.074	1.804
10.0	292.9	2.95	1.94	7.84	2.147	1.909
11.0	300.7	2.65	2.06	8.45	2.185	2.010
12.0	309.9	2.36	2.20	9.05	2.189	2.105
13.0	320.0	2.10	2.35	9.63	2.165	2.191
14.0	331.0	1.86	2.51	10.20	2.118	2.269
15.0	342.6	1.64	2.67	10.77	2.055	2.338
16.0	354.6	1.44	2.83	11.33	1.982	2.397
17.0	366.9	1.27	3.00	11.89	1.904	2.447
18.0	379.3	1.12	3.16	12.45	1.825	2.489
19.0	391.7	0.98	3.32	13.02	1.747	2.524
20.0	404.0	0.86	3.47	13.59	1.672	2.552
21.0	416.3	0.75	3.62	14.16	1.600	2.574
22.0	428.3	0.66	3.76	14.74	1.533	2.591
23.0	440.1	0.57	3.90	15.33	1.471	2.604
24.0	451.8	0.49	4.03	15.92	1.413	2.612
25.0	463.1	0.42	4.15	16.53	1.359	2.617
26.0	474.3	0.36	4.27	17.14	1.309	2.620
27.0	485.1	0.30	4.39	17.77	1.263	2.619
28.0	495.7	0.25	4.49	18.41	1.220	2.617
29.0	506.1	0.20	4.60	19.05	1.181	2.613
30.0	516.3	0.16	4.69	19.71	1.144	2.607
35.0	563.6	—0.01	5.11	23.22	0.994	2.561
40.0	606.1	—0.12	5.43	27.10	0.885	2.501
45.0	644.7	—0.20	5.69	31.42	0.802	2.436
50.0	680.3	—0.27	5.89	36.24	0.737	2.370
55.0	713.3	—0.32	6.06	41.61	0.684	2.306
60.0	744.3	—0.35	6.21	47.60	0.639	2.245
65.0	773.6	—0.39	6.33	54.28	0.602	2.187
70.0	801.5	—0.41	6.44	61.73	0.569	2.133
75.0	828.2	—0.44	6.53	70.03	0.541	2.082
80.0	854.0	—0.45	6.62	79.28	0.516	2.035
85.0	878.9	—0.47	6.70	89.56	0.493	1.990
90.0	903.1	—0.49	6.78	101.01	0.473	1.949
95.0	926.8	—0.50	6.85	113.74	0.455	1.911
100.0	949.9	—0.51	6.92	127.88	0.438	1.875
			$T=200$ K			
0.1	283.4	5.00	1.40	0.10	1.009	1.006
0.5	282.6	4.93	1.41	0.49	1.044	1.032
1.0	281.8	4.84	1.42	0.98	1.090	1.065
1.5	281.0	4.75	1.43	1.45	1.138	1.099
2.0	280.4	4.66	1.44	1.91	1.187	1.133

Table II.12 (*Continued*)

p	w	μ	k	f	α/α_0	γ/γ_0
			$T=200$ K			
2.5	280.0	4.57	1.45	2.36	1.238	1.169
3.0	279.6	4.49	1.46	2.79	1.290	1.206
3.5	279.4	4.40	1.48	3.22	1.344	1.244
4.0	279.4	4.31	1.49	3.64	1.400	1.283
4.5	279.5	4.22	1.51	4.05	1.458	1.323
5.0	279.8	4.12	1.53	4.45	1.516	1.364
6.0	280.9	3.91	1.58	5.23	1.635	1.450
7.0	283.1	3.68	1.64	5.97	1.751	1.540
8.0	286.3	3.44	1.71	6.68	1.859	1.632
9.0	290.9	3.17	1.79	7.37	1.950	1.725
10.0	296.7	2.90	1.89	8.03	2.019	1.818
11.0	303.8	2.63	1.99	8.68	2.060	1.908
12.0	312.0	2.37	2.11	9.30	2.075	1.993
13.0	321.1	2.12	2.24	9.92	2.066	2.072
14.0	331.1	1.89	2.38	10.53	2.036	2.144
15.0	341.6	1.68	2.52	11.13	1.991	2.208
16.0	352.6	1.50	2.67	11.72	1.935	2.265
17.0	364.0	1.33	2.81	12.32	1.872	2.314
18.0	375.5	1.18	2.96	12.91	1.806	2.357
19.0	387.1	1.04	3.10	13.51	1.739	2.393
20.0	398.8	0.92	3.24	14.11	1.672	2.423
21.0	410.4	0.81	3.38	14.72	1.608	2.447
22.0	421.9	0.71	3.51	15.33	1.546	2.467
23.0	433.2	0.62	3.64	15.95	1.487	2.482
24.0	444.4	0.54	3.76	16.57	1.432	2.494
25.0	455.4	0.47	3.88	17.21	1.380	2.503
26.0	466.2	0.40	3.99	17.86	1.332	2.508
27.0	476.9	0.34	4.10	18.51	1.286	2.511
28.0	487.2	0.29	4.21	19.18	1.244	2.512
29.0	497.4	0.24	4.31	19.86	1.205	2.511
30.0	507.4	0.20	4.40	20.55	1.168	2.508
35.0	554.2	0.02	4.82	24.21	1.017	2.477
40.0	596.6	—0.10	5.14	28.25	0.906	2.429
45.0	635.2	—0.19	5.41	32.74	0.821	2.373
50.0	670.8	—0.26	5.62	37.73	0.755	2.316
55.0	703.8	—0.31	5.80	43.28	0.700	2.258
60.0	734.8	—0.35	5.95	49.45	0.655	2.203
65.0	764.1	—0.38	6.08	56.33	0.617	2.150
70.0	792.0	—0.41	6.19	63.97	0.583	2.099
75.0	818.7	—0.43	6.30	72.47	0.554	2.052
80.0	844.4	—0.45	6.39	81.92	0.528	2.007
85.0	869.1	—0.47	6.47	92.41	0.505	1.965
90.0	893.2	—0.48	6.55	104.06	0.484	1.925
95.0	916.6	—0.50	6.62	116.99	0.466	1.888
100.0	939.4	—0.51	6.69	131.32	0.448	1.853

Table II.12 (*Continued*)

p	w	μ	k	f	α/α_0	γ/γ_0
			$T=210$ K			
0.1	290.5	4.60	1.40	0.10	1.008	1.006
0.5	289.9	4.53	1.41	0.49	1.039	1.028
1.0	289.3	4.44	1.42	0.98	1.078	1.057
1.5	288.9	4.35	1.43	1.46	1.119	1.087
2.0	288.5	4.27	1.44	1.92	1.160	1.117
2.5	288.3	4.19	1.45	2.38	1.202	1.148
3.0	288.2	4.10	1.46	2.83	1.245	1.179
3.5	288.2	4.02	1.47	3.27	1.289	1.211
4.0	288.4	3.94	1.49	3.70	1.333	1.244
4.5	288.7	3.86	1.51	4.13	1.379	1.278
5.0	289.1	3.77	1.52	4.54	1.424	1.312
6.0	290.5	3.60	1.56	5.36	1.516	1.382
7.0	292.6	3.40	1.61	6.14	1.605	1.454
8.0	295.7	3.20	1.67	6.90	1.688	1.528
9.0	299.6	2.99	1.73	7.64	1.760	1.602
10.0	304.6	2.76	1.81	8.36	1.818	1.676
11.0	310.6	2.54	1.89	9.06	1.859	1.749
12.0	317.4	2.32	1.99	9.75	1.883	1.818
13.0	325.1	2.11	2.09	10.43	1.890	1.884
14.0	333.4	1.91	2.19	11.10	1.881	1.945
15.0	342.4	1.73	2.31	11.77	1.860	2.001
16.0	351.8	1.55	2.42	12.43	1.829	2.052
17.0	361.5	1.40	2.54	13.08	1.789	2.098
18.0	371.5	1.25	2.65	13.74	1.745	2.138
19.0	381.7	1.12	2.77	14.40	1.697	2.174
20.0	392.0	1.00	2.89	15.06	1.647	2.205
21.0	402.4	0.89	3.00	15.73	1.596	2.231
22.0	412.8	0.80	3.12	16.40	1.546	2.254
23.0	423.2	0.71	3.23	17.08	1.497	2.273
24.0	433.4	0.62	3.34	17.77	1.449	2.289
25.0	443.7	0.55	3.44	18.47	1.403	2.302
26.0	453.8	0.48	3.54	19.17	1.359	2.312
27.0	463.7	0.42	3.64	19.89	1.318	2.320
28.0	473.6	0.36	3.74	20.62	1.278	2.325
29.0	483.3	0.31	3.83	21.36	1.241	2.329
30.0	492.8	0.26	3.92	22.11	1.205	2.331
35.0	538.2	0.07	4.32	26.07	1.056	2.323
40.0	579.9	—0.07	4.65	30.42	0.943	2.295
45.0	618.2	—0.16	4.92	35.22	0.856	2.257
50.0	653.7	—0.23	5.14	40.53	0.787	2.213
55.0	686.7	—0.29	5.33	46.41	0.731	2.168
60.0	717.7	—0.33	5.50	52.93	0.684	2.123
65.0	747.0	—0.37	5.64	60.15	0.644	2.078
70.0	774.8	—0.40	5.76	68.15	0.610	2.035
75.0	801.4	—0.42	5.87	77.01	0.579	1.994

Table II.12 (*Continued*)

p	w	μ	k	f	α/α_0	γ/γ_0
			$T=210$ K			
80.0	826.8	−0.44	5.96	86.82	0.553	1.954
85.0	851.3	−0.46	6.05	97.68	0.528	1.916
90.0	875.0	−0.48	6.13	109.68	0.507	1.880
95.0	898.0	−0.49	6.21	122.95	0.487	1.846
100.0	920.4	−0.50	6.28	137.60	0.469	1.813
			$T=220$ K			
0.1	297.4	4.24	1.40	0.10	1.007	1.005
0.5	297.0	4.17	1.41	0.50	1.034	1.025
1.0	296.6	4.08	1.42	0.98	1.068	1.051
1.5	296.4	4.00	1.43	1.46	1.103	1.077
2.0	296.3	3.92	1.44	1.93	1.138	1.104
2.5	296.3	3.85	1.45	2.40	1.173	1.130
3.0	296.4	3.77	1.46	2.86	1.209	1.158
3.5	296.6	3.69	1.47	3.31	1.246	1.185
4.0	296.9	3.62	1.49	3.75	1.282	1.213
4.5	297.3	3.54	1.50	4.19	1.319	1.242
5.0	297.9	3.47	1.52	4.62	1.356	1.271
6.0	299.5	3.31	1.55	5.46	1.428	1.329
7.0	301.7	3.15	1.59	6.29	1.499	1.389
8.0	304.5	2.98	1.64	7.09	1.564	1.450
9.0	308.2	2.80	1.70	7.87	1.622	1.511
10.0	312.6	2.61	1.76	8.64	1.671	1.572
11.0	317.8	2.42	1.83	9.39	1.708	1.632
12.0	323.8	2.24	1.90	10.14	1.733	1.689
13.0	330.4	2.06	1.98	10.87	1.746	1.745
14.0	337.6	1.88	2.07	11.59	1.748	1.797
15.0	345.4	1.72	2.16	12.31	1.740	1.845
16.0	353.5	1.57	2.25	13.03	1.723	1.890
17.0	362.1	1.42	2.35	13.75	1.699	1.931
18.0	370.9	1.29	2.44	14.46	1.670	1.968
19.0	379.9	1.17	2.54	15.18	1.637	2.002
20.0	389.0	1.05	2.64	15.90	1.600	2.032
21.0	398.3	0.95	2.73	16.63	1.562	2.058
22.0	407.7	0.85	2.83	17.36	1.523	2.082
23.0	417.1	0.76	2.92	18.09	1.483	2.102
24.0	426.5	0.68	3.02	18.84	1.444	2.120
25.0	435.9	0.61	3.11	19.59	1.405	2.135
26.0	445.2	0.54	3.20	20.35	1.367	2.147
27.0	454.5	0.47	3.29	21.13	1.330	2.158
28.0	463.7	0.42	3.38	21.91	1.295	2.167
29.0	472.8	0.36	3.46	22.71	1.260	2.174
30.0	481.9	0.31	3.54	23.51	1.228	2.179
35.0	525.4	0.11	3.91	27.76	1.085	2.187
40.0	566.0	−0.03	4.23	32.39	0.974	2.175
45.0	603.7	−0.14	4.50	37.48	0.886	2.150
50.0	638.9	−0.21	4.73	43.08	0.815	2.119

Table II.12 (*Continued*)

p	w	μ	k	f	α/α_0	γ/γ_0
			$T=220$ K			
55.0	671.8	—0.27	4.93	49.27	0.758	2.084
60.0	702.7	—0.32	5.10	56.09	0.710	2.048
65.0	731.9	—0.36	5.25	63.62	0.669	2.011
70.0	759.6	—0.39	5.38	71.94	0.633	1.975
75.0	786.0	—0.41	5.49	81.11	0.602	1.939
80.0	811.3	—0.44	5.59	91.23	0.575	1.905
85.0	835.6	—0.45	5.69	102.38	0.550	1.871
90.0	859.1	—0.47	5.77	114.68	0.528	1.838
95.0	881.7	—0.49	5.85	128.21	0.507	1.807
100.0	903.7	—0.50	5.92	143.12	0.489	1.776
			$T=230$ K			
0.1	304.1	3.92	1.40	0.10	1.006	1.005
0.5	303.9	3.85	1.41	0.50	1.030	1.023
1.0	303.7	3.77	1.42	0.99	1.060	1.046
1.5	303.7	3.69	1.43	1.47	1.090	1.069
2.0	303.7	3.62	1.44	1.95	1.120	1.092
2.5	303.9	3.54	1.45	2.42	1.150	1.116
3.0	304.2	3.47	1.46	2.88	1.181	1.140
3.5	304.5	3.40	1.47	3.34	1.211	1.164
4.0	305.0	3.33	1.48	3.79	1.242	1.188
4.5	305.6	3.27	1.50	4.24	1.272	1.213
5.0	306.2	3.20	1.51	4.68	1.302	1.238
6.0	307.9	3.06	1.54	5.55	1.362	1.288
7.0	310.2	2.91	1.58	6.41	1.419	1.339
8.0	313.0	2.76	1.62	7.24	1.472	1.390
9.0	316.4	2.61	1.67	8.07	1.519	1.442
10.0	320.5	2.45	1.72	8.88	1.560	1.493
11.0	325.2	2.29	1.78	9.67	1.592	1.543
12.0	330.5	2.13	1.84	10.46	1.616	1.591
13.0	336.4	1.98	1.91	11.24	1.631	1.638
14.0	342.8	1.83	1.98	12.02	1.638	1.683
15.0	349.6	1.68	2.05	12.79	1.636	1.725
16.0	356.9	1.54	2.13	13.55	1.628	1.764
17.0	364.4	1.41	2.21	14.32	1.614	1.801
18.0	372.3	1.29	2.29	15.09	1.595	1.835
19.0	380.4	1.18	2.37	15.86	1.572	1.866
20.0	388.6	1.07	2.45	16.63	1.546	1.894
21.0	397.0	0.97	2.54	17.41	1.517	1.919
22.0	405.5	0.88	2.62	18.19	1.487	1.942
23.0	414.0	0.80	2.70	18.98	1.456	1.962
24.0	422.6	0.72	2.78	19.78	1.424	1.980
25.0	431.2	0.65	2.86	20.59	1.392	1.996
26.0	439.8	0.58	2.94	21.40	1.360	2.010
27.0	448.5	0.52	3.02	22.23	1.328	2.022
28.0	457.0	0.46	3.09	23.07	1.297	2.032
29.0	465.6	0.40	3.17	23.91	1.267	2.041

Table II.12 (*Continued*)

p	w	μ	k	f	α/α_0	γ/γ_0
			$T=230$ K			
30.0	474.0	0.35	3.24	24.77	1.238	2.049
35.0	515.4	0.14	3.59	29.28	1.105	2.068
40.0	554.6	0.00	3.89	34.18	0.997	2.067
45.0	591.5	—0.11	4.15	39.53	0.910	2.053
50.0	626.2	—0.19	4.38	45.40	0.840	2.032
55.0	658.8	—0.26	4.58	51.86	0.781	2.006
60.0	689.5	—0.31	4.75	58.95	0.732	1.978
65.0	718.6	—0.35	4.91	66.76	0.691	1.948
70.0	746.2	—0.38	5.04	75.34	0.655	1.918
75.0	772.5	—0.41	5.16	84.78	0.623	1.888
80.0	797.7	—0.43	5.27	95.16	0.595	1.858
85.0	821.8	—0.45	5.37	106.57	0.570	1.828
90.0	845.0	—0.47	5.45	119.09	0.548	1.799
95.0	867.4	—0.48	5.53	132.84	0.527	1.771
100.0	889.0	—0.50	5.61	147.93	0.508	1.743
			$T=240$ K			
0.1	310.7	3.63	1.40	0.10	1.005	1.004
0.5	310.6	3.56	1.41	0.50	1.026	1.021
1.0	310.6	3.49	1.42	0.99	1.053	1.041
1.5	310.7	3.41	1.43	1.47	1.079	1.062
2.0	310.9	3.34	1.43	1.95	1.105	1.083
2.5	311.3	3.27	1.44	2.43	1.131	1.104
3.0	311.6	3.21	1.46	2.90	1.157	1.125
3.5	312.1	3.14	1.47	3.37	1.183	1.146
4.0	312.7	3.08	1.48	3.83	1.209	1.168
4.5	313.4	3.02	1.49	4.29	1.234	1.189
5.0	314.2	2.95	1.51	4.74	1.260	1.211
6.0	316.0	2.83	1.53	5.63	1.309	1.254
7.0	318.3	2.70	1.57	6.51	1.356	1.298
8.0	321.0	2.57	1.60	7.38	1.400	1.342
9.0	324.3	2.43	1.65	8.23	1.439	1.386
10.0	328.1	2.30	1.69	9.08	1.474	1.430
11.0	332.5	2.16	1.74	9.91	1.502	1.473
12.0	337.3	2.02	1.79	10.74	1.523	1.514
13.0	342.6	1.88	1.85	11.56	1.538	1.555
14.0	348.4	1.75	1.91	12.38	1.547	1.593
15.0	354.6	1.62	1.97	13.19	1.549	1.630
16.0	361.2	1.50	2.04	14.01	1.546	1.665
17.0	368.0	1.38	2.11	14.82	1.538	1.697
18.0	375.1	1.27	2.17	15.63	1.525	1.728
19.0	382.4	1.17	2.24	16.45	1.509	1.756
20.0	389.9	1.07	2.31	17.27	1.490	1.782
21.0	397.5	0.98	2.39	18.10	1.469	1.805
22.0	405.2	0.89	2.46	18.93	1.446	1.827
23.0	413.1	0.81	2.53	19.77	1.421	1.847
24.0	421.0	0.74	2.60	20.61	1.396	1.865

p	w	μ	k	f	α/α_0	γ/γ_0
			$T=240$ K			
25.0	428.9	0.67	2.67	21.47	1.369	1.880
26.0	436.9	0.60	2.74	22.33	1.343	1.895
27.0	444.8	0.54	2.81	23.21	1.316	1.907
28.0	452.8	0.48	2.88	24.09	1.290	1.919
29.0	460.8	0.43	2.94	24.99	1.264	1.928
30.0	468.7	0.38	3.01	25.90	1.238	1.937
35.0	507.9	0.17	3.32	30.64	1.117	1.964
40.0	545.6	0.02	3.61	35.78	1.015	1.971
45.0	581.4	—0.09	3.86	41.38	0.930	1.966
50.0	615.4	—0.18	4.08	47.49	0.860	1.952
55.0	647.6	—0.24	4.28	54.19	0.801	1.934
60.0	678.0	—0.30	4.45	61.53	0.752	1.912
65.0	706.9	—0.34	4.61	69.57	0.710	1.889
70.0	734.4	—0.37	4.75	78.39	0.674	1.864
75.0	760.6	—0.40	4.87	88.06	0.642	1.839
80.0	785.6	—0.42	4.98	98.65	0.614	1.813
85.0	809.6	—0.44	5.08	110.26	0.589	1.787
90.0	832.6	—0.46	5.17	122.97	0.566	1.762
95.0	854.8	—0.48	5.25	136.89	0.545	1.737
100.0	876.2	—0.49	5.33	152.10	0.526	1.712
			$T=250$ K			
0.1	317.1	3.36	1.40	0.10	1.005	1.004
0.5	317.2	3.30	1.41	0.50	1.023	1.019
1.0	317.3	3.23	1.42	0.99	1.047	1.037
1.5	317.6	3.16	1.42	1.48	1.070	1.056
2.0	317.9	3.10	1.43	1.96	1.093	1.075
2.5	318.3	3.03	1.44	2.44	1.115	1.094
3.0	318.9	2.97	1.45	2.92	1.138	1.113
3.5	319.5	2.91	1.46	3.39	1.160	1.131
4.0	320.1	2.85	1.48	3.86	1.182	1.150
4.5	320.9	2.79	1.49	4.32	1.204	1.169
5.0	321.7	2.73	1.50	4.79	1.225	1.189
6.0	323.7	2.62	1.53	5.70	1.267	1.227
7.0	326.0	2.50	1.56	6.60	1.306	1.265
8.0	328.8	2.39	1.59	7.49	1.343	1.303
9.0	331.9	2.27	1.63	8.37	1.376	1.342
10.0	335.6	2.15	1.67	9.25	1.405	1.379
11.0	339.6	2.03	1.71	10.12	1.429	1.416
12.0	344.1	1.91	1.76	10.98	1.449	1.452
13.0	349.1	1.79	1.81	11.84	1.463	1.488
14.0	354.4	1.67	1.86	12.69	1.472	1.521
15.0	360.0	1.55	1.91	13.54	1.476	1.554
16.0	366.0	1.44	1.97	14.40	1.475	1.584
17.0	372.3	1.34	2.03	15.25	1.471	1.614
18.0	378.8	1.24	2.09	16.11	1.463	1.641
19.0	385.5	1.14	2.15	16.97	1.452	1.666

Table II.12 (*Continued*)

p	w	μ	k	f	α/α_0	γ/γ_0
			$T=250$ K			
20.0	392.3	1.05	2.21	17.83	1.438	1.690
21.0	399.3	0.97	2.27	18.70	1.422	1.712
22.0	406.4	0.89	2.33	19.57	1.404	1.732
23.0	413.7	0.81	2.39	20.45	1.384	1.751
24.0	421.0	0.74	2.46	21.34	1.364	1.768
25.0	428.3	0.68	2.52	22.24	1.343	1.784
26.0	435.7	0.61	2.58	23.15	1.321	1.798
27.0	443.1	0.55	2.64	24.07	1.298	1.810
28.0	450.6	0.50	2.70	25.00	1.276	1.822
29.0	458.0	0.45	2.76	25.94	1.253	1.832
30.0	465.5	0.40	2.82	26.89	1.231	1.841
35.0	502.4	0.19	3.11	31.86	1.123	1.873
40.0	538.5	0.04	3.37	37.22	1.027	1.886
45.0	573.2	—0.08	3.61	43.04	0.946	1.887
50.0	606.3	—0.16	3.82	49.37	0.877	1.880
55.0	637.9	—0.23	4.02	56.28	0.819	1.867
60.0	668.0	—0.29	4.19	63.83	0.769	1.851
65.0	696.7	—0.33	4.35	72.08	0.727	1.833
70.0	724.0	—0.37	4.49	81.10	0.691	1.813
75.0	750.0	—0.40	4.61	90.96	0.659	1.792
80.0	774.9	—0.42	4.72	101.74	0.630	1.771
85.0	798.8	—0.44	4.83	113.51	0.605	1.749
90.0	821.7	—0.46	4.92	126.36	0.582	1.726
95.0	843.7	—0.47	5.00	140.39	0.561	1.704
100.0	864.9	—0.49	5.08	155.70	0.542	1.682
			$T=260$ K			
0.1	323.4	3.12	1.40	0.10	1.004	1.003
0.5	323.6	3.07	1.41	0.50	1.021	1.017
1.0	323.8	3.00	1.42	0.99	1.042	1.034
1.5	324.2	2.93	1.42	1.48	1.062	1.051
2.0	324.7	2.87	1.43	1.97	1.082	1.068
2.5	325.2	2.81	1.44	2.45	1.102	1.085
3.0	325.8	2.75	1.45	2.93	1.121	1.102
3.5	326.5	2.70	1.46	3.41	1.141	1.119
4.0	327.3	2.64	1.47	3.88	1.160	1.136
4.5	328.1	2.59	1.48	4.36	1.178	1.153
5.0	329.0	2.54	1.50	4.83	1.197	1.170
6.0	331.1	2.43	1.52	5.76	1.232	1.204
7.0	333.4	2.33	1.55	6.68	1.266	1.237
8.0	336.2	2.22	1.58	7.59	1.297	1.271
9.0	339.3	2.11	1.61	8.50	1.325	1.305
10.0	342.8	2.01	1.65	9.40	1.350	1.338
11.0	346.6	1.90	1.69	10.29	1.371	1.370
12.0	350.9	1.79	1.73	11.18	1.388	1.402
13.0	355.5	1.69	1.77	12.07	1.401	1.433
14.0	360.4	1.58	1.82	12.96	1.410	1.463

Table II.12 (*Continued*)

p	w	μ	k	i	α/α_0	γ/γ_0
			$T=260$ K			
15.0	365.7	1.48	1.86	13.85	1.414	1.491
16.0	371.2	1.38	1.91	14.74	1.416	1.519
17.0	377.0	1.28	1.96	15.63	1.413	1.545
18.0	383.0	1.19	2.02	16.52	1.408	1.569
19.0	389.2	1.11	2.07	17.41	1.400	1.592
20.0	395.6	1.02	2.12	18.32	1.390	1.614
21.0	402.0	0.94	2.18	19.22	1.377	1.634
22.0	408.7	0.87	2.23	20.14	1.363	1.653
23.0	415.4	0.80	2.29	21.06	1.348	1.671
24.0	422.1	0.73	2.34	21.99	1.331	1.687
25.0	429.0	0.67	2.40	22.93	1.314	1.702
26.0	435.9	0.61	2.45	23.87	1.296	1.715
27.0	442.8	0.56	2.51	24.83	1.277	1.728
28.0	449.8	0.50	2.56	25.80	1.258	1.739
29.0	456.8	0.45	2.62	26.78	1.238	1.750
30.0	463.8	0.41	2.67	27.77	1.219	1.759
35.0	498.7	0.21	2.93	32.94	1.123	1.793
40.0	533.1	0.05	3.17	38.51	1.035	1.810
45.0	566.5	—0.06	3.40	44.52	0.957	1.816
50.0	598.8	—0.15	3.60	51.05	0.890	1.814
55.0	629.7	—0.22	3.79	58.15	0.833	1.806
60.0	659.4	—0.28	3.96	65.89	0.784	1.795
65.0	687.7	—0.33	4.12	74.32	0.742	1.781
70.0	714.8	—0.36	4.26	83.51	0.706	1.765
75.0	740.7	—0.39	4.38	93.53	0.674	1.748
80.0	765.5	—0.42	4.50	104.45	0.645	1.730
85.0	789.2	—0.44	4.60	116.35	0.620	1.711
90.0	812.0	—0.46	4.69	129.31	0.597	1.692
95.0	833.9	—0.47	4.78	143.41	0.576	1.673
100.0	855.0	—0.49	4.86	158.77	0.557	1.653
			$T=270$ K			
0.1	329.6	2.90	1.40	0.10	1.004	1.003
0.5	329.8	2.85	1.41	0.50	1.019	1.016
1.0	330.2	2.79	1.42	0.99	1.037	1.031
1.5	330.7	2.72	1.42	1.49	1.055	1.046
2.0	331.2	2.67	1.43	1.98	1.073	1.062
2.5	331.9	2.61	1.44	2.46	1.090	1.077
3.0	332.6	2.56	1.45	2.95	1.107	1.093
3.5	333.3	2.50	1.46	3.43	1.124	1.108
4.0	334.2	2.45	1.47	3.91	1.141	1.123
4.5	335.1	2.40	1.48	4.38	1.157	1.138
5.0	336.0	2.35	1.49	4.86	1.173	1.154
6.0	338.2	2.26	1.52	5.80	1.204	1.184
7.0	340.6	2.16	1.54	6.74	1.232	1.214
8.0	343.3	2.07	1.57	7.67	1.259	1.244
9.0	346.4	1.97	1.60	8.60	1.283	1.274

Table II.12 (*Continued*)

p	w	μ	k	f	α/α_0	γ/γ_0
			$T=270$ K			
10.0	349.8	1.87	1.63	9.52	1.304	1.303
11.0	353.5	1.78	1.67	10.44	1.323	1.332
12.0	357.5	1.68	1.70	11.36	1.338	1.360
13.0	361.8	1.59	1.74	12.28	1.349	1.387
14.0	366.5	1.49	1.78	13.19	1.357	1.414
15.0	371.4	1.40	1.82	14.11	1.363	1.439
16.0	376.6	1.31	1.87	15.03	1.365	1.464
17.0	382.0	1.22	1.91	15.95	1.364	1.487
18.0	387.6	1.14	1.96	16.88	1.360	1.510
19.0	393.4	1.06	2.01	17.81	1.354	1.531
20.0	399.3	0.99	2.06	18.74	1.347	1.550
21.0	405.4	0.91	2.10	19.68	1.337	1.569
22.0	411.6	0.84	2.15	20.63	1.326	1.587
23.0	417.8	0.78	2.20	21.59	1.313	1.603
24.0	424.2	0.72	2.25	22.55	1.300	1.618
25.0	430.6	0.66	2.30	23.52	1.285	1.632
26.0	437.1	0.60	2.35	24.51	1.270	1.645
27.0	443.6	0.55	2.40	25.50	1.254	1.657
28.0	450.1	0.50	2.45	26.51	1.238	1.668
29.0	456.7	0.45	2.50	27.52	1.221	1.679
30.0	463.3	0.41	2.54	28.55	1.204	1.688
35.0	496.3	0.21	2.78	33.91	1.120	1.723
40.0	529.2	0.06	3.01	39.65	1.039	1.743
45.0	561.4	—0.05	3.22	45.84	0.966	1.753
50.0	592.6	—0.14	3.41	52.54	0.901	1.754
55.0	622.8	—0.22	3.59	59.81	0.846	1.751
60.0	651.9	—0.27	3.76	67.71	0.797	1.743
65.0	679.9	—0.32	3.91	76.30	0.755	1.733
70.0	706.7	—0.36	4.05	85.63	0.719	1.720
75.0	732.4	—0.39	4.18	95.78	0.687	1.707
80.0	757.0	—0.42	4.29	106.82	0.658	1.692
85.0	780.7	—0.44	4.40	118.81	0.633	1.676
90.0	803.4	—0.46	4.49	131.85	0.610	1.660
95.0	825.2	—0.47	4.58	146.00	0.589	1.643
100.0	846.2	—0.49	4.66	161.37	0.570	1.626
			$T=280$ K			
0.1	335.6	2.70	1.40	0.10	1.003	1.003
0.5	335.9	2.65	1.41	0.50	1.017	1.014
1.0	336.4	2.59	1.42	0.99	1.033	1.028
1.5	337.0	2.53	1.42	1.49	1.049	1.043
2.0	337.6	2.48	1.43	1.98	1.065	1.057
2.5	338.4	2.43	1.44	2.47	1.080	1.071
3.0	339.1	2.38	1.45	2.96	1.095	1.084
3.5	340.0	2.33	1.46	3.44	1.110	1.098
4.0	340.8	2.28	1.47	3.93	1.125	1.112
4.5	341.8	2.23	1.48	4.41	1.139	1.126

Table II.12 (*Continued*)

p	w	μ	k	j	α/α_0	γ/γ_0
			$T=280$ K			
5.0	342.8	2.19	1.49	4.89	1.153	1.140
6.0	345.0	2.10	1.51	5.85	1.179	1.167
7.0	347.5	2.01	1.54	6.80	1.204	1.194
8.0	350.2	1.92	1.56	7.75	1.227	1.221
9.0	353.2	1.84	1.59	8.69	1.248	1.248
10.0	356.5	1.75	1.62	9.63	1.266	1.274
11.0	360.1	1.66	1.65	10.58	1.282	1.299
12.0	364.0	1.58	1.68	11.52	1.295	1.325
13.0	368.1	1.49	1.72	12.46	1.306	1.349
14.0	372.5	1.41	1.75	13.40	1.313	1.373
15.0	377.2	1.32	1.79	14.34	1.318	1.396
16.0	382.1	1.24	1.83	15.29	1.321	1.418
17.0	387.2	1.16	1.87	16.24	1.321	1.439
18.0	392.4	1.09	1.91	17.19	1.319	1.459
19.0	397.9	1.01	1.96	18.15	1.314	1.478
20.0	403.5	0.94	2.00	19.11	1.308	1.496
21.0	409.2	0.88	2.04	20.08	1.301	1.514
22.0	415.0	0.81	2.09	21.06	1.292	1.530
23.0	420.9	0.75	2.13	22.05	1.281	1.545
24.0	426.9	0.69	2.18	23.05	1.270	1.559
25.0	432.9	0.64	2.22	24.05	1.258	1.573
26.0	439.0	0.59	2.26	25.07	1.245	1.585
27.0	445.2	0.54	2.31	26.09	1.232	1.596
28.0	451.3	0.49	2.35	27.13	1.218	1.607
29.0	457.5	0.44	2.40	28.18	1.203	1.617
30.0	463.8	0.40	2.44	29.24	1.189	1.626
35.0	495.1	0.21	2.66	34.76	1.113	1.662
40.0	526.5	0.07	2.86	40.66	1.040	1.684
45.0	557.4	—0.05	3.06	47.01	0.971	1.696
50.0	587.7	—0.14	3.25	53.87	0.910	1.700
55.0	617.1	—0.21	3.42	61.29	0.856	1.700
60.0	645.6	—0.27	3.58	69.33	0.808	1.695
65.0	673.1	—0.32	3.73	78.05	0.767	1.688
70.0	699.6	—0.36	3.87	87.50	0.730	1.678
75.0	725.1	—0.39	3.99	97.76	0.698	1.667
80.0	749.5	—0.41	4.11	108.88	0.670	1.655
85.0	773.1	—0.44	4.21	120.95	0.644	1.642
90.0	795.7	—0.46	4.31	134.03	0.622	1.628
95.0	817.5	—0.47	4.40	148.20	0.601	1.614
100.0	838.4	—0.49	4.48	163.55	0.582	1.599
			$T=290$ K			
0.1	341.5	2.51	1.40	0.10	1.003	1.003
0.5	341.9	2.47	1.41	0.50	1.015	1.013
1.0	342.5	2.41	1.41	1.00	1.030	1.026
1.5	343.2	2.36	1.42	1.49	1.044	1.039
2.0	343.9	2.31	1.43	1.98	1.058	1.052

Table II.12 (*Continued*)

p	w	μ	k	f	α/α_0	γ/γ_0
			$T=290$ K			
2.5	344.7	2.26	1.44	2.48	1.072	1.065
3.0	345.5	2.21	1.45	2.97	1.085	1.077
3.5	346.4	2.17	1.46	3.46	1.098	1.090
4.0	347.3	2.12	1.47	3.94	1.111	1.103
4.5	348.3	2.08	1.48	4.43	1.123	1.115
5.0	349.4	2.04	1.49	4.92	1.135	1.128
6.0	351.6	1.95	1.51	5.88	1.159	1.152
7.0	354.1	1.87	1.53	6.85	1.180	1.177
8.0	356.9	1.79	1.55	7.81	1.200	1.201
9.0	359.9	1.71	1.58	8.77	1.218	1.225
10.0	363.1	1.63	1.61	9.73	1.234	1.249
11.0	366.6	1.55	1.64	10.69	1.248	1.272
12.0	370.3	1.48	1.67	11.65	1.260	1.294
13.0	374.3	1.40	1.70	12.61	1.269	1.316
14.0	378.5	1.32	1.73	13.57	1.276	1.338
15.0	383.0	1.24	1.76	14.54	1.281	1.358
16.0	387.6	1.17	1.80	15.51	1.283	1.378
17.0	392.4	1.10	1.84	16.48	1.284	1.398
18.0	397.4	1.03	1.87	17.46	1.282	1.416
19.0	402.6	0.96	1.91	18.45	1.279	1.434
20.0	407.9	0.90	1.95	19.44	1.274	1.450
21.0	413.3	0.83	1.99	20.44	1.268	1.466
22.0	418.8	0.78	2.03	21.44	1.260	1.481
23.0	424.3	0.72	2.07	22.46	1.252	1.495
24.0	430.0	0.66	2.11	23.48	1.242	1.509
25.0	435.7	0.61	2.15	24.51	1.232	1.521
26.0	441.5	0.56	2.19	25.56	1.221	1.533
27.0	447.3	0.52	2.23	26.61	1.210	1.544
28.0	453.2	0.47	2.27	27.68	1.197	1.554
29.0	459.1	0.43	2.31	28.76	1.185	1.563
30.0	465.0	0.39	2.35	29.85	1.172	1.572
35.0	494.9	0.21	2.55	35.51	1.105	1.608
40.0	524.8	0.07	2.74	41.56	1.038	1.631
45.0	554.5	—0.04	2.93	48.05	0.975	1.644
50.0	583.8	—0.13	3.10	55.05	0.916	1.651
55.0	612.4	—0.21	3.27	62.60	0.864	1.653
60.0	640.3	—0.27	3.42	70.76	0.818	1.651
65.0	667.3	—0.32	3.57	79.59	0.777	1.646
70.0	693.4	—0.36	3.70	89.14	0.741	1.639
75.0	718.6	—0.39	3.82	99.48	0.709	1.631
80.0	742.9	—0.42	3.94	110.67	0.680	1.621
85.0	766.3	—0.44	4.04	122.78	0.655	1.610
90.0	788.8	—0.46	4.14	135.88	0.632	1.599
95.0	810.5	—0.47	4.23	150.05	0.611	1.586
100.0	831.4	—0.49	4.31	165.36	0.592	1.573

Table II.12 (*Continued*)

p	w	μ	k	f	α/α_0	γ/γ_0
			$T=300$ K			
0.1	347.4	2.34	1.40	0.10	1.003	1.002
0.5	347.8	2.30	1.41	0.50	1.014	1.012
1.0	348.5	2.25	1.41	1.00	1.027	1.024
1.5	349.2	2.20	1.42	1.49	1.040	1.036
2.0	350.0	2.15	1.43	1.99	1.052	1.048
2.5	350.8	2.10	1.44	2.48	1.064	1.059
3.0	351.7	2.06	1.45	2.98	1.076	1.071
3.5	352.6	2.02	1.46	3.47	1.088	1.083
4.0	353.6	1.97	1.46	3.96	1.099	1.094
4.5	354.7	1.93	1.47	4.45	1.110	1.106
5.0	355.8	1.90	1.48	4.94	1.121	1.117
6.0	358.1	1.82	1.50	5.92	1.141	1.140
7.0	360.6	1.74	1.52	6.89	1.160	1.162
8.0	363.4	1.67	1.55	7.87	1.177	1.184
9.0	366.3	1.60	1.57	8.84	1.193	1.206
10.0	369.5	1.52	1.60	9.81	1.207	1.227
11.0	372.9	1.45	1.62	10.79	1.219	1.248
12.0	376.5	1.38	1.65	11.77	1.229	1.268
13.0	380.4	1.31	1.68	12.75	1.238	1.288
14.0	384.4	1.24	1.71	13.73	1.244	1.308
15.0	388.7	1.17	1.74	14.71	1.248	1.326
16.0	393.1	1.10	1.77	15.70	1.251	1.345
17.0	397.7	1.03	1.81	16.70	1.251	1.362
18.0	402.5	0.97	1.84	17.70	1.250	1.379
19.0	407.4	0.91	1.88	18.71	1.248	1.395
20.0	412.4	0.85	1.91	19.72	1.244	1.410
21.0	417.5	0.79	1.95	20.74	1.239	1.425
22.0	422.8	0.74	1.99	21.78	1.233	1.439
23.0	428.1	0.68	2.02	22.81	1.225	1.452
24.0	433.5	0.63	2.06	23.86	1.217	1.465
25.0	438.9	0.58	2.10	24.92	1.208	1.476
26.0	444.4	0.54	2.13	25.99	1.199	1.487
27.0	450.0	0.49	2.17	27.07	1.189	1.498
28.0	455.6	0.45	2.21	28.16	1.178	1.507
29.0	461.2	0.41	2.24	29.27	1.167	1.517
30.0	466.8	0.37	2.28	30.38	1.156	1.525
35.0	495.3	0.20	2.46	36.18	1.096	1.560
40.0	524.0	0.07	2.64	42.35	1.035	1.583
45.0	552.6	—0.04	2.81	48.97	0.976	1.598
50.0	580.8	—0.13	2.98	56.09	0.921	1.607
55.0	608.6	—0.21	3.13	63.75	0.871	1.610
60.0	635.8	—0.27	3.28	72.01	0.826	1.610
65.0	662.3	—0.31	3.42	80.93	0.786	1.608
70.0	688.0	—0.35	3.55	90.57	0.750	1.603
75.0	712.9	—0.39	3.67	100.97	0.718	1.596

Table II.12 (*Continued*)

p	w	μ	k	f	α/α_0	γ/γ_0
			$T=300$ K			
80.0	737.0	—0.42	3.79	112.21	0.690	1.589
85.0	760.2	—0.44	3.89	124.35	0.665	1.580
90.0	782.6	—0.46	3.99	137.45	0.642	1.570
95.0	804.3	—0.47	4.08	151.59	0.621	1.560
100.0	825.1	—0.49	4.16	166.85	0.602	1.549
			$T=350$ K			
0.1	375.0	1.65	1.40	0.10	1.002	1.002
0.5	375.7	1.62	1.40	0.50	1.008	1.008
1.0	376.6	1.58	1.41	1.00	1.016	1.017
1.5	377.6	1.54	1.42	1.50	1.024	1.025
2.0	378.6	1.51	1.42	2.00	1.031	1.033
2.5	379.6	1.48	1.43	2.50	1.038	1.041
3.0	380.7	1.44	1.44	3.00	1.045	1.049
3.5	381.8	1.41	1.45	3.51	1.051	1.057
4.0	382.9	1.38	1.45	4.01	1.058	1.064
4.5	384.1	1.35	1.46	4.51	1.064	1.072
5.0	385.3	1.33	1.47	5.02	1.070	1.080
6.0	387.8	1.27	1.49	6.02	1.081	1.095
7.0	390.4	1.22	1.50	7.04	1.091	1.109
8.0	393.2	1.17	1.52	8.05	1.100	1.124
9.0	396.1	1.12	1.54	9.08	1.109	1.138
10.0	399.1	1.07	1.56	10.10	1.116	1.152
11.0	402.2	1.02	1.58	11.13	1.123	1.166
12.0	405.5	0.97	1.60	12.17	1.128	1.179
13.0	408.9	0.92	1.62	13.21	1.132	1.192
14.0	412.4	0.87	1.64	14.26	1.136	1.205
15.0	416.1	0.83	1.66	15.31	1.138	1.217
16.0	419.8	0.78	1.68	16.37	1.139	1.229
17.0	423.7	0.74	1.71	17.44	1.140	1.241
18.0	427.7	0.69	1.73	18.52	1.139	1.252
19.0	431.8	0.65	1.76	19.61	1.138	1.263
20.0	436.0	0.61	1.78	20.70	1.136	1.274
21.0	440.2	0.57	1.81	21.81	1.133	1.284
22.0	444.5	0.53	1.83	22.92	1.130	1.294
23.0	448.9	0.49	1.86	24.05	1.126	1.303
24.0	453.3	0.45	1.88	25.18	1.121	1.312
25.0	457.8	0.42	1.91	26.33	1.116	1.321
26.0	462.4	0.39	1.93	27.49	1.111	1.329
27.0	466.9	0.35	1.96	28.66	1.105	1.337
28.0	471.5	0.32	1.98	29.84	1.099	1.344
29.0	476.2	0.29	2.01	31.03	1.092	1.351
30.0	480.8	0.26	2.04	32.24	1.085	1.358
35.0	504.4	0.13	2.17	38.49	1.048	1.387
40.0	528.3	0.02	2.29	45.11	1.008	1.409
45.0	552.2	—0.07	2.42	52.16	0.967	1.426
50.0	576.2	—0.15	2.54	59.68	0.927	1.438

Table II.12 (*Continued*)

p	w	μ	k	f	α/α_0	γ/γ_0
			$T=350$ K			
55.0	600.2	—0.22	2.66	67.70	0.888	1.446
60.0	624.0	—0.27	2.78	76.28	0.850	1.451
65.0	647.7	—0.32	2.89	85.46	0.816	1.454
70.0	671.0	—0.36	3.00	95.28	0.783	1.455
75.0	694.0	—0.40	3.11	105.79	0.754	1.455
80.0	716.6	—0.43	3.21	117.05	0.726	1.454
85.0	738.7	—0.45	3.31	129.10	0.702	1.452
90.0	760.2	—0.47	3.40	142.00	0.679	1.448
95.0	781.2	—0.49	3.49	155.81	0.658	1.445
100.0	801.6	—0.51	3.57	170.58	0.640	1.440
			$T=400$ K			
0.1	400.5	1.16	1.40	0.10	1.001	1.001
0.5	401.3	1.13	1.40	0.50	1.005	1.006
1.0	402.4	1.10	1.41	1.00	1.010	1.012
1.5	403.5	1.08	1.41	1.50	1.014	1.018
2.0	404.6	1.05	1.42	2.01	1.019	1.024
2.5	405.7	1.03	1.43	2.51	1.023	1.029
3.0	406.9	1.00	1.43	3.02	1.027	1.035
3.5	408.1	0.98	1.44	3.53	1.031	1.041
4.0	409.3	0.96	1.44	4.03	1.035	1.046
4.5	410.5	0.94	1.45	4.54	1.038	1.052
5.0	411.8	0.91	1.46	5.06	1.041	1.057
6.0	414.4	0.87	1.47	6.08	1.048	1.068
7.0	417.0	0.83	1.49	7.12	1.053	1.078
8.0	419.8	0.80	1.50	8.15	1.058	1.089
9.0	422.6	0.76	1.52	9.20	1.063	1.099
10.0	425.5	0.72	1.53	10.25	1.067	1.109
11.0	428.5	0.69	1.55	11.31	1.070	1.118
12.0	431.6	0.65	1.56	12.38	1.073	1.128
13.0	434.8	0.62	1.58	13.45	1.075	1.137
14.0	438.0	0.59	1.60	14.54	1.076	1.146
15.0	441.3	0.55	1.61	15.63	1.077	1.155
16.0	444.7	0.52	1.63	16.73	1.077	1.164
17.0	448.2	0.49	1.65	17.84	1.077	1.172
18.0	451.8	0.46	1.67	18.95	1.076	1.180
19.0	455.4	0.43	1.68	20.08	1.075	1.188
20.0	459.1	0.40	1.70	21.22	1.073	1.196
21.0	462.8	0.37	1.72	22.37	1.071	1.203
22.0	466.6	0.34	1.74	23.53	1.069	1.210
23.0	470.4	0.31	1.76	24.70	1.066	1.217
24.0	474.3	0.28	1.78	25.88	1.062	1.224
25.0	478.2	0.25	1.80	27.07	1.058	1.231
26.0	482.2	0.23	1.82	28.27	1.055	1.237
27.0	486.2	0.20	1.84	29.49	1.050	1.243
28.0	490.2	0.18	1.86	30.72	1.046	1.249
29.0	494.3	0.16	1.88	31.96	1.041	1.254

Table II.12 (*Continued*)

p	w	μ	k	f	α/α_0	γ/γ_0
			$T=400$ K			
30.0	498.4	0.13	1.90	33.21	1.036	1.260
35.0	519.0	0.03	2.00	39.69	1.009	1.283
40.0	539.8	—0.06	2.10	46.53	0.980	1.302
45.0	560.7	—0.13	2.19	53.78	0.950	1.317
50.0	581.7	—0.20	2.29	61.46	0.919	1.329
55.0	602.7	—0.25	2.38	69.62	0.888	1.338
60.0	623.7	—0.30	2.48	78.28	0.858	1.344
65.0	644.8	—0.34	2.57	87.49	0.830	1.349
70.0	665.7	—0.38	2.66	97.27	0.802	1.353
75.0	686.6	—0.41	2.75	107.68	0.775	1.355
80.0	707.3	—0.44	2.83	118.75	0.751	1.357
85.0	727.8	—0.47	2.92	130.51	0.727	1.357
90.0	748.1	—0.49	3.00	143.03	0.706	1.357
95.0	768.0	—0.51	3.08	156.33	0.686	1.356
100.0	787.6	—0.53	3.16	170.48	0.667	1.355
			$T=450$ K			
0.1	424.2	0.79	1.39	0.10	1.001	1.001
0.5	425.1	0.77	1.40	0.50	1.003	1.005
1.0	426.2	0.75	1.40	1.00	1.006	1.009
1.5	427.4	0.73	1.41	1.51	1.009	1.013
2.0	428.5	0.71	1.41	2.01	1.011	1.018
2.5	429.7	0.69	1.42	2.52	1.014	1.022
3.0	431.0	0.67	1.42	3.03	1.016	1.026
3.5	432.2	0.65	1.43	3.54	1.018	1.031
4.0	433.4	0.64	1.44	4.05	1.020	1.035
4.5	434.7	0.62	1.44	4.56	1.022	1.039
5.0	436.0	0.60	1.45	5.08	1.024	1.043
6.0	438.6	0.57	1.46	6.11	1.028	1.051
7.0	441.3	0.54	1.47	7.16	1.031	1.059
8.0	444.0	0.52	1.48	8.21	1.033	1.066
9.0	446.8	0.49	1.50	9.27	1.035	1.074
10.0	449.6	0.46	1.51	10.33	1.037	1.081
11.0	452.5	0.44	1.52	11.41	1.039	1.089
12.0	455.4	0.41	1.54	12.49	1.040	1.096
13.0	458.4	0.38	1.55	13.58	1.040	1.103
14.0	461.5	0.36	1.56	14.68	1.040	1.109
15.0	464.6	0.33	1.58	15.79	1.040	1.116
16.0	467.8	0.31	1.59	16.91	1.040	1.122
17.0	471.0	0.29	1.61	18.04	1.039	1.129
18.0	474.3	0.26	1.62	19,18	1.038	1.135
19.0	477.6	0.24	1.64	20,33	1.036	1.141
20.0	481.0	0.22	1.65	21.49	1.035	1.147
21.0	484.4	0.19	1.67	22.66	1.032	1.152
22.0	487.8	0.17	1.68	23.84	1.030	1.158
23.0	491.3	0.15	1.70	25.03	1.027	1.163
24.0	494.8	0.13	1.71	26.23	1.025	1.168

Table II.12 (*Continued*)

p	w	μ	k	f	α/α_0	γ/γ_0
			$T=450$ K			
25.0	498.4	0.11	1.73	27.45	1.021	1.173
26.0	502.0	0.09	1.75	28.67	1.018	1.178
27.0	505.6	0.07	1.76	29.91	1.015	1.183
28.0	509.2	0.05	1.78	31.16	1.011	1.188
29.0	512.9	0.03	1.79	32.43	1.007	1.192
30.0	516.6	0.01	1.81	33.70	1.003	1.196
35.0	535.2	—0.07	1.89	40.28	0.981	1.216
40.0	554.0	—0.14	1.97	47.21	0.958	1.232
45.0	572.9	—0.20	2.05	54.52	0.934	1.245
50.0	591.8	—0.25	2.13	62.24	0.909	1.255
55.0	610.7	—0.30	2.20	70.40	0.884	1.264
60.0	629.6	—0.34	2.28	79.02	0.860	1.271
65.0	648.5	—0.38	2.35	88.13	0.835	1.276
70.0	667.5	—0.41	2.43	97.77	0.812	1.280
75.0	686.4	—0.44	2.50	107.97	0.789	1.283
80.0	705.3	—0.46	2.58	118.76	0.767	1.286
85.0	724.2	—0.49	2.65	130.17	0.746	1.288
90.0	743.0	—0.51	2.72	142.23	0.726	1.289
95.0	761.7	—0.53	2.79	154.99	0.707	1.290
100.0	780.2	—0.54	2.86	168.49	0.689	1.290
			$T=500$ K			
0.1	446.4	0.51	1.39	0.10	1.000	1.001
0.5	447.3	0.49	1.39	0.50	1.002	1.003
1.0	448.5	0.48	1.40	1.00	1.003	1.007
1.5	449.7	0.46	1.40	1.51	1.005	1.010
2.0	450.9	0.45	1.41	2.01	1.006	1.014
2.5	452.1	0.43	1.41	2.52	1.008	1.017
3.0	453.3	0.42	1.42	3.03	1.009	1.020
3.5	454.6	0.40	1.42	3.54	1.010	1.024
4.0	455.8	0.39	1.43	4.06	1.011	1.027
4.5	457.1	0.38	1.43	4.57	1.012	1.030
5.0	458.4	0.37	1.44	5.09	1.013	1.033
6.0	461.0	0.34	1.45	6.13	1.015	1.039
7.0	463.6	0.32	1.46	7.18	1.016	1.045
8.0	466.3	0.30	1.47	8.23	1.017	1.051
9.0	469.0	0.28	1.48	9.30	1.018	1.057
10.0	471.8	0.25	1.49	10.37	1.018	1.063
11.0	474.6	0.23	1.50	11.46	1.019	1.068
12.0	477.4	0.21	1.52	12.55	1.019	1.074
13.0	480.3	0.20	1.53	13.65	1.018	1.079
14.0	483.3	0.18	1.54	14.76	1.018	1.084
15.0	486.2	0.16	1.55	15.88	1.017	1.090
16.0	489.2	0.14	1.56	17.00	1.016	1.095
17.0	492.2	0.12	1.58	18.14	1.015	1.100
18.0	495.3	0.10	1.59	19.29	1.013	1.104
19.0	498.4	0.08	1.60	20.45	1.011	1.109

Table II.12 (*Continued*)

p	w	μ	k	f	α/α_0	γ/γ_0
			$T=500$ K			
20.0	501.6	0.07	1.61	21.62	1.009	1.114
21.0	504.8	0.05	1.63	22.80	1.007	1.118
22.0	508.0	0.03	1.64	23.99	1.005	1.122
23.0	511.2	0.01	1.65	25.19	1.002	1.127
24.0	514.5	—0.00	1.67	26.40	1.000	1.131
25.0	517.8	—0.02	1.68	27.63	0.997	1.135
26.0	521.1	—0.03	1.69	28.86	0.994	1.139
27.0	524.4	—0.05	1.71	30.11	0.991	1.143
28.0	527.8	—0.06	1.72	31.37	0.988	1.146
29.0	531.2	—0.08	1.73	32.64	0.984	1.150
30.0	534.5	—0.09	1.75	33.92	0.981	1.153
35.0	551.7	—0.16	1.81	40.53	0.962	1.169
40.0	569.0	—0.22	1.88	47.47	0.942	1.183
45.0	586.5	—0.27	1.95	54.77	0.921	1.194
50.0	603.9	—0.31	2.01	62.45	0.900	1.203
55.0	621.2	—0.35	2.08	70.54	0.879	1.211
60.0	638.6	—0.38	2.14	79.05	0.858	1.218
65.0	656.0	—0.41	2.21	88.02	0.838	1.223
70.0	673.3	—0.44	2.27	97.46	0.817	1.227
75.0	690.7	—0.46	2.33	107.41	0.798	1.231
80.0	708.1	—0.49	2.39	117.89	0.778	1.234
85.0	725.5	—0.51	2.46	128.92	0.759	1.236
90.0	742.9	—0.53	2.52	140.54	0.741	1.238
95.0	760.3	—0.55	2.58	152.77	0.723	1.239
100.0	777.6	—0.56	2.64	165.65	0.706	1.240
			$T=550$ K			
0.1	467.3	0.29	1.38	0.10	1.000	1.001
0.5	468.2	0.28	1.39	0.50	1.001	1.003
1.0	469.4	0.27	1.39	1.00	1.002	1.005
1.5	470.6	0.25	1.39	1.51	1.003	1.008
2.0	471.8	0.24	1.40	2.01	1.003	1.011
2.5	473.1	0.23	1.40	2.52	1.004	1.013
3.0	474.3	0.22	1.41	3.03	1.004	1.016
3.5	475.5	0.21	1.41	3.55	1.005	1.019
4.0	476.8	0.20	1.42	4.06	1.005	1.021
4.5	478.1	0.19	1.42	4.58	1.006	1.024
5.0	479.3	0.18	1.43	5.09	1.006	1.026
6.0	481.9	0.16	1.44	6.14	1.006	1.031
7.0	484.5	0.14	1.45	7.19	1.006	1.036
8.0	487.2	0.12	1.46	8.25	1.006	1.041
9.0	489.8	0.11	1.47	9.32	1.006	1.045
10.0	492.5	0.09	1.48	10.39	1.006	1.050
11.0	495.2	0.07	1.49	11.48	1.005	1.054
12.0	498.0	0.06	1.50	12.57	1.005	1.059
13.0	500.8	0.04	1.51	13.68	1.004	1.063
14.0	503.6	0.03	1.52	14.79	1.003	1.067

Table II.12 (*Continued*)

p	w	μ	k	f	α/α_0	γ/γ_0
			$T=550$ K			
15.0	506.4	0.01	1.53	15.91	1.001	1.071
16.0	509.3	—0.00	1.54	17.04	1.000	1.075
17.0	512.2	—0.02	1.55	18.19	0.998	1.079
18.0	515.1	—0.03	1.56	19.34	0.997	1.083
19.0	518.1	—0.04	1.57	20.50	0.995	1.087
20.0	521.0	—0.06	1.58	21.67	0.993	1.090
21.0	524.0	—0.07	1.59	22.85	0.990	1.094
22.0	527.1	—0.08	1.60	24.05	0.988	1.097
23.0	530.1	—0.10	1.62	25.25	0.986	1.101
24.0	533.2	—0.11	1.63	26.46	0.983	1.104
25.0	536.3	—0.12	1.64	27.69	0.980	1.107
26.0	539.4	—0.14	1.65	28.93	0.978	1.111
27.0	542.5	—0.15	1.66	30.17	0.975	1.114
28.0	545.7	—0.16	1.67	31.43	0.972	1.117
29.0	548.8	—0.17	1.68	32.70	0.969	1.120
30.0	552.0	—0.18	1.70	33.99	0.965	1.123
35.0	568.0	—0.24	1.75	40.58	0.949	1.136
40.0	584.3	—0.29	1.81	47.49	0.931	1.147
45.0	600.5	—0.33	1.87	54.74	0.912	1.157
50.0	616.8	—0.36	1.93	62.35	0.894	1.165
55.0	633.0	—0.39	1.98	70.33	0.875	1.173
60.0	649.2	—0.42	2.04	78.71	0.857	1.178
65.0	665.4	—0.45	2.10	87.50	0.839	1.183
70.0	681.5	—0.47	2.15	96.73	0.821	1.188
75.0	697.6	—0.49	2.20	106.42	0.803	1.191
80.0	713.8	—0.51	2.26	116.58	0.786	1.194
85.0	729.9	—0.53	2.31	127.25	0.769	1.196
90.0	746.1	—0.55	2.37	138.44	0.753	1.198
95.0	762.3	—0.56	2.42	150.18	0.737	1.200
100.0	778.5	—0.58	2.47	162.50	0.721	1.201
			$T=600$ K			
0.1	487.0	0.12	1.38	0.10	1.000	1.000
0.5	488.0	0.11	1.38	0.50	1.000	1.002
1.0	489.2	0.10	1.38	1.00	1.001	1.004
1.5	490.4	0.09	1.39	1.51	1.001	1.007
2.0	491.6	0.08	1.39	2.02	1.001	1.009
2.5	492.8	0.07	1.40	2.52	1.001	1.011
3.0	494.1	0.06	1.40	3.03	1.001	1.013
3.5	495.3	0.05	1.41	3.55	1.001	1.015
4.0	496.6	0.04	1.41	4.06	1.001	1.017
4.5	497.8	0.04	1.41	4.58	1.001	1.019
5.0	499.1	0.03	1.42	5.10	1.001	1.021
6.0	501.6	0.01	1.43	6.14	1.000	1.025
7.0	504.2	—0.00	1.44	7.19	1.000	1.029
8.0	506.8	—0.02	1.44	8.25	0.999	1.033
9.0	509.4	—0.03	1.45	9.32	0.998	1.037

Table II.12 (*Continued*)

p	w	μ	k	f	α/α_0	γ/γ_0
			$T=600$ K			
10.0	512.0	—0.04	1.46	10.40	0.997	1.040
11.0	514.7	—0.05	1.47	11.48	0.996	1.044
12.0	517.3	—0.07	1.48	12.58	0.995	1.047
13.0	520.0	—0.08	1.49	13.68	0.994	1.051
14.0	522.7	—0.09	1.50	14.80	0.992	1.054
15.0	525.5	—0.10	1.51	15.92	0.991	1.058
16.0	528.2	—0.11	1.52	17.05	0.989	1.061
17.0	531.0	—0.13	1.53	18.19	0.987	1.064
18.0	533.8	—0.14	1.54	19.35	0.985	1.067
19.0	536.6	—0.15	1.55	20.51	0.983	1.070
20.0	539.5	—0.16	1.56	21.68	0.981	1.073
21.0	542.3	—0.17	1.57	22.86	0.979	1.076
22.0	545.2	—0.18	1.58	24.05	0.977	1.079
23.0	548.1	—0.19	1.59	25.25	0.974	1.082
24.0	551.0	—0.20	1.60	26.47	0.972	1.085
25.0	554.0	—0.21	1.61	27.69	0.969	1.087
26.0	556.9	—0.22	1.62	28.92	0.966	1.090
27.0	559.9	—0.23	1.63	30.17	0.964	1.093
28.0	562.9	—0.24	1.64	31.42	0.961	1.095
29.0	565.8	—0.25	1.65	32.69	0.958	1.098
30.0	568.8	—0.26	1.66	33.96	0.955	1.100
35.0	584.0	—0.31	1.71	40.53	0.939	1.111
40.0	599.3	—0.35	1.76	47.39	0.923	1.121
45.0	614.7	—0.38	1.81	54.56	0.906	1.130
50.0	630.1	—0.41	1.86	62.08	0.889	1.137
55.0	645.4	—0.44	1.91	69.94	0.873	1.143
60.0	660.7	—0.46	1.96	78.17	0.856	1.149
65.0	675.9	—0.48	2.01	86.78	0.840	1.153
70.0	691.0	—0.50	2.06	95.79	0.824	1.157
75.0	706.2	—0.52	2.11	105.22	0.808	1.161
80.0	721.3	—0.54	2.15	115.09	0.792	1.164
85.0	736.4	—0.55	2.20	125.41	0.777	1.166
90.0	751.6	—0.57	2.25	136.21	0.762	1.168
95.0	766.7	—0.58	2.29	147.50	0.747	1.170
100.0	781.9	—0.59	2.34	159.31	0.733	1.171
			$T=650$ K			
0.1	505.9	—0.02	1.37	0.10	1.000	1.000
0.5	506.8	—0.03	1.37	0.50	1.000	1.002
1.0	508.0	—0.04	1.38	1.00	1.000	1.004
1.5	509.2	—0.04	1.38	1.51	1.000	1.005
2.0	510.4	—0.05	1.39	2.02	0.999	1.007
2.5	511.6	—0.06	1.39	2.52	0.999	1.009
3.0	512.9	—0.07	1.39	3.03	0.999	1.011
3.5	514.1	—0.07	1.40	3.55	0.999	1.012
4.0	515.3	—0.08	1.40	4.06	0.998	1.014
4.5	516.5	—0.09	1.41	4.58	0.998	1.016

Table II.12 (*Continued*)

p	w	μ	k	i	α/α_0	γ/γ_0
			$T=650$ K			
5.0	517.8	—0.09	1.41	5.10	0.997	1.017
6.0	520.3	—0.11	1.42	6.14	0.996	1.021
7.0	522.8	—0.12	1.43	7.19	0.995	1.024
8.0	525.3	—0.13	1.43	8.25	0.994	1.027
9.0	527.9	—0.14	1.44	9.32	0.993	1.030
10.0	530.5	—0.15	1.45	10.40	0.992	1.033
11.0	533.0	—0.16	1.46	11.48	0.990	1.036
12.0	535.6	—0.17	1.47	12.58	0.989	1.039
13.0	538.3	—0.18	1.47	13.68	0.987	1.042
14.0	540.9	—0.19	1.48	14.79	0.985	1.045
15.0	543.6	—0.20	1.49	15.91	0.983	1.047
16.0	546.2	—0.21	1.50	17.04	0.982	1.050
17.0	548.9	—0.22	1.51	18.18	0.980	1.053
18.0	551.6	—0.23	1.52	19.33	0.978	1.055
19.0	554.3	—0.24	1.53	20.49	0.975	1.058
20.0	557.0	—0.24	1.54	21.66	0.973	1.060
21.0	559.8	—0.25	1.54	22.84	0.971	1.063
22.0	562.5	—0.26	1.55	24.02	0.969	1.065
23.0	565.3	—0.27	1.56	25.22	0.966	1.068
24.0	568.1	—0.28	1.57	26.43	0.964	1.070
25.0	570.9	—0.29	1.58	27.65	0.961	1.072
26.0	573.7	—0.30	1.59	28.88	0.959	1.075
27.0	576.6	—0.30	1.60	30.11	0.956	1.077
28.0	579.4	—0.31	1.61	31.36	0.953	1.079
29.0	582.2	—0.32	1.62	32.62	0.950	1.081
30.0	585.1	—0.33	1.62	33.89	0.948	1.083
35.0	599.5	—0.37	1.67	40.41	0.933	1.093
40.0	614.1	—0.40	1.72	47.21	0.918	1.101
45.0	628.7	—0.43	1.76	54.30	0.902	1.109
50.0	643.3	—0.45	1.81	61.71	0.887	1.115
55.0	657.9	—0.48	1.85	69.45	0.871	1.121
60.0	672.4	—0.50	1.90	77.53	0.856	1.126
65.0	686.9	—0.51	1.94	85.96	0.841	1.130
70.0	701.3	—0.53	1.98	94.76	0.826	1.134
75.0	715.6	—0.55	2.03	103.95	0.812	1.137
80.0	730.0	—0.56	2.07	113.54	0.798	1.140
85.0	744.3	—0.57	2.11	123.54	0.784	1.142
90.0	758.5	—0.58	2.15	133.98	0.770	1.144
95.0	772.9	—0.60	2.20	144.87	0.756	1.146
100.0	787.2	—0.61	2.24	156.22	0.743	1.147
			$T=700$ K			
0.1	523.9	—0.13	1.37	0.10	1.000	1.000
0.5	524.8	—0.14	1.37	0.50	1.000	1.002
1.0	526.0	—0.14	1.37	1.00	0.999	1.003
1.5	527.2	—0.15	1.38	1.51	0.999	1.004
2.0	528.4	—0.16	1.38	2.02	0.998	1.006

Table II.12 (*Continued*)

p	w	μ	k	f	α/α_0	γ/γ_0
			$T=700$ K			
2.5	529.6	—0.16	1.38	2.52	0.998	1.007
3.0	530.8	—0.17	1.39	3.03	0.997	1.009
3.5	532.0	—0.18	1.39	3.55	0.997	1.010
4.0	533.2	—0.18	1.39	4.06	0.996	1.012
4.5	534.4	—0.19	1.40	4.58	0.996	1.013
5.0	535.6	—0.19	1.40	5.10	0.995	1.014
6.0	538.1	—0.20	1.41	6.14	0.994	1.017
7.0	540.6	—0.21	1.42	7.19	0.992	1.020
8.0	543.1	—0.22	1.42	8.25	0.991	1.022
9.0	545.5	—0.23	1.43	9.32	0.989	1.025
10.0	548.1	—0.24	1.44	10.39	0.987	1.028
11.0	550.6	—0.25	1.45	11.47	0.986	1.030
12.0	553.1	—0.26	1.45	12.57	0.984	1.033
13.0	555.7	—0.26	1.46	13.67	0.982	1.035
14.0	558.2	—0.27	1.47	14.78	0.980	1.037
15.0	560.8	—0.28	1.48	15.90	0.978	1.040
16.0	563.4	—0.29	1.48	17.02	0.976	1.042
17.0	566.0	—0.29	1.49	18.16	0.974	1.044
18.0	568.6	—0.30	1.50	19.31	0.972	1.046
19.0	571.2	—0.31	1.51	20.46	0.970	1.048
20.0	573.8	—0.32	1.52	21.62	0.968	1.051
21.0	576.5	—0.32	1.52	22.80	0.965	1.053
22.0	579.2	—0.33	1.53	23.98	0.963	1.055
23.0	581.8	—0.34	1.54	25.17	0.961	1.057
24.0	584.5	—0.34	1.55	26.37	0.958	1.059
25.0	587.2	—0.35	1.56	27.59	0.956	1.061
26.0	589.9	—0.36	1.56	28.81	0.953	1.063
27.0	592.6	—0.36	1.57	30.04	0.951	1.064
28.0	595.3	—0.37	1.58	31.28	0.948	1.066
29.0	598.1	—0.38	1.59	32.53	0.945	1.068
30.0	600.8	—0.38	1.60	33.79	0.943	1.070
35.0	614.6	—0.42	1.64	40.26	0.929	1.078
40.0	628.5	—0.44	1.68	46.99	0.914	1.085
45.0	642.5	—0.47	1.72	54.00	0.900	1.092
50.0	656.5	—0.49	1.76	61.31	0.885	1.098
55.0	670.5	—0.51	1.80	68.92	0.871	1.103
60.0	684.4	—0.53	1.84	76.85	0.857	1.107
65.0	698.2	—0.54	1.88	85.11	0.843	1.111
70.0	712.0	—0.56	1.92	93.72	0.829	1.115
75.0	725.7	—0.57	1.96	102.68	0.815	1.118
80.0	739.3	—0.58	2.00	112.01	0.802	1.120
85.0	752.9	—0.59	2.04	121.72	0.789	1.123
90.0	766.5	—0.60	2.08	131.83	0.776	1.125
95.0	780.1	—0.61	2.11	142.36	0.764	1.126
100.0	793.7	—0.62	2.15	153.30	0.752	1.128

Table II.12 (*Continued*)

p	w	μ	k	f	α/α_0	γ/γ_0
			$T=750$ K			
0.1	541.2	—0.22	1.36	0.10	1.000	1.000
0.5	542.1	—0.23	1.36	0.50	0.999	1.001
1.0	543.3	—0.23	1.37	1.00	0.999	1.003
1.5	544.4	—0.24	1.37	1.51	0.998	1.004
2.0	545.6	—0.24	1.37	2.01	0.998	1.005
2.5	546.8	—0.25	1.38	2.52	0.997	1.006
3.0	548.0	—0.25	1.38	3.03	0.996	1.007
3.5	549.2	—0.26	1.38	3.55	0.995	1.009
4.0	550.4	—0.26	1.39	4.06	0.995	1.010
4.5	551.6	—0.27	1.39	4.58	0.994	1.011
5.0	552.8	—0.27	1.39	5.09	0.993	1.012
6.0	555.2	—0.28	1.40	6.14	0.992	1.014
7.0	557.6	—0.29	1.41	7.19	0.990	1.017
8.0	560.0	—0.30	1.41	8.24	0.988	1.019
9.0	562.5	—0.30	1.42	9.31	0.986	1.021
10.0	564.9	—0.31	1.43	10.38	0.985	1.023
11.0	567.4	—0.32	1.43	11.46	0.983	1.025
12.0	569.9	—0.33	1.44	12.55	0.981	1.027
13.0	572.3	—0.33	1.45	13.65	0.979	1.030
14.0	574.8	—0.34	1.46	14.76	0.977	1.032
15.0	577.3	—0.34	1.46	15.88	0.975	1.033
16.0	579.9	—0.35	1.47	17.00	0.973	1.035
17.0	582.4	—0.36	1.48	18.13	0.971	1.037
18.0	584.9	—0.36	1.48	19.27	0,968	1.039
19.0	587.4	—0.37	1.49	20.42	0.966	1.041
20.0	590.0	—0.38	1.50	21.58	0.964	1.043
21.0	592.6	—0.38	1.51	22.75	0.962	1.045
22.0	595.1	—0.39	1.51	23.93	0.959	1.046
23.0	597.7	—0.39	1.52	25.11	0.957	1.048
24.0	600.3	—0.40	1.53	26.31	0.954	1.050
25.0	602.9	—0.40	1.54	27.51	0.952	1.052
26.0	605.5	—0.41	1.54	28.73	0.949	1.053
27.0	608.1	—0.42	1.55	29.95	0.947	1.055
28.0	610.7	—0.42	1.56	31.18	0.944	1.056
29.0	613.4	—0.43	1.57	32.43	0.942	1.058
30.0	616.0	—0.43	1.57	33.68	0.939	1.059
35.0	629.3	—0.46	1.61	40.09	0.926	1.067
40.0	642.6	—0.48	1.65	46.75	0.912	1.073
45.0	656.1	—0.50	1.69	53.68	0.898	1.079
50.0	669.5	—0.52	1.73	60.89	0.885	1.084
55.0	682.9	—0.54	1.76	68.38	0.871	1.089
60.0	696.3	—0.55	1.80	76.17	0.858	1.093
65.0	709.6	—0.57	1.84	84.27	0.844	1.096
70.0	722.9	—0.58	1.87	92.69	0.831	1.099
75.0	736.0	—0.59	1.91	101.44	0.819	1.102

Table II.12 (*Continued*)

p	w	μ	k	f	α/α_0	γ/γ_0
			$T=750$ K			
80.0	749.1	—0.60	1.94	110.54	0.806	1.105
85.0	762.2	—0.61	1.98	119.99	0.794	1.107
90.0	775.2	—0.62	2.01	129.80	0.782	1.109
95.0	788.2	—0.63	2.05	139.99	0.771	1.110
100.0	801.1	—0.63	2.08	150.58	0.759	1.112
			$T=800$ K			
0.1	557.8	—0.30	1.35	0.10	1.000	1.000
0.5	558.7	—0.30	1.36	0.50	0.999	1.001
1.0	559.9	—0.31	1.36	1.00	0.999	1.002
1.5	561.0	—0.31	1.36	1.51	0.998	1.003
2.0	562.2	—0.32	1.37	2.01	0.997	1.004
2.5	563.4	—0.32	1.37	2.52	0.996	1.005
3.0	564.5	—0.32	1.37	3.03	0.995	1.006
3.5	565.7	—0.33	1.38	3.54	0.995	1.007
4.0	566.9	—0.33	1.38	4.06	0.994	1.008
4.5	568.1	—0.34	1.38	4.57	0.993	1.009
5.0	569.2	—0.34	1.39	5.09	0.992	1.010
6.0	571.6	—0.35	1.39	6.13	0.990	1.012
7.0	574.0	—0.35	1.40	7.18	0.988	1.014
8.0	576.4	—0.36	1.40	8.24	0.986	1.016
9.0	578.8	—0.37	1.41	9.30	0.985	1.018
10.0	581.2	—0.37	1.42	10.37	0.983	1.020
11.0	583.6	—0.38	1.42	11.45	0.981	1.022
12.0	586.0	—0.38	1.43	12.54	0.979	1.023
13.0	588.4	—0.39	1.44	13.63	0.977	1.025
14.0	590.8	—0.40	1.44	14.74	0.974	1.027
15.0	593.3	—0.40	1.45	15.85	0.972	1.029
16.0	595.7	—0.41	1.46	16.97	0.970	1.030
17.0	598.2	—0.41	1.46	18.10	0.968	1.032
18.0	600.6	—0.42	1.47	19.24	0.966	1.034
19.0	603.1	—0.42	1.48	20.38	0.963	1.035
20.0	605.6	—0.43	1.48	21.54	0.961	1.037
21.0	608.1	—0.43	1.49	22.70	0.959	1.038
22.0	610.6	—0.44	1.50	23.87	0.957	1.040
23.0	613.1	—0.44	1.50	25.05	0.954	1.041
24.0	615.6	—0.45	1.51	26.24	0.952	1.043
25.0	618.1	—0.45	1.52	27.43	0.949	1.044
26.0	620.6	—0.45	1.53	28.64	0.947	1.046
27.0	623.1	—0.46	1.53	29.86	0.945	1.047
28.0	625.7	—0.46	1.54	31.08	0.942	1.048
29.0	628.2	—0.47	1.55	32.32	0.940	1.050
30.0	630.7	—0.47	1.55	33.56	0.937	1.051
35.0	643.5	—0.50	1.59	39.91	0.924	1.057
40.0	656.4	—0.52	1.62	46.51	0.911	1.063
45.0	669.4	—0.53	1.66	53.36	0.898	1.068
50.0	682.4	—0.55	1.69	60.47	0.885	1.073

p	w	μ	k	f	α/α_0	γ/γ_0
			$T=800$ K			
55.0	695.3	—0.57	1.73	67.85	0.872	1.077
60.0	708.2	—0.58	1.76	75.51	0.859	1.081
65.0	721.1	—0.59	1.79	83.46	0.846	1.084
70.0	733.8	—0.60	1.83	91.71	0.834	1.087
75.0	746.5	—0.61	1.86	100.27	0.822	1.090
80.0	759.2	—0.62	1.89	109.14	0.810	1.092
85.0	771.7	—0.63	1.93	118.35	0.799	1.094
90.0	784.3	—0.63	1.96	127.90	0.788	1.096
95.0	796.7	—0.64	1.99	137.79	0.777	1.097
100.0	809.2	—0.65	2.02	148.05	0.766	1.099
			$T=850$ K			
0.1	573.9	—0.36	1.35	0.10	1.000	1.000
0.5	574.8	—0.36	1.35	0.50	0.999	1.001
1.0	576.0	—0.37	1.35	1.00	0.998	1.002
1.5	577.1	—0.37	1.36	1.51	0.997	1.003
2.0	578.3	—0.38	1.36	2.01	0.997	1.004
2.5	579.4	—0.38	1.36	2.52	0.996	1.005
3.0	580.5	—0.38	1.37	3.03	0.995	1.005
3.5	581.7	—0.39	1.37	3.54	0.994	1.006
4.0	582.9	—0.39	1.37	4.06	0.993	1.007
4.5	584.0	—0.39	1.38	4.57	0.992	1.008
5.0	585.2	—0.40	1.38	5.09	0.991	1.009
6.0	587.5	—0.40	1.38	6.13	0.989	1.011
7.0	589.8	—0.41	1.39	7.18	0.987	1.012
8.0	592.2	—0.41	1.40	8.23	0.985	1.014
9.0	594.5	—0.42	1.40	9.29	0.983	1.016
10.0	596.8	—0.42	1.41	10.36	0.981	1.017
11.0	599.2	—0.43	1.42	11.44	0.979	1.019
12.0	601.6	—0.43	1.42	12.52	0.977	1.020
13.0	603.9	—0.44	1.43	13.62	0.975	1.022
14.0	606.3	—0.44	1.43	14.72	0.973	1.023
15.0	608.7	—0.45	1.44	15.82	0.971	1.025
16.0	611.1	—0.45	1.45	16.94	0.968	1.026
17.0	613.5	—0.46	1.45	18.07	0.966	1.028
18.0	615.9	—0.46	1.46	19.20	0.964	1.029
19.0	618.3	—0.46	1.46	20.34	0.962	1.030
20.0	620.7	—0.47	1.47	21.49	0.959	1.032
21.0	623.1	—0.47	1.48	22.64	0.957	1.033
22.0	625.5	—0.48	1.48	23.81	0.955	1.034
23.0	627.9	—0.48	1.49	24.98	0.952	1.036
24.0	630.4	—0.49	1.50	26.16	0.950	1.037
25.0	632.8	—0.49	1.50	27.35	0.948	1.038
26.0	635.3	—0.49	1.51	28.55	0.945	1.039
27.0	637.7	—0.50	1.52	29.76	0.943	1.041
28.0	640.2	—0.50	1.52	30.98	0.941	1.042
29.0	642.6	—0.50	1.53	32.20	0.938	1.043

p	w	μ	k	$\bar{i}$	α/α_0	γ/γ_0
			$T=850$ K			
30.0	645.1	—0.51	1.53	33.44	0.936	1.044
35.0	657.5	—0.53	1.57	39.74	0.923	1.050
40.0	670.0	—0.54	1.60	46.27	0.911	1.055
45.0	682.5	—0.56	1.63	53.05	0.898	1.059
50.0	695.0	—0.57	1.66	60.07	0.886	1.064
55.0	707.6	—0.59	1.70	67.34	0.873	1.067
60.0	720.1	—0.60	1.73	74.88	0.861	1.071
65.0	732.5	—0.61	1.76	82.69	0.849	1.074
70.0	744.9	—0.62	1.79	90.78	0.837	1.077
75.0	757.2	—0.63	1.82	99.16	0.826	1.079
80.0	769.4	—0.63	1.85	107.83	0.814	1.081
85.0	781.6	—0.64	1.88	116.82	0.804	1.083
90.0	793.7	—0.65	1.91	126.12	0.793	1.085
95.0	805.7	—0.65	1.94	135.75	0.782	1.086
100.0	817.7	—0.66	1.97	145.71	0.772	1.088
			$T=900$ K			
0.1	589.6	—0.41	1.34	0.10	1.000	1.000
0.5	590.5	—0.42	1.35	0.50	0.999	1.001
1.0	591.6	—0.42	1.35	1.00	0.998	1.002
1.5	592.7	—0.42	1.35	1.51	0.997	1.002
2.0	593.8	—0.43	1.36	2.01	0.996	1.003
2.5	594.9	—0.43	1.36	2.52	0.995	1.004
3.0	596.1	—0.43	1.36	3.03	0.994	1.005
3.5	597.2	—0.44	1.36	3.54	0.993	1.005
4.0	598.3	—0.44	1.37	4.06	0.992	1.006
4.5	599.5	—0.44	1.37	4.57	0.991	1.007
5.0	600.6	—0.44	1.37	5.09	0.990	1.008
6.0	602.9	—0.45	1.38	6.12	0.988	1.009
7.0	605.2	—0.45	1.38	7.17	0.986	1.011
8.0	607.5	—0.46	1.39	8.22	0.984	1.012
9.0	609.8	—0.46	1.40	9.28	0.982	1.013
10.0	612.1	—0.47	1.40	10.35	0.980	1.015
11.0	614.4	—0.47	1.41	11.42	0.978	1.016
12.0	616.7	—0.48	1.41	12.51	0.976	1.018
13.0	619.0	—0.48	1.42	13.60	0.974	1.019
14.0	621.3	—0.48	1.42	14.69	0.972	1.020
15.0	623.6	—0.49	1.43	15.80	0.969	1.021
16.0	626.0	—0.49	1.44	16.91	0.967	1.023
17.0	628.3	—0.49	1.44	18.03	0.965	1.024
18.0	630.7	—0.50	1.45	19.16	0.963	1.025
19.0	633.0	—0.50	1.45	20.29	0.961	1.026
20.0	635.3	—0.51	1.46	21.44	0.958	1.028
21.0	637.7	—0.51	1.47	22.59	0.956	1.029
22.0	640.1	—0.51	1.47	23.75	0.954	1.030
23.0	642.4	—0.52	1.48	24.92	0.951	1.031
24.0	644.8	—0.52	1.48	26.09	0.949	1.032

p	w	μ	k	f	α/α_0	γ/γ_0
			T=900 K			
25.0	647.2	—0.52	1.49	27.27	0.947	1.033
26.0	649.5	—0.53	1.49	28.47	0.944	1.034
27.0	651.9	—0.53	1.50	29.67	0.942	1.036
28.0	654.3	—0.53	1.51	30.88	0.940	1.037
29.0	656.7	—0.54	1.51	32.09	0.937	1.038
30.0	659.1	—0.54	1.52	33.32	0.935	1.039
35.0	671.1	—0.55	1.55	39.57	0.923	1.044
40.0	683.2	—0.57	1.58	46.04	0.911	1.048
45.0	695.4	—0.58	1.61	52.74	0.899	1.052
50.0	707.5	—0.60	1.64	59.68	0.887	1.056
55.0	719.7	—0.61	1.67	66.85	0.875	1.059
60.0	731.8	—0.62	1.70	74.27	0.863	1.063
65.0	743.9	—0.63	1.73	81.95	0.851	1.065
70.0	755.9	—0.63	1.76	89.90	0.840	1.068
75.0	767.8	—0.64	1.79	98.11	0.829	1.070
80.0	779.7	—0.65	1.81	106.61	0.818	1.072
85.0	791.5	—0.65	1.84	115.39	0.808	1.074
90.0	803.2	—0.66	1.87	124.47	0.798	1.076
95.0	814.9	—0.66	1.90	133.86	0.788	1.077
100.0	826.6	—0.67	1.93	143.56	0.778	1.078
			T=950 K			
0.1	604.8	—0.46	1.34	0.10	1.000	1.000
0.5	605.6	—0.46	1.34	0.50	0.999	1.001
1.0	606.7	—0.46	1.35	1.00	0.998	1.001
1.5	607.8	—0.47	1.35	1.51	0.997	1.002
2.0	608.9	—0.47	1.35	2.01	0.996	1.003
2.5	610.0	—0.47	1.35	2.52	0.995	1.003
3.0	611.2	—0.47	1.36	3.03	0.994	1.004
3.5	612.3	—0.48	1.36	3.54	0.993	1.005
4.0	613.4	—0.48	1.36	4.05	0.992	1.005
4.5	614.5	—0.48	1.36	4.57	0.991	1.006
5.0	615.6	—0.48	1.37	5.08	0.990	1.007
6.0	617.8	—0.49	1.37	6.12	0.988	1.008
7.0	620.1	—0.49	1.38	7.16	0.986	1.009
8.0	622.3	—0.50	1.38	8.22	0.984	1.011
9.0	624.6	—0.50	1.39	9.27	0.982	1.012
10.0	626.8	—0.50	1.39	10.34	0.980	1.013
11.0	629.1	—0.51	1.40	11.41	0.977	1.014
12.0	631.4	—0.51	1.40	12.49	0.975	1.015
13.0	633.6	—0.52	1.41	13.58	0.973	1.017
14.0	635.9	—0.52	1.42	14.67	0.971	1.018
15.0	638.2	—0.52	1.42	15.77	0.969	1.019
16.0	640.5	—0.52	1.43	16.88	0.967	1.020
17.0	642.8	—0.53	1.43	18.00	0.964	1.021
18.0	645.0	—0.53	1.44	19.12	0.962	1.022
19.0	647.3	—0.53	1.44	20.25	0.960	1.023

Table II.12 (*Continued*)

p	w	μ	k	f	α/α_0	γ/γ_0
			$T=950$ K			
20.0	649.6	—0,54	1.45	21.39	0.958	1.024
21.0	651.9	—0,54	1.45	22.54	0.955	1.025
22.0	654.2	—0,54	1.46	23.69	0.953	1.026
23.0	656.5	—0,55	1.47	24.85	0.951	1.027
24.0	658.8	—0,55	1.47	26.02	0.949	1.028
25.0	661.2	—0,55	1.48	27.20	0.946	1.029
26.0	663.5	—0,55	1.48	28.38	0.944	1.030
27.0	665.8	—0,56	1.49	29.57	0.942	1.031
28.0	668.1	—0,56	1.49	30.78	0.939	1.032
29.0	670.4	—0,56	1.50	31.98	0.937	1.033
30.0	672.8	—0,57	1.50	33.20	0.935	1.034
35.0	684.5	—0,58	1.53	39.41	0.923	1.038
40.0	696.2	—0,59	1.56	45.82	0.911	1.042
45.0	708.0	—0,60	1.59	52.45	0.900	1.046
50.0	719.9	—0,61	1.62	59.30	0.888	1.050
55.0	731.7	—0,62	1.65	66.39	0.876	1.053
60.0	743.5	—0,63	1.67	73.70	0.865	1.056
65.0	755.2	—0,64	1.70	81.26	0.854	1.058
70.0	766.9	—0,65	1.73	89.07	0.843	1.061
75.0	778.5	—0,65	1.76	97.14	0.832	1.063
80.0	790.1	—0,66	1.78	105.47	0.822	1.065
85.0	801.5	—0,67	1.81	114.07	0.812	1.066
90.0	813.0	—0,67	1.84	122.95	0.802	1.068
95.0	824.3	—0,67	1.86	132.11	0.792	1.069
100.0	835.6	—0,68	1.89	141.57	0.783	1.070
			$T=1000$ K			
0.1	619.6	—0,50	1.34	0.10	1.000	1.000
0.5	620.4	—0,50	1.34	0.50	0.999	1.001
1.0	621.5	—0,50	1.34	1.00	0.998	1.001
1.5	622.6	—0,50	1.34	1.51	0.997	1.002
2.0	623.7	—0,51	1.35	2.01	0.996	1.002
2.5	624.8	—0,51	1.35	2.52	0.995	1.003
3.0	625.9	—0,51	1.35	3.03	0.994	1.004
3.5	626.9	—0,51	1.35	3.54	0.993	1.004
4.0	628.0	—0,51	1.36	4.05	0.992	1.005
4.5	629.1	—0,52	1.36	4.57	0.991	1.005
5.0	630.2	—0,52	1.36	5.08	0.990	1.006
6.0	632.4	—0,52	1.37	6.12	0.988	1.007
7.0	634.6	—0,53	1.37	7.16	0.986	1.008
8.0	636.8	—0,53	1.38	8.21	0.983	1.009
9.0	639.0	—0,53	1.38	9.27	0.981	1.010
10.0	641.3	—0,54	1.39	10.33	0.979	1.011
11.0	643.5	—0,54	1.39	11.40	0.977	1.013
12.0	645.7	—0,54	1.40	12.47	0.975	1.014
13.0	647.9	—0,55	1.40	13.56	0.973	1.015
14.0	650.1	—0,55	1.41	14.65	0.971	1.016

Table II.12 (*Continued*)

p	w	μ	k	f	α/α_0	γ/γ_0
			$T=1000$ K			
15.0	652.4	—0.55	1.41	15.75	0.968	1.017
16.0	654.6	—0.55	1.42	16.85	0.966	1.018
17.0	656.8	—0.56	1.42	17.96	0.964	1.019
18.0	659.1	—0.56	1.43	19.08	0.962	1.020
19.0	661.3	—0.56	1.43	20.21	0.960	1.020
20.0	663.6	—0.56	1.44	21.34	0.957	1.021
21.0	665.8	—0.57	1.44	22.48	0.955	1.022
22.0	668.1	—0.57	1.45	23.63	0.953	1.023
23.0	670.3	—0.57	1.45	24.79	0.951	1.024
24.0	672.6	—0.57	1.46	25.95	0.948	1.025
25.0	674.8	—0.58	1.47	27.12	0.946	1.026
26.0	677.1	—0.58	1.47	28.30	0.944	1.027
27.0	679.4	—0.58	1.48	29.49	0.942	1.028
28.0	681.6	—0.58	1.48	30.68	0.939	1.028
29.0	683.9	—0.59	1.49	31.88	0.937	1.029
30.0	686.2	—0.59	1.49	33.09	0.935	1.030
35.0	697.6	—0.60	1.52	39.25	0.924	1.034
40.0	709.0	—0.61	1.54	45.61	0.912	1.038
45.0	720.5	—0.62	1.57	52.18	0.901	1.041
50.0	732.0	—0.63	1.60	58.95	0.889	1.044
55.0	743.5	—0.64	1.62	65.95	0.878	1.047
60.0	755.0	—0.65	1.65	73.17	0.867	1.050
65.0	766.4	—0.66	1.68	80.62	0.857	1.052
70.0	777.8	—0.66	1.70	88.30	0.846	1.054
75.0	789.2	—0.67	1.73	96.23	0.836	1.056
80.0	800.4	—0.67	1.75	104.41	0.826	1.058
85.0	811.6	—0.68	1.78	112.84	0.816	1.060
90.0	822.8	—0.68	1.80	121.53	0.806	1.061
95.0	833.8	—0.68	1.83	130.49	0.797	1.062
100.0	844.8	—0.69	1.85	139.73	0.788	1.064
			$T=1050$ K			
0.1	634.0	--0.53	1.33	0.10	1.000	1.000
0.5	634.8	—0.53	1.33	0.50	0.999	1.001
1.0	635.9	—0.53	1.34	1.00	0.998	1.001
1.5	637.0	—0.53	1.34	1.51	0.997	1.002
2.0	638.1	—0.54	1.34	2.01	0.996	1.002
2.5	639.1	—0.54	1.34	2.52	0.995	1.003
3.0	640.2	—0.54	1.35	3.03	0.994	1.003
3.5	641.3	—0.54	1.35	3.54	0.993	1.004
4.0	642.3	—0.55	1.35	4.05	0.992	1.004
4.5	643.4	—0.55	1.35	4.56	0.991	1.005
5.0	644.5	—0.55	1.36	5.08	0.990	1.005
6.0	646.7	—0.55	1.36	6.11	0.988	1.006
7.0	648.8	—0.56	1.37	7.15	0.985	1.007
8.0	651.0	—0.56	1.37	8.20	0.983	1.008
9.0	653.2	—0.56	1.38	9.26	0.981	1.009

Table II.12 (*Continued*)

p	w	μ	k	f	α/α_0	γ/γ_0
			$T=1050$ K			
10.0	655.3	—0.56	1.38	10.32	0.979	1.010
11.0	657.5	—0.57	1.39	11.39	0.977	1.011
12.0	659.7	—0.57	1.39	12.46	0.975	1.012
13.0	661.9	—0.57	1.40	13.54	0.973	1.013
14.0	664.0	—0.57	1.40	14.63	0.970	1.014
15.0	666.2	—0.58	1.41	15.72	0.968	1.015
16.0	668.4	—0.58	1.41	16.82	0.966	1.016
17.0	670.6	—0.58	1.42	17.93	0.964	1.016
18.0	672.8	—0.58	1.42	19.05	0.962	1.017
19.0	675.0	—0.59	1.43	20.17	0.959	1.018
20.0	677.2	—0.59	1.43	21.30	0.957	1.019
21.0	679.4	—0.59	1.44	22.43	0.955	1.020
22.0	681.6	—0.59	1.44	23.58	0.953	1.021
23.0	683.8	—0.59	1.45	24.73	0.951	1.021
24.0	686.0	—0.60	1.45	25.89	0.949	1.022
25.0	688.2	—0.60	1.46	27.05	0.946	1.023
26.0	690.4	—0.60	1.46	28.22	0.944	1.024
27.0	692.6	—0.60	1.47	29.40	0.942	1.025
28.0	694.8	—0.60	1.47	30.59	0.940	1.025
29.0	697.1	—0.61	1.48	31.78	0.937	1.026
30.0	699.3	—0.61	1.48	32.98	0.935	1.027
35.0	710.4	—0.62	1.51	39.10	0.924	1.030
40.0	721.6	—0.63	1.53	45.41	0.913	1.034
45.0	732.8	—0.64	1.56	51.91	0.902	1.037
50.0	744.0	—0.65	1.58	58.62	0.891	1.040
55.0	755.2	—0.65	1.61	65.54	0.880	1.042
60.0	766.4	—0.66	1.63	72.66	0.870	1.045
65.0	777.6	—0.67	1.66	80.01	0.859	1.047
70.0	788.7	—0.67	1.68	87.58	0.849	1.049
75.0	799.8	—0.68	1.70	95.38	0.839	1.051
80.0	810.8	—0.68	1.73	103.42	0.829	1.052
85.0	821.7	—0.69	1.75	111.69	0.820	1.054
90.0	832.6	—0.69	1.78	120.22	0.811	1.055
95.0	843.4	—0.69	1.80	129.00	0.802	1.057
100.0	854.1	—0.69	1.82	138.03	0.793	1.058
			$T=1100$ K			
0,1	648.1	—0.56	1.33	0.10	1.000	1.000
0,5	649.0	—0.56	1.33	0.50	0.999	1.000
1,0	650.0	—0.56	1.33	1.00	0.998	1.001
1,5	651.0	—0.56	1.34	1.51	0.997	1.001
2,0	652.1	—0.56	1.34	2.01	0.996	1.002
2,5	653.2	—0.57	1.34	2.52	0.995	1.002
3,0	654.2	—0.57	1.34	3.03	0.994	1.003
3,5	655.3	—0.57	1.35	3.54	0.993	1.003
4,0	656.3	—0.57	1.35	4.05	0.992	1.004
4,5	657.4	—0.57	1.35	4.56	0.991	1.004

Table II.12 (*Continued*)

p	w	μ	k	f	α/α_0	γ/γ_0
			$T=1100$ K			
5.0	658.4	—0.57	1.35	5.08	0.990	1.005
6.0	660.6	—0.58	1.36	6.11	0.987	1.006
7.0	662.7	—0.58	1.36	7.15	0.985	1.006
8.0	664.8	—0.58	1.37	8.20	0.983	1.007
9.0	669.9	—0.59	1.37	9.25	0.981	1.008
10.0	669.1	—0.59	1.38	10.31	0.979	1.009
11.0	671.2	—0.59	1.38	11.37	0.977	1.010
12.0	673.4	—0.59	1.38	12.44	0.975	1.011
13.0	675.5	—0.60	1.39	13.52	0.972	1.012
14.0	677.6	—0.60	1.39	14.61	0.970	1.012
15.0	679.8	—0.60	1.40	15.70	0.968	1.013
16.0	681.9	—0.60	1.40	16.80	0.966	1.014
17.0	684.1	—0.60	1.41	17.90	0.964	1.015
18.0	686.2	—0.61	1.41	19.01	0.962	1.015
19.0	688.4	—0.61	1.42	20.13	0.960	1.016
20.0	690.5	—0.61	1.42	21.26	0.957	1.017
21.0	692.7	—0.61	1.43	22.39	0.955	1.018
22.0	694.8	—0.61	1.43	23.53	0.953	1.018
23.0	697.0	—0.61	1.44	24.67	0.951	1.019
24.0	699.1	—0.62	1.44	25.82	0.949	1.020
25.0	701.3	—0.62	1.45	26.98	0.947	1.021
26.0	703.5	—0.62	1.45	28.15	0.944	1.021
27.0	705.6	—0.62	1.46	29.32	0.942	1.022
28.0	707.8	—0.62	1.46	30.50	0.940	1.023
29.0	710.0	—0.63	1.47	31.69	0.938	1.023
30.0	712.1	—0.63	1.47	32.88	0.936	1.024
35.0	723.0	—0.64	1.49	38.96	0.925	1.027
40.0	733.9	—0.64	1.52	45.22	0.914	1.030
45.0	744.9	—0.65	1.54	51.67	0.904	1.033
50.0	755.8	—0.66	1.57	58.31	0.893	1.036
55.0	766.8	—0.67	1.59	65.15	0.882	1.038
60.0	777.7	—0.67	1.61	72.19	0.872	1.040
65.0	788.7	—0.68	1.64	79.44	0.862	1.042
70.0	799.5	—0.68	1.66	86.91	0.852	1.044
75.0	810.3	—0.69	1.68	94.59	0.842	1.046
80.0	821.1	—0.69	1.71	102.49	0.833	1.048
85.0	831.8	—0.69	1.73	110.63	0.824	1.049
90.0	842.4	—0.70	1.75	119.00	0.815	1.050
95.0	853.0	—0.70	1.77	127.61	0.806	1.052
100.0	863.5	—0.70	1.79	136.46	0.797	1.053
			$T=1150$ K			
0.1	661.9	—0.58	1.33	0.10	1.000	1.000
0.5	662.7	—0.58	1.33	0.50	0.999	1.000
1.0	663.8	—0.58	1.33	1.00	0.998	1.001
1.5	664.8	—0.59	1.33	1.51	0.997	1.001
2.0	665.8	—0.59	1.34	2.01	0.996	1.002

Table II.12 (*Continued*)

p	w	μ	k	f	α/α_0	γ/γ_0
			$T=1150$ K			
2.5	666.9	—0.59	1.34	2.52	0.995	1.002
3.0	667.9	—0.59	1.34	3.03	0.994	1.003
3.5	669.0	—0.59	1.34	3.54	0.993	1.003
4.0	670.0	—0.59	1.34	4.05	0.992	1.003
4.5	671.0	—0.60	1.35	4.56	0.991	1.004
5.0	672.1	—0.60	1.35	5.07	0.990	1.004
6.0	674.2	—0.60	1.35	6.11	0.987	1.005
7.0	676.3	—0.60	1.36	7.15	0.985	1.006
8.0	678.3	—0.60	1.36	8.19	0.983	1.007
9.0	680.4	—0.61	1.37	9.24	0.981	1.007
10.0	682.5	—0.61	1.37	10.30	0.979	1.008
11.0	684.6	—0.61	1.38	11.36	0.977	1.009
12.0	686.7	—0.61	1.38	12.43	0.975	1.010
13.0	688.8	—0.62	1.38	13.51	0.973	1.010
14.0	690.9	—0.62	1.39	14.59	0.970	1.011
15.0	693.0	—0.62	1.39	15.68	0.968	1.012
16.0	695.2	—0.62	1.40	16.77	0.966	1.012
17.0	697.3	—0.62	1.40	17.87	0.964	1.013
18.0	699.4	—0.62	1.41	18.98	0.962	1.014
19.0	701.5	—0.63	1.41	20.09	0.960	1.015
20.0	703.6	—0.63	1.42	21.21	0.958	1.015
21.0	705.7	—0.63	1.42	22.34	0.956	1.016
22.0	707.8	—0.63	1.42	23.48	0.953	1.017
23.0	709.9	—0.63	1.43	24.62	0.951	1.017
24.0	712.0	—0.63	1.43	25.76	0.949	1.018
25.0	714.2	—0.64	1.44	26.92	0.947	1.019
26.0	716.3	—0.64	1.44	28.08	0.945	1.019
27.0	718.4	—0.64	1.45	29.24	0.943	1.020
28.0	720.5	—0.64	1.45	30.42	0.941	1.020
29.0	722.7	—0.64	1.46	31.60	0.939	1.021
30.0	724.8	—0.64	1.46	32.78	0.937	1.022
35.0	735.4	—0.65	1.48	38.82	0.926	1.025
40.0	746.1	—0.66	1.51	45.04	0.916	1.027
45.0	756.8	—0.66	1.53	51.43	0.905	1.030
50.0	767.5	—0.67	1.55	58.01	0.895	1.032
55.0	778.2	—0.68	1.57	64.78	0.885	1.034
60.0	788.9	—0.68	1.60	71.75	0.875	1.036
65.0	799.6	—0.69	1.62	78.91	0.865	1.038
70.0	810.3	—0.69	1.64	86.28	0.855	1.040
75.0	820.9	—0.70	1.66	93.85	0.846	1.042
80.0	831.4	—0.70	1.68	101.63	0.836	1.043
85.0	841.9	—0.70	1.71	109.64	0.827	1.045
90.0	852.3	—0.70	1.73	117.86	0.819	1.046
95.0	862.6	—0.71	1.75	126.32	0.810	1.047
100.0	872.9	—0.71	1.77	135.00	0.802	1.048

Table II.12 (*Continued*)

p	ω	μ	k	f	α/α_0	γ/γ_0
			$T=1200$ K			
0.1	675.4	—0.60	1.32	0.10	1.000	1.000
0.5	676.3	—0.60	1.33	0.50	0.999	1.000
1.0	677.3	—0.61	1.33	1.00	0.998	1.001
1.5	678.3	—0.61	1.33	1.51	0.997	1.001
2.0	679.3	—0.61	1.33	2.01	0.996	1.002
2.5	680.3	—0.61	1.33	2.52	0.995	1.002
3.0	681.4	—0.61	1.34	3.03	0.994	1.002
3.5	682.4	—0.61	1.34	3.53	0.993	1.003
4.0	683.4	—0.61	1.34	4.05	0.992	1.003
4.5	684.4	—0.62	1.34	4.56	0.991	1.003
5.0	685.4	—0.62	1.34	5.07	0.990	1.004
6.0	687.5	—0.62	1.35	6.10	0.988	1.004
7.0	689.6	—0.62	1.35	7.14	0.985	1.005
8.0	691.6	—0.62	1.36	8.18	0.983	1.006
9.0	693.7	—0.63	1.36	9.23	0.981	1.007
10.0	695.7	—0.63	1.37	10.29	0.979	1.007
11.0	697.8	—0.63	1.37	11.35	0.977	1.008
12.0	699.9	—0.63	1.37	12.42	0.975	1.009
13.0	701.9	—0.63	1.38	13.49	0.973	1.009
14.0	704.0	—0.63	1.38	14.57	0.971	1.010
15.0	706.1	—0.64	1.39	15.66	0.969	1.011
16.0	708.1	—0.64	1.39	16.75	0.966	1.011
17.0	710.2	—0.64	1.40	17.85	0.964	1.012
18.0	712.3	—0.64	1.40	18.95	0.962	1.013
19.0	714.3	—0.64	1.40	20.06	0.960	1.013
20.0	716.4	—0.64	1.41	21.18	0.958	1.014
21.0	718.5	—0.65	1.41	22.30	0.956	1.014
22.0	720.6	—0.65	1.42	23.43	0.954	1.015
23.0	722.6	—0.65	1.42	24.56	0.952	1.016
24.0	724.7	—0.65	1.43	25.71	0.950	1.016
25.0	726.8	—0.65	1.43	26.85	0.948	1.017
26.0	728.9	—0.65	1.43	28.01	0.946	1.017
27.0	730.9	—0.65	1.44	29.17	0.944	1.018
28.0	733.0	—0.65	1.44	30.34	0.942	1.018
29.0	735.1	—0.66	1.45	31.51	0.939	1.019
30.0	737.2	—0.66	1.45	32.69	0.937	1.020
35.0	747.6	—0.66	1.47	38.70	0.927	1.022
40.0	758.1	—0.67	1.49	44.87	0.917	1.025
45.0	768.6	—0.68	1.52	51.21	0.907	1.027
50.0	779.1	—0.68	1.54	57.74	0.897	1.029
55.0	789.6	—0.69	1.56	64.44	0.887	1.031
60.0	800.0	—0.69	1.58	71.33	0.877	1.033
65.0	810.5	—0.70	1.60	78.41	0.867	1.035
70.0	820.9	—0.70	1.62	85.68	0.858	1.037
75.0	831.3	—0.70	1.64	93.16	0.849	1.038

Table II.12 (*Continued*)

p	w	μ	k	f	α/α_0	γ/γ_0
			T=1200 K			
80.0	841.6	—0.71	1.67	100.83	0.840	1.040
85.0	851.9	—0.71	1.69	108.71	0.831	1.041
90.0	862.1	—0.71	1.71	116.81	0.822	1.042
95.0	872.3	—0.71	1.73	125.12	0.814	1.043
100.0	882.4	—0.72	1.75	133.65	0.806	1.044
			T=1250 K			
0.1	688.7	—0.62	1.32	0.10	1.000	1.000
0.5	689.5	—0.62	1.32	0.50	0.999	1.000
1.0	690.5	—0.62	1.32	1.00	0.998	1.001
1.5	691.5	—0.63	1.33	1.51	0.997	1.001
2.0	692.5	—0.63	1.33	2.01	0.996	1.001
2.5	693.5	—0.63	1.33	2.52	0.995	1.002
3.0	694.5	—0.63	1.33	3.02	0.994	1.002
3.5	695.5	—0.63	1.34	3.53	0.993	1.002
4.0	696.5	—0.63	1.34	4.04	0.992	1.003
4.5	697.5	—0.63	1.34	4.56	0.991	1.003
5.0	698.6	—0.63	1.34	5.07	0.990	1.003
6.0	700.6	—0.64	1.35	6.10	0.988	1.004
7.0	702.6	—0.64	1.35	7.14	0.986	1.005
8.0	704.6	—0.64	1.35	8.18	0.983	1.005
9.0	706.6	—0.64	1.36	9.23	0.981	1.006
10.0	708.7	—0.64	1.36	10.28	0.979	1.007
11.0	710.7	—0.65	1.37	11.34	0.977	1.007
12.0	712.7	—0.65	1.37	12.40	0.975	1.008
13.0	714.8	—0.65	1.37	13.48	0.973	1.008
14.0	716.8	—0.65	1.38	14.55	0.971	1.009
15.0	718.8	—0.65	1.38	15.64	0.969	1.010
16.0	720.9	—0.65	1.39	16.72	0.967	1.010
17.0	722.9	—0.65	1.39	17.82	0.965	1.011
18.0	724.9	—0.66	1.39	18.92	0.963	1.011
19.0	727.0	—0.66	1.40	20.03	0.961	1.012
20.0	729.0	—0.66	1.40	21.14	0.959	1.012
21.0	731.0	—0.66	1.41	22.26	0.957	1.013
22.0	733.1	—0.66	1.41	23.38	0.955	1.014
23.0	735.1	—0.66	1.42	24.51	0.952	1.014
24.0	737.1	—0.66	1.42	25.65	0.950	1.015
25.0	739.2	—0.66	1.42	26.79	0.948	1.015
26.0	741.2	—0.67	1.43	27.94	0.946	1.016
27.0	743.3	—0.67	1.43	29.10	0.944	1.016
28.0	745.3	—0.67	1.44	30.26	0.942	1.017
29.0	747.4	—0.67	1.44	31.43	0.940	1.017
30.0	749.4	—0.67	1.44	32.61	0.938	1.018
35.0	759.6	—0.68	1.46	38.58	0.928	1.020
40.0	769.9	—0.68	1.49	44.71	0.918	1.022
45.0	780.2	—0.69	1.51	51.01	0.908	1.025
50.0	790.4	—0.69	1.53	57.48	0.898	1.027

Table II.12 (*Continued*)

p	w	μ	k	f	α/α_0	γ/γ_0
			$T=1250$ K			
55.0	800.7	—0.70	1.55	64.12	0.889	1.029
60.0	811.0	—0.70	1.57	70.94	0.879	1.030
65.0	821.2	—0.70	1.59	77.94	0.870	1.032
70.0	831.5	—0.71	1.61	85.13	0.861	1.034
75.0	841.6	—0.71	1.63	92.51	0.852	1.035
80.0	851.8	—0.71	1.65	100.08	0.843	1.036
85.0	861.9	—0.72	1.67	107.85	0.834	1.037
90.0	871.9	—0.72	1.69	115.82	0.826	1.039
95.0	881.9	—0.72	1.71	124.00	0.818	1.040
100.0	891.8	—0.72	1.73	132.39	0.810	1.041
			$T=1300$ K			
0.1	701.7	—0.64	1.32	0.10	1.000	1.000
0.5	702.5	—0.64	1.32	0.50	0.999	1.000
1.0	703.5	—0.64	1.32	1.00	0.998	1.001
1.5	704.5	—0.64	1.32	1.51	0.997	1.001
2.0	705.5	—0.64	1.33	2.01	0.996	1.001
2.5	706.5	—0.65	1.33	2.52	0.995	1.002
3.0	707.4	—0.65	1.33	3.02	0.994	1.002
3.5	708.4	—0.65	1.33	3.53	0.993	1.002
4.0	709.4	—0.65	1.33	4.04	0.992	1.002
4.5	710.4	—0.65	1.34	4.55	0.991	1.003
5.0	711.4	—0.65	1.34	5.07	0.990	1.003
6.0	713.4	—0.65	1.34	6.10	0.988	1.004
7.0	715.4	—0.65	1.35	7.13	0.986	1.004
8.0	717.4	—0.66	1.35	8.17	0.984	1.005
9.0	719.4	—0.66	1.35	9.22	0.982	1.005
10.0	721.4	—0.66	1.36	10.27	0.979	1.006
11.0	723.4	—0.66	1.36	11.33	0.977	1.007
12.0	725.4	—0.66	1.37	12.39	0.975	1.007
13.0	727.4	—0.66	1.37	13.46	0.973	1.008
14.0	729.4	—0.66	1.37	14.54	0.971	1.008
15.0	731.4	—0.67	1.38	15.62	0.969	1.009
16.0	733.4	—0.67	1.38	16.70	0.967	1.009
17.0	735.4	—0.67	1.39	17.79	0.965	1.010
18.0	737.4	—0.67	1.39	18.89	0.963	1.010
19.0	739.4	—0.67	1.39	19.99	0.961	1.011
20.0	741.4	—0.67	1.40	21.10	0.959	1.011
21.0	743.4	—0.67	1.40	22.22	0.957	1.012
22.0	745.4	—0.67	1.41	23.34	0.955	1.012
23.0	747.4	—0.67	1.41	24.47	0.953	1.013
24.0	749.4	—0.68	1.41	25.60	0.951	1.013
25.0	751.4	—0.68	1.42	26.74	0.949	1.014
26.0	753.4	—0.68	1.42	27.88	0.947	1.014
27.0	755.4	—0.68	1.42	29.03	0.945	1.015
28.0	757.4	—0.68	1.43	30.19	0.943	1.015
29.0	759.4	—0.68	1.43	31.35	0.941	1.016

Table II.12 (*Continued*)

p	w	μ	k	f	α/α_0	γ/γ_0
			$T = 1300$ K			
30.0	761.4	—0.68	1.44	32.52	0.939	1.016
35.0	771.5	—0.69	1.46	38.46	0.929	1.018
40.0	781.5	—0.69	1.48	44.56	0.920	1.020
45.0	791.6	—0.70	1.50	50.81	0.910	1.022
50.0	801.7	—0.70	1.52	57.23	0.900	1.024
55.0	811.8	—0.70	1.54	63.82	0.891	1.026
60.0	821.8	—0.71	1.56	70.57	0.882	1.028
65.0	831.9	—0.71	1.57	77.50	0.873	1.029
70.0	841.9	—0.72	1.59	84.61	0.864	1.031
75.0	851.9	—0.72	1.61	91.90	0.855	1.032
80.0	861.9	—0.72	1.63	99.38	0.846	1.033
85.0	871.8	—0.72	1.65	107.05	0.838	1.035
90.0	881.6	—0.72	1.67	114.91	0.830	1.036
95.0	891.4	—0.73	1.69	122.96	0.822	1.037
100.0	901.2	—0.73	1.71	131.22	0.814	1.038
			$T = 1350$ K			
0.1	714.5	—0.65	1.32	0.10	1.000	1.000
0.5	715.3	—0.65	1.32	0.50	0.999	1.000
1.0	716.2	—0.66	1.32	1.00	0.998	1.001
1.5	717.2	—0.66	1.32	1.51	0.997	1.001
2.0	718.2	—0.66	1.32	2.01	0.996	1.001
2.5	719.2	—0.66	1.33	2.52	0.995	1.001
3.0	720.1	—0.66	1.33	3.02	0.994	1.002
3.5	721.1	—0.66	1.33	3.53	0.993	1.002
4.0	722.1	—0.66	1.33	4.04	0.992	1.002
4.5	723.1	—0.66	1.33	4.55	0.991	1.003
5.0	724.0	—0.66	1.34	5.07	0.990	1.003
6.0	726.0	—0.67	1.34	6.09	0.988	1.003
7.0	728.0	—0.67	1.34	7.13	0.986	1.004
8.0	729.9	—0.67	1.35	8.17	0.984	1.004
9.0	731.9	—0.67	1.35	9.21	0.982	1.005
10.0	733.8	—0.67	1.35	10.26	0.980	1.005
11.0	735.8	—0.67	1.36	11.32	0.978	1.006
12.0	737.8	—0.67	1.36	12.38	0.976	1.006
13.0	739.7	—0.68	1.37	13.45	0.974	1.007
14.0	741.7	—0.68	1.37	14.52	0.972	1.007
15.0	743.7	—0.68	1.37	15.60	0.970	1.008
16.0	745.6	—0.68	1.38	16.68	0.968	1.008
17.0	747.6	—0.68	1.38	17.77	0.966	1.009
18.0	749.6	—0.68	1.38	18.86	0.964	1.009
19.0	751.5	—0.68	1.39	19.96	0.962	1.010
20.0	753.5	—0.68	1.39	21.07	0.960	1.010
21.0	755.5	—0.68	1.40	22.18	0.958	1.011
22.0	757.4	—0.69	1.40	23.30	0.956	1.011
23.0	759.4	—0.69	1.40	24.42	0.954	1.012
24.0	761.4	—0.69	1.41	25.55	0.952	1.012

Table II.12 (*Continued*)

p	w	μ	k	f	α/α_0	γ/γ_0
			$T=1350$ K			
25.0	763.4	—0.69	1.41	26.69	0.950	1.013
26.0	765.3	—0.69	1.41	27.83	0.948	1.013
27.0	767.3	—0.69	1.42	28.97	0.946	1.013
28.0	769.3	—0.69	1.42	30.12	0.944	1.014
29.0	771.2	—0.69	1.43	31.28	0.942	1.014
30.0	773.2	—0.69	1.43	32.45	0.940	1.015
35.0	783.1	—0.70	1.45	38.35	0.930	1.017
40.0	793.0	—0.70	1.47	44.41	0.921	1.019
45.0	802.9	—0.70	1.49	50.63	0.912	1.021
50.0	812.8	—0.71	1.51	57.00	0.902	1.022
55.0	822.7	—0.71	1.52	63.53	0.893	1.024
60.0	832.6	—0.72	1.54	70.23	0.884	1.026
65.0	842.4	—0.72	1.56	77.09	0.875	1.027
70.0	852.3	—0.72	1.58	84.13	0.866	1.028
75.0	862.1	—0.72	1.60	91.34	0.858	1.030
80.0	871.9	—0.73	1.62	98.72	0.849	1.031
85.0	881.6	—0.73	1.64	106.29	0.841	1.032
90.0	891.3	—0.73	1.66	114.05	0.833	1.033
95.0	900.9	—0.73	1.67	121.99	0.825	1.034
100.0	910.5	—0.73	1.69	130.12	0.818	1.035
			$T=1400$ K			
0.1	727.0	—0.67	1.31	0.10	1.000	1.000
0.5	727.8	—0.67	1.32	0.50	0.999	1.000
1.0	728.7	—0.67	1.32	1.00	0.998	1.001
1.5	729.7	—0.67	1.32	1.51	0.997	1.001
2.0	730.7	—0.67	1.32	2.01	0.996	1.001
2.5	731.6	—0.67	1.32	2.52	0.995	1.001
3.0	732.6	—0.67	1.33	3.02	0.994	1.002
3.5	733.5	—0.67	1.33	3.53	0.993	1.002
4.0	734.5	—0.68	1.33	4.04	0.992	1.002
4.5	735.5	—0.68	1.33	4.55	0.991	1.002
5.0	736.4	—0.68	1.33	5.06	0.990	1.003
6.0	738.4	—0.68	1.34	6.09	0.988	1.003
7.0	740.3	—0.68	1.34	7.12	0.986	1.004
8.0	742.2	—0.68	1.34	8.16	0.984	1.004
9.0	744.2	—0.68	1.35	9.21	0.982	1.004
10.0	746.1	—0.68	1.35	10.26	0.980	1.005
11.0	748.0	—0.69	1.35	11.31	0.978	1.005
12.0	750.0	—0.69	1.36	12.37	0.976	1.006
13.0	751.9	—0.69	1.36	13.43	0.974	1.006
14.0	753.8	—0.69	1.37	14.50	0.972	1.007
15.0	755.8	—0.69	1.37	15.58	0.970	1.007
16.0	757.7	—0.69	1.37	16.66	0.968	1.008
17.0	759.6	—0.69	1.38	17.75	0.966	1.008
18.0	761.6	—0.69	1.38	18.84	0.964	1.009
19.0	763.5	—0.69	1.38	19.94	0.962	1.009

Table II.12 (*Continued*)

p	w	μ	k	f	α/α_0	γ/γ_0
			$T=1400$ K			
20.0	765.5	—0.69	1.39	21.04	0.960	1.009
21.0	767.4	—0.69	1.39	22.15	0.958	1.010
22.0	769.3	—0.70	1.39	23.26	0.956	1.010
23.0	771.3	—0.70	1.40	24.38	0.955	1.011
24.0	773.2	—0.70	1.40	25.51	0.953	1.011
25.0	775.2	—0.70	1.41	26.64	0.951	1.012
26.0	777.1	—0.70	1.41	27.77	0.949	1.012
27.0	779.0	—0.70	1.41	28.91	0.947	1.012
28.0	781.0	—0.70	1.42	30.06	0.945	1.013
29.0	782.9	—0.70	1.42	31.21	0.943	1.013
30.0	784.9	—0.70	1.42	32.37	0.941	1.014
35.0	794.6	—0.71	1.44	38.25	0.932	1.015
40.0	804.3	—0.71	1.46	44.28	0.922	1.017
45.0	814.0	—0.71	1.48	50.45	0.913	1.019
50.0	823.7	—0.72	1.50	56.78	0.904	1.021
55.0	833.5	—0.72	1.52	63.26	0.895	1.022
60.0	843.2	—0.72	1.53	69.90	0.886	1.024
65.0	852.9	—0.73	1.55	76.70	0.877	1.025
70.0	862.5	—0.73	1.57	83.67	0.869	1.026
75.0	872.2	—0.73	1.59	90.80	0.861	1.027
80.0	881.8	—0.73	1.61	98.11	0.852	1.028
85.0	891.4	—0.73	1.62	105.58	0.844	1.029
90.0	900.9	—0.73	1.64	113.24	0.836	1.030
95.0	910.4	—0.74	1.66	121.07	0.829	1.031
100.0	919.8	—0.74	1.67	129.09	0.821	1.032
			$T=1450$ K			
0.1	739.3	—0.68	1.31	0.10	1.000	1.000
0.5	740.1	—0.68	1.31	0.50	0.999	1.000
1.0	741.0	—0.68	1.32	1.00	0.998	1.000
1.5	742.0	—0.68	1.32	1.51	0.997	1.001
2.0	742.9	—0.68	1.32	2.01	0.996	1.001
2.5	743.9	—0.68	1.32	2.52	0.995	1.001
3.0	744.8	—0.68	1.32	3.02	0.994	1.001
3.5	745.8	—0.69	1.32	3.53	0.993	1.002
4.0	746.7	—0.69	1.33	4.04	0.992	1.002
4.5	747.7	—0.69	1.33	4.55	0.991	1.002
5.0	748.6	—0.69	1.33	5.06	0.990	1.002
6.0	750.5	—0.69	1.33	6.09	0.988	1.003
7.0	752.4	—0.69	1.34	7.12	0.986	1.003
8.0	754.3	—0.69	1.34	8.16	0.984	1.004
9.0	756.2	—0.69	1.34	9.20	0.982	1.004
10.0	758.1	—0.69	1.35	10.25	0.980	1.005
11.0	760.1	—0.70	1.35	11.30	0.978	1.005
12.0	762.0	—0.70	1.35	12.36	0.976	1.005
13.0	763.9	—0.70	1.36	13.42	0.974	1.006
14.0	765.8	—0.70	1.36	14.49	0.973	1.006

Table II.12 (*Continued*)

p	w	μ	k	f	α/α_0	γ/γ_0
			$T = 1450$ K			
15.0	767.7	—0.70	1.37	15.56	0.971	1.007
16.0	769.6	—0.70	1.37	16.64	0.969	1.007
17.0	771.5	—0.70	1.37	17.73	0.967	1.007
18.0	773.4	—0.70	1.38	18.82	0.965	1.008
19.0	775.3	—0.70	1.38	19.91	0.963	1.008
20.0	777.2	—0.70	1.38	21.01	0.961	1.009
21.0	779.1	—0.70	1.39	22.11	0.959	1.009
22.0	781.0	—0.70	1.39	23.22	0.957	1.009
23.0	782.9	—0.71	1.39	24.34	0.955	1.010
24.0	784.9	—0.71	1.40	25.46	0.953	1.010
25.0	786.8	—0.71	1.40	26.59	0.951	1.011
26.0	788.7	—0.71	1.40	27.72	0.950	1.011
27.0	790.6	—0.71	1.41	28.86	0.948	1.011
28.0	792.5	—0.71	1.41	30.00	0.946	1.012
29.0	794.4	—0.71	1.41	31.15	0.944	1.012
30.0	796.3	—0.71	1.42	32.30	0.942	1.012
35.0	805.9	—0.71	1.44	38.16	0.933	1.014
40.0	815.4	—0.72	1.45	44.15	0.924	1.016
45.0	825.0	—0.72	1.47	50.29	0.915	1.017
50.0	834.5	—0.72	1.49	56.57	0.906	1.019
55.0	844.1	—0.73	1.51	63.01	0.897	1.020
60.0	853.7	—0.73	1.52	69.60	0.888	1.022
65.0	863.2	—0.73	1.54	76.34	0.880	1.023
70.0	872.7	—0.73	1.56	83.24	0.871	1.024
75.0	882.2	—0.74	1.58	90.30	0.863	1.025
80.0	891.7	—0.74	1.59	97.53	0.855	1.026
85.0	901.1	—0.74	1.61	104.92	0.847	1.027
90.0	910.5	—0.74	1.63	112.48	0.840	1.028
95.0	919.8	—0.74	1.64	120.22	0.832	1.029
100.0	929.1	—0.74	1.66	128.13	0.825	1.030
			$T = 1500$ K			
0.1	751.5	—0.69	1.31	0.10	1.000	1.000
0.5	752.2	—0.69	1.31	0.50	0.999	1.000
1.0	753.1	—0.69	1.31	1.00	0.998	1.000
1.5	754.1	—0.69	1.32	1.51	0.997	1.001
2.0	755.0	—0.69	1.32	2.01	0.996	1.001
2.5	755.9	—0.69	1.32	2.51	0.995	1.001
3.0	756.9	—0.70	1.32	3.02	0.994	1.001
3.5	757.8	—0.70	1.32	3.53	0.993	1.002
4.0	758.8	—0.70	1.32	4.04	0.992	1.002
4.5	759.7	—0.70	1.33	4.55	0.991	1.002
5.0	760.6	—0.70	1.33	5.06	0.990	1.002
6.0	762.5	—0.70	1.33	6.09	0.988	1.003
7.0	764.4	—0.70	1.33	7.12	0.986	1.003
8.0	766.3	—0.70	1.34	8.15	0.984	1.003
9.0	768.1	—0.70	1.34	9.20	0.983	1.004

Table II.12 (*Continued*)

p	w	μ	k	f	α/α_0	γ/γ_0
			$T=1500$ K			
10.0	770.0	—0.70	1.34	10.24	0.981	1.004
11.0	771.9	—0.71	1.35	11.29	0.979	1.005
12.0	773.8	—0.71	1.35	12.35	0.977	1.005
13.0	775.6	—0.71	1.35	13.41	0.975	1.005
14.0	777.5	—0.71	1.36	14.48	0.973	1.006
15.0	779.4	—0.71	1.36	15.55	0.971	1.006
16.0	781.3	—0.71	1.36	16.62	0.969	1.007
17.0	783.2	—0.71	1.37	17.71	0.967	1.007
18.0	785.0	—0.71	1.37	18.79	0.965	1.007
19.0	786.9	—0.71	1.38	19.88	0.963	1.008
20.0	788.8	—0.71	1.38	20.98	0.962	1.008
21.0	790.7	—0.71	1.38	22.08	0.960	1.008
22.0	792.6	—0.71	1.39	23.19	0.958	1.009
23.0	794.4	—0.71	1.39	24.30	0.956	1.009
24.0	796.3	—0.71	1.39	25.42	0.954	1.009
25.0	798.2	—0.71	1.40	26.54	0.952	1.010
26.0	800.1	—0.72	1.40	27.67	0.950	1.010
27.0	802.0	—0.72	1.40	28.80	0.949	1.010
28.0	803.8	—0.72	1.41	29.94	0.947	1.011
29.0	805.7	—0.72	1.41	31.09	0.945	1.011
30.0	807.6	—0.72	1.41	32.24	0.943	1.011
35.0	817.0	—0.72	1.43	38.07	0.934	1.013
40.0	826.4	—0.72	1.45	44.03	0.925	1.015
45.0	835.8	—0.73	1.46	50.13	0.916	1.016
50.0	845.2	—0.73	1.48	56.38	0.908	1.018
55.0	854.6	—0.73	1.50	62.77	0.899	1.019
60.0	864.0	—0.73	1.51	69.31	0.890	1.020
65.0	873.4	—0.74	1.53	75.99	0.882	1.021
70.0	882.8	—0.74	1.55	82.83	0.874	1.022
75.0	892.1	—0.74	1.56	89.83	0.866	1.024
80.0	901.4	—0.74	1.58	96.98	0.858	1.025
85.0	910.7	—0.74	1.60	104.30	0.850	1.025
90.0	919.9	—0.74	1.61	111.77	0.843	1.026
95.0	929.1	—0.74	1.63	119.42	0.835	1.027
100.0	938.3	—0.75	1.65	127.23	0.828	1.028

REFERENCES

1. Altunin, V. v. Teplofizicheskiye svoystva dvuokisi ugleroda (Thermophysical Properties of Carbon Dioxide). Standards Publishing House, Moscow, 1975, 557 pp.
2. Anisimov, M. A., Koval'chuk, B. A., Rabinovich, V. A., Smirnov, V. A. Results of an experimental study of specific heat c_v of single- and two-phase argon. Teplofiz. Svoistva Veshchestv Mater., 1977, No. 12, pp. 86–106.
3. Blagoy, Yu. P., Rudenko, N. S. Density of solutions of liquefied N_2-O_2 and Ar-O_2. Izv. Vuzov. Fiz., 1958, v. 1, No. 6, pp. 145–151.
4. Brilliantov, N. A. Investigation of the Joule–Thomson effect in air and oxygen at low pressures. Zh. Tekh. Fiz., 1948, v. 18, No. 9, pp. 1113–1122.
5. Vasserman, A. A. Concerning incorporation of specifics of thermodynamic behavior of a binary gas mixture in compiling the equation of state. Kholod. Tekh. Tekhnol., 1975, No. 21, pp. 88–94.
6. Vasserman, A. A., Golovskiy, Ye. A., Mitsevich, E. P., Tsymarny, V. A. Measurement of the density of air at temperatures from 78 to 190 K and pressures to 600 bar. VINITI Deposition No. 2953/76.
7. Vasserman, A. A., Kazavchinskiy, Ya. Z., Rabinovich, V. A. Thermophysical properties of air and air components. New York, Halsted Press, John Wiley & Sons, 1971; original published in Moscow by Nauka Press, 1966; also available in English from NTIS, Springfield, Va., as TT 70-50095, A18.
8. Vasserman, A. A., Kreyzerova, A. Ya. A computer technique for compiling an equation of state for a real gas and a binary mixture from experimental thermal data. Teplof. Vys. Temp., 1976, v. 14, No. 3, p. 67.
9. Vasserman, A. A., Rabinovich, V. A. Thermophysical properties of liquid air and its components. New York, Halsted Press, John Wiley & Sons, 1970; originally published in Moscow by Standard Publishing House, 1968; also available in English from NTIS, Springfield, as TT 69-55092, A12.
10. Vukalovich, M. P., Zubarev, V. A., Aleksandrov, A. A., Kozlov, A. D. Experimental study of specific volumes of air. Teploenergetika, No. 1, pp. 70–73, 1968.
11. Dodge, B. F. Chemical engineering thermodynamics. New York, McGraw-Hill, 1950.

12. Ishkin, I. P., Kaganer, M. G. Investigation of thermodynamic properties of air and nitrogen at low temperatures under pressure. Part I. Determination of the isothermal Joule–Thomson effect of air and nitrogen. Zh. Tekh. Fiz., 1956, v. 26, pp. 2329–2337.
13. Ishkin, I. P., Kaganer, M. G. Investigation of thermodynamic properties of air and nitrogen at low temperatures under pressure. Part II. Thermodynamic diagrams of state of air and nitrogen. Zh. Tekh. Fiz., 1956, v. 26, pp. 2338–2347.
14. Kirillin, V. A., Sheindlin, A. Ye. Termodinamika rastvorov (Thermodynamics of solutions). Moscow, Gosenergoizdat, 1956.
15. Kozlov, A. D. An experimental study of the specific volumes of air at temperatures from 20 to 600 °C and pressures of 20 to 700 bar. Author's abstract of Candidate thesis. Moscow Energetics Institute, Moscow, 1968.
16. Narinskiy, G. B. Thermodynamic principles of the separation of air by low-temperature rectification. Author's abstract of Ph.D. thesis. Institute of Physical Problems, USSR Academy of Sciences, Moscow, 1970.
17. Prikhod'ko, A., Yavnel', A. Investigation of solid oxygen–nitrogen mixtures. Acta Physikochim. (USSR), 1939, v. 11, p. 783,196.
18. Rogovaya, I. A., Kaganer, M. G. A facility for determining the compressibility of gases at pressures to 200 atm and temperatures from 0 to —200 °C. Zh. Fiz. Khim., 1960, v. 34, no. 9, pp. 1933–1937.
19. Spiridonov, G. A., Kozlov, A. D., Sychev, V. V. Determination of thermodynamic functions of gases from *pvT* data by mathematical computer experiment. Teplofiz. Svoistva Veshchestv Mater., 1976, No. 10, pp. 35–53.
20. Sychev, V. V., Kozlov, A. D., Spiridonov, G. A. Thermodynamic properties of air at temperatures from 150 to 1000 K and pressures to 1000 bar. Teplofiz. Svoistva Veshchestv Mater., 1972, No. 5, pp. 4–20.
22. Glushko, V. P. Termodinamicheskiye svoystva individual'nykh veshchestv (Thermodynamic properties of individual substances), v. 1, USSR Academy of Sciences, Moscow, 1962.
23. Himmelblau, D. M. Process analysis by statistical methods. New York, John Wiley & Sons, 1969.
24. Chashkin, Yu. R., Gorbunova, V. G., Voronel', A. V. Effect of impurities on the behavior of the thermodynamic potential in the liquid–vapor critical point. Zh. Eksp. Tekhn. Fiz., 1965, v. 49, no. 2, pp. 433–437, 1965.
25. Amagat, E. H. Memoir sur la compressibilite des gaz a des pressions elevees. Ann. Chim. et Phys., 1880, v. 19, pp. 345–385.
26. Amagat, E. H. Compressibilite des gaz oxygene, hydragene, azote et air jusqu'a 300 atm. Compt. Rend. Acad. Sci., Paris, 1888, v. 197, pp. 522–524.
27. Amagat, E. H. Memoires sur l'elasticite et la dilatobilite des fluides jusqu aux tres hautes pressions. Ann. Chim. et Phys., 1893, v. 29, pp. 68–136.
28. Andrussow, L. Relation cntre les coefficients de ses phénomènes et l'équation de Maxwell. J. Chim. Phys., 1955, v. 52, No. 4, pp. 295–306.
29. Armstrong, G. T., Goldstein, J. M., Roberts, D. E. Liquid-vapor phase equilibrium solutions of oxygen and nitrogen at pressures below one atmosphere. J. Res. NBS, 1955, v. 55, pp. 265–277.
30. Baehr, H. D., Schwier, K. Die thermodynamischen Eigenschaften der Luft im Temperaturbereich zwischen —210°C und +1250°C bis zu Drucken von 4500 bar. Berlin—Göttingen—Heidelberg, Springer Verlag, 1961, 136 s.
31. Baker, H. D. The Joule-Effect in air. Phys. Rev., 1943, v. 64, No. 9–10, pp. 302–311.
32. Behn, U. Über die Sublimationswärme der Kohlensäure und die Verdampfungswärme der Luft. Ann. Phys., (4) 1900, B. 1, s. 270–274.
33. Bennewitz, K., Schulze O. Eine neue Methode zur Bestimmung der spezifischen Wärme von Gasen und Dämpfen. Z. Phys. Chemie, 1940, A 186, s. 299–313.
34. Blanke, W. Messung der thermischen Zustandsgrossen von Luft im Zweiphasengebiet und seiner Umgebung. Diss. vorgelegt von Dipl. Physiker Walter Blanke. Bochum, 1973, 151 s.

35. Bradley, W. P., Hale, C. F. The nozzle expansion of air at high pressure. Phys. Rev., 1909, v. 29, pp. 258–292.
36. Brinkworth, J. H. The measurement of the ratio of the specific heats using small volumes of gas. The ratios of the specific heats of air and of hydrogen at atmospheric pressure and at temperatures between 20°C and —183°C. Proc. Roy. Soc. (London), 1925, v. A107, pp. 510–543.
37. Burnett, E. S. Compressibility determinations without volume measurements. J. Appl. Mech., 1936, v. 3, No. 4, pp. A136–A140.
38. Cockett, A. H. The binary system nitrogen-oxygen at 1,3158 atm. Proc. Roy. Soc. (London), 1957, v. A239, pp. 76–92.
39. Colwell, R. C., Friend, A. W., McGraw, D. A. The velocity of sound in air. J. Franklin Inst., 1938, v. 225, No. 5, pp. 579–583.
40. Cooks, S. R. On the velocity of sound in gases and the ratio of specific heats at the temperature of liquid air. Phys. Rev., 1906, v. 23, pp. 212–237.
41. Dailey, B. P., Felsing, W. A. Heat capacities at higher temperatures of ethane and propane. J. Am. Chem. Soc., 1943, v. 65, No. 1, pp. 42–44.
42. Dalton, J. P. Researches on the Joule-Kelvin-effect, especially at low temperatures. Commun. Phys. Lab. Univ. Leiden, 1909, N 109a, 109c.
43. Dana, L. I. The latent heat of vaporization of liquid oxygen-nitrogen mixtures. Proc. Amer. Acad. Arts. Sci., 1925, v. 60, pp. 241–267.
44. Dawe, R. A., Snowdon, P. N. Experimental enthalpies for gaseous air in the range 1 to 100 bar and 222.00 to 366.45 K. Physica, 1974, v. 76, No. 1, pp. 167–171.
45. Dewar, J. The liquefaction of air and research at low temperatures. Chem. News, 1896, v. 73, pp. 40–46.
46. Din, F. Thermodynamic functions of gases, v. 2. London, Butterworths, 1956, pp. 1–55.
47. Din, F. The liquid-vapour equilibrium of the system nitrogen + oxygen at pressures up to 10 atm. Trans. Faraday Soc., 1960, v. 56, pp. 668–681.
48. Dixon, H. B., Campball, C., Parker, A. The velocity of sound in gases at high temperatures and the ratio of the specific heats. Proc. Roy. Soc. (London), 1921, v. A100, pp. 1–26.
49. Dixon, H. B., Greenwood, G. On the velocity of sound in gases and vapours and the ratio of the specific heats. Proc. Roy. Soc. (London), 1924, v. A105, pp. 199–220.
50. Dodge B. F., Dunbar, A. K. On investigation of the coexisting liquid and vapor phases of solutions of oxygen and nitrogen. J. Am. Chem. Soc., 1927, v. 49, No. 3, pp. 591–610.
51. Eucken, A., Hauck, F. Die spezifischen Wärmen Cp und Cr einiger Stoffe im festen Flüssigen und hyeprkritischen Gebiet zwischen 90° und 320° abs. Z. Phys. Chem., 1928, B. 134, S. 161–177.
52. Eucken, A., von Lüde, K. Die spezifische Wärme der Gase bei mittleren und hohen Temperaturen. I. Die spezifische Wärme der Gase: Luft, Stickstoff, Sauerstoff, Kohlenoxyd, Kohlensäure, Stickoxydul und Methan zwischen 0° und 220°C. Z. Phys. Chem., 1929, B 5, S. 413–441.
53. Fenner, R. C., Richtmyer, F. K. Heat of vaporization of liquid air. Phys. Rev., 1905, v. 20, pp. 77–84.
54. Furukawa, G. T., McCoskey, R. E. The condensation line of air and the heats of vaporization of oxygen and nitrogen. Techn. Note, U.S. NACA No. 2969, 1953.
55. Giacomini, F. A. The temperature dependency of the molecular heats of gases, especially of ammonia, methane and hydrogen at low temperatures. Phil. Mag. (6), 1925, v. 50, pp. 146–156.
56. Glassman, I., Bonilla, C. F. Thermal conductivity and Prandtl number of air at high temperatures. Chem. Eng. Progress, Symp. Series, 1953, v. 49, No. 5, pp. 153–162.
57. Hall, L. A. A bibliography of thermophysical properties of air from 0 to 360 K. NBS Techn. Note, No. 383, 1969, 121 pp.
58. Hardy, H. C., Telfair, D., Pielemeier, W. H. The velocity of sound in air. J. Acoustic. Soc. Amer., 1942, v. 13, pp. 226–233.

59. Hausen, H. Der Thomson-Joule Effect und die Zustandsgrossen der Luft bei Drucken bis zu 200 At und Temperaturen zwischen +10° und —175°C. Forsch. Geb. Ing., 1926, H. 247, S. 1-57.
60. Heck, R. C. H. The new specific heats, addenda. Mech. Eng., 1941, v. 63, pp. 126-135.
61. Henry, P. S. H. The specific heats of air, oxygen and nitrogen from 20° to 370°C. Proc. Roy. Soc. (London), 1931, v. A 133, pp. 492-506.
62. Hodge, A. H. An experimental determination of ultrasonic velocity in several gases at pressures between one and one hundred atmospheres. J. Chem. Phys., 1937, v. 5, No. 12, pp. 974-977.
63. Holborn, L., Jakob, M. Die spezifische Wärme Cp. der Luft bei 60°C und 1 bis 300 At. Forsch. Geb. Ing., 1916, H. 187/188, s. 3-54.
64. Holborn, L., Otto, J. Über die isothermen von Stickstoff, Sauerstoff, und Helium. Z. Physik, 1922, B. 10, S. 367-376.
65. Holborn, L., Otto, J. Über die Isothermen einiger Gase bis 400 grad und ihre Bedeutung für das Gasthermometer. Z. Physik, 1924, B. 23, S. 77-94.
66. Holborn, L., Otto, J. Über die Isothermen einiger Gas zwischen 400° und —183°. Z. Physik, 1925, B. 33, S. 1-11.
67. Holborn, L., Schultze, H. Über die Druckwage und die Isothermen von Luft, Argon und Helium zwischen 0 und 200°. Ann. Physik, 1915 (4), B. 47, S. 1089-1111.
68. Hoxton, L. G. The Joule-Thomson effect for air at moderate temperatures and pressures. Phys. Rev., 1919, v. 13, No. 6, pp. 438-479.
69. Hubbard, J. C., hodge, A. H. Ratio of specific heats of air, N_2 and CO_2 as a function of pressure by the ultrasonic method. J. Chem. Phys., 1937, v. 5, No. 12, pp. 978-979.
70. Itterbeek, A. van, Dael, W. van. Measurements on the velocity of sound in liquid oxygen and nitrogen and mixtures of nitrogen and oxygen under high pressures. Bull Inst. Int. du Froid., 1958, Annexe 1, pp. 295-306.
71. Itterbeek, A. van, Rop, W. de. Isotherms on the velocity of sound in air under pressures up to 20 atm. combined with thermal diffusion. Appl. Sci. Res., 1956, v. A6, pp. 21-28.
72. Jakob, M. Die spezifische Wärme der Luft im Bereich von 0 bis 200 At und von —80 bis 250°. Z. Techn. Phys., 1923, No. 12, S. 460-480.
73. Kiyama, R. Ultrapressure. VII. The compressibility of the air under ultra pressure. Rev. Phys. Chem. Jap., 1945, v. 19, pp. 38-42.
74. Kaye, G. W. C., Sherratt, G. G. The velocity of sound in gases in tubes. Proc. Roy. Soc. (London), 1933, v. A141, pp. 123-143.
75. Kistiakowsky, G. B., Rice, W. W. Gaseous heat capacities. 1. The method and the heat capacities of C_2H_6 and C_2D_6. J. Chem. Phys., 1939, v. 7, No. 5, pp. 281-288.
76. Koch, P. P. Über das Verhältnis der spezifischen Wärmen Cp/Cv = K in trockener, kohlensäurefrier atmosphärischer Luft als Funktion des Druckes bei der Temepraturen 0° und — 79.3°C. Ann. Phys., 1908, B. 26, s. 551-579.
77. Koch, P. P. Über das Verhaltnis der spezifischen Wärmen Cp/Cv = K in trockener kohlensäurefreier atmosphärischer Luft als Funktion des Druckes bei des Temperaturen 0° und — 79.3°. Ann. Phys., (4), 1908, v. 27, No. 2, S. 311-345.
78. Kucnen, J. P., Clark, A. L. Critical point, critical phenomena and a few condensation-constants of air. Commun. Phys. Lab. Univ. Leiden, 1917, No. 150, s. 124.
79. Kuenen, J. P., Verschoyle, T., Urk, A. van. The critical curve of oxygen-nitrogen mixtures, the critical phenomena and some isotherms of two mixtures with 50% and 75% by volume of oxygen in the neighborhood of the critical point. Commun. Phys. Lab. univ. Leiden, 1922, No. 161, p. 48.
80. Ladenburg, A., Krügel, C. Über die spezifischen Gewichte einiger verflüssigter Gase. Ber. Deut. Chem. Gesellschaft, 1899, B. 32, No. 2, s. 1415-1418.
81. Landolt, H., Börnstein, R. Zahlenwerte und Funktionen aus Physik, Chemie, Astronomie, Geophysik und Technik. B. 2, Berlin, Springer Verlag, 1961, s. 231.

82. Lourie, H. Chaleur specifique des gas. Chaleur Ind., 1930, v. 11, pp. 423–435.
83. Masi, I. F. Survey of experimental determination of heat capacity of ten technically important gases. Trans. ASME, 1954, v. 76, pp. 1067–1975.
84. Michels, A., Wassenaar, T., Levelt, J. M., Graaff, W. de. Compressibility isotherms of air at temperatures between —25°C and —155°C and at densities up to 560 Amagats (pressures up to 1000 atmospheres). Appl. Sci. Res., 1954, A4, No. 5-6, pp. 381-392.
85. Michels, A., Wassenaar, T., Seventer, W. van. Isothermen of air between 0°C and 75°C and at pressures up to 2200 atm. Appl. Sci. Res., 1953, A4, pp. 52–56.
86. Noell, F. Die Abhängigkeit des Thomson-Joule-Effektes für Luft von Druck und Temperatur bei Drücken bis 150 At und Temperaturen von —55° bis +250°C. Forschungsarb., 1916, H. 184, s. 3–46.
87. Occhialini, A., Bodareu, E. The dielectric constants of air up to 350 atmospheres. Ann. Physik, 1913, v. 42, pp. 67–93.
88. Olszewski, K. Temperature of inversion of the Joule-Kelvin effect for air and nitrogen. Phil. Mag., (6), 1907, v. 13, pp. 722-724.
89. Partington, J. R. The ratio of the specific heats of air and of carbon dioxide. Proc. Roy. Soc., 1921, v. A100, pp. 27–49.
90. Partington, J. R., Shilling, W. G. The variation of the specific heat of air with temperature. Trans. Faraday Soc., 1923, v. 18, pp. 386-393.
91. Penning, F. M. Isotherms of diatomic substances and their binary mixtures. XXIV. Isochores of air and several other gases. Commun. Phys. Lab. Univ. Leiden, 1923, No. 166, pp. 3–37.
92. Quigley, T. H. An experimental determination of the velocity of sound in dry, CO, free air and methane at temperatures below the ice point. Phys. Rev., 1945, v. 67, No. 9–10, pp. 298–303.
93. Roebuck, J. R. The Joule-Thomson-effect in air. II. Proc. Amer. Acad. Arts Sci., 1925, v. 60, pp. 537–596.
94. Roebuck, J. R. The Joule-Thomson-effect in air. II. Proc. Amer. Acad. Arts Sci., 1930, v. 64, pp. 287-334.
95. Roebuck, J. R., Murrel, T. A. The Kelvin scale from the gas scales by use of Joule-Thomson data. In: Temperature. Its measurement and control in science and industry. Reinhold Publ. Co., New York, 1941, pp. 60–73.
96. Romberg, H. Neue Messungen der thermischen Zustandsgrössen der Luft bei tiefen. Temperaturen und die Berechnung der kalorischen Zustandsgrössen mit Hilfe des Kihara—Potentials. VDI—Forschungsheft, 1971, No. 543, 35 s.
97. Ruhemann, M., Lichter, A., Komarow, P. Zustandsdiagramme niedrig schmelzeuder gemische. II. Das Schmelzdiagramm Sauerstoff—Stickstoff und das Zustandsdiagramm Stickstoff—Kohlenoxyd. Phys. Z. Sowjetunion, 1935, B. 8, H. 3, S. 326-336.
98. Scheel, K., Heuse W. Die spezifische Wärme Cp der Luft bei Zimmertemperatur und tiefen Temperaturen. Phys. Z., 1911, B. 12, S. 1074–1076.
99. Scheel, K., Heuse W. Die spezifische Wärme der Luft bei Zimmertemperatur und tiefen Temperaturen. Ann. Phys., (4), 1912, B. 37, s. 79–95.
100. Scheel, K., Heuse, W. Die spezifische Wärme von Helium und einiger Zweiatomigen Gasen. Ann. Phys., (4), 1913, B. 40, s. 473–492.
101. Shearer, J. S. Heat vaporization of oxygen, nitrogen and air. Phys. Rev., 1903, v. 17, pp. 469–475.
102. Shilling, W. G., Partington, J. R. Measurements of the velocity of sound in air, nitrogen, and oxygen with special reference to the temperature coefficients of molecular heats. Phil. Mag., (7), 1928, v. 6, No. 38, pp. 920–939.
103. Smith, D. H., Harlow, r. G. The velocity of sound in air, nitrogen, and argon. Brit. J. Appl. Phys., 1963, v. 14, pp. 102–106.
104. Tables of thermal properties of gases. Hilsenrath J., Beckett, C. W., Benedict, W. S., Fano,

L., Hoge, H. J., Masi, J. F., Nuttall, R. L., Touloukian, Y. S., Wodley, H. W., NBS Circular No. 564, 1955, 437 pp.

105. Thermodynamische Funktionen idealer Gase für Temperaturen bis 6000 K. Baher, H. D., Hartmann, H., Pohl, H. C., Schomächer, H., Springer-Verlag, Berlin, 1968, 415 s.
106. Tucker, W. S. The determination of velocity of sound by the employment of closed resonators and the hot-wire microphone. Phil. Mag., (7), 1943, v. 34, pp. 217–235.
107. Vogel, E. Über die Temperaturänderung von Luft und Sauerstoff beim Strömen durch eine Drosselstelle bei 10°C und Drücken bis zu 150 atmosphären. Forschungsarb., 1911, H. 108/109, s. 1–31.
108. Walker, G., Christian, W. J., Budenholzer, R. A. The vapor pressure of dry air at low temperatures. Adv. Cryog. Engng, 1966, v. 11, pp. 372–378.
109. Williams, V. C. Written discussion on the thermodynamic properties of air at low temperatures. Trans. Amer. Inst. Chem. Eng., 1943, v. 39, No. 1, pp. 433–438.
110. Witkowski, A. W. Thermodynamic properties of air. Phil. Mag., (5), 1896, v. 41, pp. 288–315.
111. Witkowski, A. W. bull. Acad. Sci., Krakow, 1898, p. 282.
112. Witt, G. Über die Verdampfungswärme flüssigen Luft. Arkiv. Mat. Astron. Fysik, 1912, B. 7, No. 32, pp. 1–13.
113. Womersley, W. D. The specific heats of air, steam and carbon dioxide. Proc. Roy. Soc. (London), 1921, v. A100, p. 483–498.